# Effective
# Writing

D0170720

# Effective Writing

## For Engineers, Managers, Scientists

### SECOND EDITION

**H. J. Tichy**

*Professor Emerita*
*Herbert H. Lehman College of the City University*
*of New York*

with

**Sylvia Fourdrinier**

*Former Managing Editor, AIChE Journal and Symposium Series*

WILEY

**A Wiley-Interscience Publication**
**JOHN WILEY & SONS**
**New York / Chichester / Brisbane / Toronto / Singapore**

Copyright © 1988 by H. J. Tichy

All rights reserved. Published simultaneously in Canada.

Reproduction or translation of any part of this work
beyond that permitted by Section 107 or 108 of the
1976 United States Copyright Act without the permission
of the copyright owner is unlawful. Requests for
permission or further information should be addressed to
the Permissions Department, John Wiley & Sons, Inc.

*Library of Congress Cataloging in Publication Data:*

Tichy, H. J. (Henrietta J.)
    Effective writing for engineers, managers, scientists / H. J. Tichy
  with Sylvia Fourdrinier.—2nd ed.
      p. cm.

    "A Wiley-Interscience publication."
    Bibliography: p.
    Includes index.
    ISBN 0-471-80708-7
    1. English language—Rhetoric. 2. English language—Technical
English. 3. English language—Business English. 4. Technical
writing. 5. Business report writing. I. Fourdrinier, Sylvia.
II. Title.

PE1475.T53 1988                      87-28576]
808'.0666—dc19                    CIP

Printed in the United States of America

10 9 8 7 6 5 4 3 2 1

# Preface

One gratification of preparing the second edition of *Effective Writing* is the opportunity to fulfill the wishes and needs expressed by readers of the first edition. They telephoned and wrote from places too numerous to list, but a few examples give a good idea of the range: Albuquerque, Anchorage, Ashdod, Beirut, Berkeley, Brooklyn, Calcutta, Capetown, Carbondale, Detroit, Dhahran, Dobbs Ferry, and so on, past Hursley Park, Kirribilli, and Poughkeepsie to Yellowstone and Yokohama, Zion and Zurich. Instructors and students wrote from community colleges, four-year colleges, university graduate courses, and even from writing classes for the members of the faculties of university departments of science and technology who were studying to improve the writing of their students as well as their own. Readers learning alone also sent comments and requests: a solitary forest worker teaching himself better ways of writing reports, a supervisor training student nurses to improve paragraphs, a lawyer in a tower fifty stories above Fifth Avenue in Manhattan who was working late to achieve briefer style, and an engineer on a tanker in mid-Atlantic who was having trouble starting to write reports.

In spite of the variety of places and people, the comments and requests are similar. "Examples, examples, and more examples," is the cry from readers who feared that in bringing the book up to date we might discard some of the illustrative paragraphs. We are glad to answer the request with a large number of illustrations of ways to begin, ways to develop paragraphs, ways to organize. We culled these from a variety of sources to offer many approaches, many styles, and many kinds of effective writing with brief comments on each. Some of the examples are paragraphs such as readers who study and apply the recommended principles may hope to write soon themselves; others are illustrations of classic excellence to serve as inspiration. Many readers of the first edition confessed that they had always been unwilling to read outside their own scientific or technological fields and the daily paper, but that some of our examples had led them to want to read more works by the same authors and to enjoy magazines, books, and journals in fields other than their own. They had discovered a main road to good writing—reading for enjoyment.

Writers in business, science, and technology who for all their working years had been writing only letters, reports, and memorandums asked for help with other forms that they are now being asked to write. We added advice about the most frequent requests: minutes, news releases, résumés, and style guides. Our own experience in large and small companies and in government convinced us that many writers there need the new section on reports, which modernizes every phase from initial organization to final selection of style. We also noted increased difficulties with diction, as did our readers; and so we provided "Appendix B: Problem Words and Phrases" to guide writers away from common errors as well as from the old familiar mistakes.

Two specific problems of usage are treated in detail to meet reader requests and our own sense of the prevailing needs. We are glad to see that the formal style previously required in some company reports, in journals, and in the theses of many graduate schools is now seldom used. Thus writers are free to choose more natural styles that permit the active voice and even, where appropriate, personal pronouns. To switch from the impersonal passive voice to the active voice requires practice and a knowledge of the appropriate uses of each, instruction which this edition supplies. Many readers tell us that they face another new difficulty, avoiding sexism in language without breaking established rules or writing awkwardly. We all share this problem and must search for the form that will serve best in each instance. The available choices discussed here should help writers to find appropriate ones and to avoid ineffective repetition, fads, and expressions that do not suit the writers' styles. Patience and constant vigilance are necessary, especially if the language grew in social climates very different from those of the present.

Another change in this edition is the expanded discussion of fallacies about writing. Well-meaning writers and editors accept and recommend to others a growing number of these fallacies. From the ubiquitous "Never begin a sentence with *because* (or whatever word the teacher detests)" to "Punctuate the way you breathe," these admonitions remain with students and are passed on long after everything else in the lower-school course has been forgotten.

A large number of pleas for punctuation rules and advice about documentation convinced us that many readers want everything in one book that they can use as a reference book, carry between home and office, study on trips, and, they tell us, read on vacation. Having added a comprehensive section on punctuation and a representative selection of modern documentation with examples, we hope that this edition will prove to be as complete as necessary.

Finally, we have kept the techniques and material that readers especially liked in the first edition, particularly the concentration on common errors and weaknesses instead of a general study of every point of grammar and rhetoric. "I couldn't face memorizing rules and lists," wrote a reader who studies by herself. "I was lucky to be able to rely on your experience to learn what I was likely to need. I enjoyed the anecdotes and stories that helped me spot errors and correct them without trying to remember pages of instruction." For readers like her we have replaced the errors no longer common enough to seem contagious with those now in vogue. Getting rid of faddish errors is like weeding; for every weed removed, new ones soon appear. But by using an up-to-date study of common errors and by carefully examining first drafts, most writers can achieve a correct style.

We are grateful to the readers of the first edition for their interest, comments, and suggestions. Their telephone calls and letters were helpful and heartening. We hope they like the results.

*Encouraging, heartening,* and *inspiring* are the epithets for the colleagues to whom we feel most indebted. The opportunity to express our appreciation to them is another gratification of preparing this second edition. Some of these colleagues we have observed closely at work, others we have seen just occasionally, and one we know only through the reputation of his work and the quality of his publications. To lift our spirits in moments of discouragement or frustration, we have only to think of these colleagues and their success in improving writing.

Managers who effectively support writing improvement and encourage their writers and editors can achieve remarkable results, particularly if they are as sincere and gifted leaders as Robert W. Boggs, of The Procter and Gamble Company; Chester Poetsch, formerly of Vicks Divisions Research, Richardson Merrill Company; and C. E. Schleckser, Jr., of Exxon Research and Development Company.

Great teachers have inspired us by the miracles that occur in classes led by superior instructors. For Siegfried Mandel of Colorado University, technical writing has been a secondary interest; yet his students in colleges and industry show the results of superior instruction, and he has written and edited stimulating books in the field. Ernest Mazzatenta has for many years guided writers at General Motors Research Laboratories in Warren, Michigan in a course that not only improves writing as well as any course we know about but also teaches most of his students to like the hard work of writing well. In addition, his service in many offices of the Society for Technical Communication, including the presidency, influenced the development of higher writing and ethical standards in that organization. F. Peter Woodford, of The Rockefeller University, is responsible for great improvement in writing in the biological sciences. The book of advice he edited for those teaching graduate students to write better reports and journal articles has had a strong influence. A respected scholar, Dr. Woodford has been a powerful voice for writing improvement on many committees on science and technology. We often wish that when we were students we had enjoyed the privilege of having such great teachers.

Over twenty-five years ago F. J. Van Antwerpen, former Secretary of the American Institute of Chemical Engineers, convinced hundreds of chemical engineers and us that engineers want to write better and can write better. From the days when urging engineers to improve their writing was not popular to the present, his abiding interest in good reading and good writing had influence in many unexpected places.

Beatrice Shube, as a senior editor at John Wiley & Sons, has been a strong force for improving writing thoughout her long career, even though she has often had to do so indirectly. Wiley authors quickly learn to trust her judgment and her honest treatment of them and so are willing to heed her advice about writing better though they might reject the advice of a less concerned, less sincere editor. Members of learned societies, where she is a familiar, respected figure, trust Miss Shube to back them in urging upon beginners the importance of good writing. It is a pleasure to work with this excellent editor, whose heart and head are always in the right places.

Because of these distinguished professionals and others like them, there has been marked progress toward better writing in industry, science, and technology in the past twenty years. That progress will continue just as long as people like them are working to improve writing and editing and are providing encouragement and inspiration for the rest of us.

H. J. TICHY
SYLVIA FOURDRINIER

*Grassy Key, Florida*
*January 1988*

# Contents

## PART TWO.  STANDARDS OF CORRECTNESS

# PART THREE. STYLE

## PART FOUR. ADVICE ON COMMON FORMS

## 15. Reports and Style Guides ....... 435

## Appendices                                                                465

## Index                                                                     563

# Effective Writing

# Slaughter of the Language: A Whodunit

*Nearly all men die of their*
*remedies, and not of their maladies.*
*Molière*

When the Lord High Executioner of the town of Titipu announces, "If I should ever be called upon to act professionally, I am happy to think that there will be no difficulty in finding plenty of people whose loss will be a distinct gain to society at large," he voices a sentiment still prevalent. With hatchet instead of snickersnee, we search for someone to blame for the faults of our society. Hollywood, we cry, is responsible for our low culture; tobacco companies are to blame for our smoking; Madison Avenue fosters materialism; purveyors of pornography inspire sexual immorality; TV is the cause of juvenile violence.

Of course, spoilsports may maintain that we are witch hunting. But the witches are more like bad fairies who visit the cradle and are thereafter held responsible for all the weaknesses of the adult. The hunting, however, is true hunting, especially when the villain is hard to find. Then with all the avidity and perseverance and little of the skill of Dupin, Sherlock Holmes, Father Brown, Miss Marple, Lord Peter Wimsey, and J. B. Fletcher, we try to detect the guilty.

High on our little list of those who must be found is the villain responsible for the death of good writing. Who, we demand, is to blame for the miserable writing in government documents, in business reports, in instructions of all kinds, in articles in learned journals, and in papers read at conventions? Like readers of detective stories, we are confronted by a number of suspicious-looking characters and a number of suspiciously innocent-looking characters. Which of them killed good writing?

Because we know that a reader of mystery stories cannot depend on the seemingly guilty to be innocent, we must view with suspicion the likely murderer of good writing, the villain accused most frequently by the other suspects—Education. Like most scoundrels, Education has a poor reputation but is attractive and desirable. Anyone who doubts the desirability should interview the parents of high school students who have been refused admission to their preferred college. Yet obviously Education is no good. Eighth-grade children cannot read, write, or count. High school students waste their time on extracurricular activities or, according to their parents, lose their health cramming for college entrance examinations. Why, even Ph.D.'s cannot write English. It must be the English teacher who kills good writing.

In the tradition of all likely murderers, the English teachers quickly produce an unshakable alibi: they were not there when many of the poorest writers of today were being educated. They cite the large number of middle-aged and elderly scientists and technologists who cannot write, and they prove that these poor writers received their degrees without taking even one college English course. If this does not convince us that they are blameless, the English professors will insist that students who have the benefit of college English courses write better than their fathers who did not. When employers of recent graduates deny that the graduates can write acceptably, each professor, insisting that they wrote well when they left college, demands of the employers, "What have you done to them?"

Business and Industry cringe at this attack, probably out of the habit of being scapegoats for the sins of society. Indeed, the names Business and Industry are aliases; every reader knows that the villains are more properly called Big Business and Conglomerate. Experts rush to testify against these villains. They submit as evidence Exhibit One—poor business reports that Business and Industry force beginners to imitate; Exhibit Two—antiquated business jargon; Exhibit Three—lists of ignorant and incompetent supervisors of business writing. Then disgruntled, unsuccessful employees testify to such frightening instances of the discouragement of imagination and originality by Big Business and Conglomerate that our story threatens to become a horror tale of prolonged sadism.

But Business and Industry restore us to sanity with their defense. They exhibit poor letters written by college graduates applying for their first jobs. Business and Industry offer in evidence their excellent training programs and tearfully describe their courses in creativity that prove how much they really love imagination and originality, at least as described and prescribed by their instructors in creativity. Then, fretted by accusations, Business and Industry return the attack of the professor of English. They have little difficulty, they say, finding competent instructors for their courses in science, mathematics, psychology, and business management. But unless training directors conduct national talent hunts or are unusually lucky, the educators hired to improve business writing are likely to be of two kinds: those who think that old-fashioned grammar is the answer to all writing problems and those newfangled instructors who encourage employees to demonstrate their imagination and informality by replacing *Dear Sir* in their letters with *Hello, how are you this lovely morning?*

An influential part of Industry, the Technical Writing Industry, accuses another character who is often held responsible for infamies—the Government. Many technical writers, editors, and publishers attribute the poor quality of their publications not to a lack of understanding of the technical material or to a lack of writing skill but to government specifications. The rules and regulations of Government are responsible, they testify, for the incomprehensibility of military instruction manuals. Our detective, who has secret longings to wear a trench coat and hunt international spies, suggests that these manuals sound like the work of saboteurs.

But readers of tax forms, the *Congressional Record*, and other government writings know that the Government does not need saboteurs to kill good writing. Readers can testify that good writing in government publications has been dead a long time. *Federal prose, gobbledygook*, and other less polite epithets for government writing are proof that the remains were beginning to smell as early as 1940, when *gobbledygook* became a popular description. The Government is the first character in our story to offer no defense. Perhaps the crime is too old to bother about, perhaps Government slaughter is legal slaughter, or perhaps the Government has been accused of so many more serious crimes that the murder of clear communication is a charge not worth noticing. It is even possible that the Government did present a defense, but as usual nobody could understand what the Government had written. In mitigation of the charges, the Government might have offered books written by some of its competent administrators to revive good writing in their departments. The Government might even have turned the attack on three of its favorite whipping boys: TV, Hollywood, and Advertising.

On TV the slaughter of the written word has been bloody. When words infrequently appear on a television screen, they often are read aloud, apparently for illiterate viewers; and as though that were not insult enough, the voice repeats or explains any difficult words. Hollywood is quick to point out that motion pictures never treat writing this way. No, indeed, motion picture audiences are expected to read. They are expected to read not English, but subtitle English, a semiforeign variety related to pidgin. Both TV and Hollywood defend themselves by offering in evidence their best dramatic presentations to prove their sponsorship of good writing. Finally, they desperately try to divert suspicion by calling attention to the faults of another popular scapegoat, Advertising. Advertising has no alibi, no plea, no defense. It so often uses words to misinform, to mislead, and to deceive that it expects accusers to overlook the great number of well-written advertisements and to think, "Aha, the murderer!"

But any detective can see that among the accusers are those who have been poisoning good writing for some time—those in colleges and uni-

3

versities, in business and industry, and in government who proliferate writings to gain promotions and prestige. The prevailing belief that recognition and advancement are awarded in direct proportion to the number of pages one publishes is frequently correct. Administrators shirk the demanding labor of evaluating publications and the task of investigating other qualifications for advancement and use quantity of writing as a test because it is easy to measure. Thus quantity defeats quality. If a dozen poor papers weigh more heavily on the scale for promotion, who in industry, government, or education will spend the energy and time required for one paper of excellence?

Even where academic pressures are not so extreme as to mean "publish or perish," they do mean write or want—want money, recognition, and reduced teaching programs. Faculty members therefore write much that should not be published. They fabricate textbooks that duplicate better ones. They send articles on the round of scholarly journals until the articles are accepted, very likely by some editor who secured one promotion for establishing a poor and unnecessary journal and will get another for editing it. And their students ape their methods unto the third and fourth generation of those that flourish by words alone.

In industry and government the situation is worse because those seeking to advance by profuse publication do not need the assistance of a misguided editor. They publish at company or government expense a flood of memorandums, reports, articles, pamphlets, booklets, and books that other employees spend company or government time reading. Thus these writers demonstrate their value and enlarge their departments, although the only increase in work is the burden of writing, reproducing, and reading their largely worthless papers.

Certainly basing promotions on pounds of publications is responsible for much nugatory writing, but it is not likely that this alone is killing excellence. A much more noxious influence is poisoning, strangling, and knifing good writing. These murderers, who destroy completely both good writers and good writing, are villains, attractive and appealing deceivers, panderers to credulity and indolence, smooth promoters disguised as friends. Members of the Write-Good-Quick-and-Easy School advance the belief that magically rapid improvement in writing can be achieved by miraculously little work. Every teacher of the school promotes one remedy for all writers and for all faults of writing, but every teacher has a different remedy—a magic average number of words for sentences, avoiding long words, beginning every paper with a statement of the problem, using only one organization, three or five or seven (choose your own magic number) fallacies. These panaceas have one feature in common: they appeal to the credulous and the lazy.

The most harmful effect of belief in the amulets and fetishes of the WGQE School is a weakening of the will to work to improve writing. Convinced that there must be quick-and-easy methods, writers disillusioned with one panacea seek another. Thus this school unfits them for

4

the work of learning to write well. Although the best authors testify that writing is hard work and that the revision essential to good writing is even harder, the dupes of the WGQE School still dream, like Walter Mitty, of magical success.

They would call anyone crazy who told them that just by holding their racket properly poor tennis players could become excellent players, but these writers hack their sentences to some prescribed length or squeeze their ideas into some ritual organization and expect to become good writers. Unfortunately they do not. Some who enroll in the WGQE School become poorer writers by following the fallacies preached to them. There is no substitute for the discipline of learning and practicing the principles of good writing. Believing that there is can kill the possibility of improvement. But fortunately writers who study and practice discover in the happy ending that good writing is not dead but only slumbering.

**Part One**

# STEPS TO BETTER WRITING

# CHAPTER 1

# Starting and Stalling

*He has half the deed done, who has made a beginning.*

*Horace*

To start writing, simple as it may seem, is difficult for many writers. They suffer anguish when they approach a blank page. They postpone, prevaricate, and procrastinate so frequently and persuasively that they often deceive everyone, even themselves. Other tasks that have been waiting for months suddenly seem urgent; neglected disagreeable duties develop immediate appeal; and plans hitherto nebulous demand instant action. Anything—daydreaming, desk-cleaning, boss-baiting—serves to postpone writing. No dodger is more artful than a reluctant writer, and the most artful dodgers of all are the occasional writers. Professional writers may waste a little time, it is true, but every working day they have to write. However, field representatives who report infrequently to the home office, scientists who turn reluctantly from their experiments to report their findings, administrators who write only a few reports a year, and other occasional writers are masters of the long delay. They can sit and stare at a blank page until some more pressing demand releases them. We have heard at least a thousand such evaders lament their procrastination. But we still sympathize each time because we too have experienced the problem and its frustrations. After all, for fifteen years we found excuses for not starting to prepare this edition.

Reluctance to start is not peculiar to writers of nonfiction, although many of them think that it is. They believe that poets and authors of prose fiction have a special inspiration that enables them to fill pages painlessly. But Anatole France declared that he grew dizzy at the sight of a blank page and might never have produced his masterpieces without Mme de Caillavet to force him to spend several hours a day at his desk— even when he was on vacation.[1] Some authors need the goad of poverty. Samuel Johnson states forthrightly, "No man but a blockhead ever wrote except for money." He composed *Rasselas* to pay for his mother's funeral. S. J. Perelman agrees, because he writes "with the grocer sitting on my shoulder." And Lawrence Durrell expresses the experience of many writers: "You see, the beauty of it is that when you are really frantic and worried about money, you find that if it's going to be a question of writing to live, why, you just damn well buckle to and do it."[2]

Writers who do not have the equivalent of Lawrence Durrell's wolf at the door or of Anatole France's mistress in the office must force themselves to write—and to write promptly. But how? The knowledge that they will write better if they allow time for revision helps, but our experience indicates that for many of them this incentive is not strong enough. Therefore with our students we sought ways of making the writing pleasanter and easier, and we found some simple methods that work for them.

Each writer is, of course, an individual problem best helped by separate consideration. Professionals can study themselves and their writing to discover which recommendations will help them. One or another of the following suggestions usually aids even the most stubborn starters to begin more readily and therefore to write better.

## EDITING AND REVISING WHILE WRITING THE FIRST DRAFT

Many reluctant writers hate to begin because they try to do too much in the first draft. They are determined to produce the final copy immediately, and for most writers this is a frustrating experience in stalling because it requires constant shifting from creation to revision.

In the middle of the first paragraph these writers stop to change a sentence. Before stopping they had the next few sentences clearly in mind and the next paragraph nearly ready. After stopping they cannot remember what they intended to say next. Hoping to pick up the thread of their thought, they reread what they have written, consult the outline, and waste time pressing to remember.

But they cannot recapture what was ready before they interrupted themselves; the words elude them. Finally they give up the attempt to recall and force themselves to write new sentences. The new sentences are unsatisfactory because the utter rightness of the words that they have forgotten haunts the writers. Some authors confessed to us that they had suffered this irritating experience two or three times for each page and accepted it as a necessary pain of writing. And pain it is for most of those who try to edit while they write. Unfavorable criticism is always dampening if not crushing; yet writers trying to edit while they write inflict discouraging self-criticism needlessly at a time when they are most sensitive.

Jim Smith, who is typical of these writers, consults his outline or the thoughts in his mind and feels ready to write two or three paragraphs. Before he finishes the first sentence, he decides that one of the words is poorly chosen. Smith tries to think of another word, consults a dictionary, may even consult a thesaurus if he is a diligent editor, and then gives up and replaces the offending word with one that is no better and may be worse. When he started the editing, Smith had two paragraphs in mind, ready to flow onto the paper. Now he does not even know how to finish his first sentence and after mulling over the possibilities decides that he does not like that sentence anyway. He crumples the paper, throws it away, and goes to the coffee machine, where he confides to a fellow worker, "I hate to write." Who wouldn't?

If someone else were to edit his work as he does, he would call that person a nitpicker or worse; yet just when Smith needs encouragement, he submits himself to the most painful of all criticism—harsh, detailed self-criticism. His readiness to write is gone, and he drives home that night wishing he had a job that did not require writing. While he is waiting at an intersection, the word he wanted pops into his head; but it is too late. He does not even plan to remember the word. Smith's will to write has gone, and the fallacy of trying to edit the first draft while writing it is responsible.

Trying to revise while writing is like straining to remember a name that has slipped one's mind. The more a person strains, the more the name eludes capture. But if the writer relaxes and thinks of something else, his memory functions painlessly and produces *Bolenciecwcz* triumphantly without conscious effort or strain. If writers do not stop to revise, their writing may flow for pages, and then the wanted correction or change, like the forgotten name, will obligingly pop into their consciousness. If it does not, they need not worry. It will appear without further summons during dinner that night or at breakfast the next morning.

Writing and revising require different frames of mind. Revision demands cool objectivity; writing, even technical writing, is a fiery, or at least a warm, procreation. Writing should be kept at a boil. When the creator turns critic, the fire dies. Then the energy that should have been spent writing must be devoted to accepting the cold criticism and laboring to start the fire again. Doing this two to four times for every page is so exasperating that it is no wonder that writers who edit while they write hate to start.

## AVOIDING OTHER INTERRUPTIONS OF THE FLOW

For quicker and better results, writers who have started should keep going, letting the words pour forth as long as the flow lasts. They must cease interrupting themselves when they are writing and must avoid other interruptions. A secretary's questions, the telephone, a friend passing the time of day—any one of these may make them forget the words they were just about to write. Hemingway said it simply and clearly: "The telephone and visitors are the work destroyers. . . . You can write any time people will leave you alone and not interrupt you."[3]

Writing without interruption produces miraculous results. Students tell us that when they are alone and uninterrupted they produce in thirty minutes what would otherwise take them hours. And their uninterrupted writing has fewer errors and better unity and coherence. When some medical writers who were used to working in open cubicles where disturbances and distractions were frequent tried working in a quiet library, they doubled their output and felt more relaxed at the end of the working day. And we are always a little amused at the end of a long course by the number of students who tell us, "Well, it was a very full course, and I learned a lot, but do you know what helped me most? It was that advice you gave us the first day to write without interruption." Even executives who were reluctant to arrange do-not-disturb periods because they had been told that executives should be available all day, soon agreed with Heinrich Heine about writing: "I need solitude for my work."

## THE PROCRASTINATORS

Some writers who cannot get started are unconsciously evading work. As long as they do not start writing, they do not have to write. The advice that spurs them best is, "Start anywhere. The section that you write first does not have to stand first in your final paper. Choose the section that interests you most or the section that you know best and write it." Once the wheels start, they keep turning, and such writers usually find that somewhere along the way they write a good beginning. Unconscious postponers, they can bemoan their inability to start—right up to their deadline. They must learn to suspect the excuse that they do not know how to begin.

Some writers cannot begin because they are not ready to write. They have not thought their subjects through, or they have not completed the research for the section they are trying to write. If they outline their papers, they crystallize their thinking before they write, or they discover that they must complete their research. Incidentally, some writers who are thought to be very slow are not sluggish at all. They have been counting as writing time the hours they spend deciding what to say. There is nothing wrong about thinking with one's pen in hand. But it is an error to confuse time spent deciding what to say with time spent expressing it and then to dislike writing because one's thinking about the subject matter is slow.

## THE FUNCTION OF THE INTRODUCTION

Some writers are problem starters because they have an artificial concept of a paper as an introduction, body, and conclusion, each part so separate from the others that often each is on a different subject. Our undergraduate writers approaching their first research papers illustrate this problem in its clearest and most elementary form. A student preparing a theme on the use of light and dark in Conrad's *Victory* has

12

completed the research, has analyzed the findings, and should be ready to write, but the student is blocked. He thinks that he has to write an introduction, and to him that means a discussion of Conrad's life and works. He is not ready to write the discussion because it is not the subject of his research, but for some reason he considers it the only way to begin his paper.

In industry the introduction devoted to another subject also obstructs some writers. A worker in a personnel department could not begin her report on three machines for processing cards. She had all the information about the machines, had organized the material, and had decided upon her recommendations. However, she was stalled by the conviction that she had to begin with a history of the use of these machines in personnel departments in the United States or at least in her company. In the same company a market research analyst could not start writing a short memorandum on three new reasons for marketing his product in the New England area because he was planning an introduction that summarized the history of the product, a history unrelated to his three reasons.

The concept of the introduction is at fault here. These writers had difficulty starting because they were trying to write the beginning of another paper, often of a book. They did not have the material, and it was not part of their subject; therefore all the forces that were blocking them were legitimate. But some writers feel that they can start only with unnecessary introductions even though their good sense tells them that they will discard the introductions later. If they are inflexible personalities, then all they can do is keep these off-the-subject introductions as brief as possible and consider writing them a warming-up period. Thus they will write shorter and shorter unnecessary introductions and, we hope, may gradually eliminate them.

Chemists and engineers often have difficulty beginning a paper for industry for similar reasons; they find it difficult to depart from the traditional organization used in their branch of science or engineering and especially difficult to start in a new way. If they have been in the habit of beginning, as some journal articles do, with detailed background or a complete history of the research in the subject that they have investigated, they may find it hard to tell a client only about the progress made in the last few months. Once they understand the situation and read some reports, most of them learn to discuss recent findings without a special introduction. But some who have been taught the fallacy too well or who are inflexible by nature will continue to waste time by writing traditional introductions even if they must later discard them.

Familiarity with only one plan makes change difficult and unpleasant for some scientists and engineers, and those just out of college may be stubborn in their insistence on a particular organization. Colleges could help by teaching students to practice using various organizations for their reports instead of demanding that they learn only one. That one, though still used in a few professional journals, may not be suitable for reports in industry or government or even for other journals.

Some supervisors try to help writers by suggesting that a paper begin with a summary or with the problem. Both are often good ways of beginning (Chapter 5). But no beginning is good for all papers. Here, as in other phases of supervision, salutary suggestions are destructive if presented as commands. An assigned method of beginning all reports places writers in straitjackets and often results in awkward introductions that seem to be the monotonous products of assembly lines. When a beginning is prescribed, writers who never had trouble starting soon develop difficulties and waste their energies fighting this cookie-cutting instruction. It is frequently the best writers who are the most annoyed and hampered by such regimentation.

## SOME PHYSICAL INFLUENCES

Another kind of regimentation that may make writing so unpleasant that writers hate to begin is the requirement of using one physical method, such as handwriting or typing or dictating. The unwillingness of a writer or a company to experiment with methods may be harmful.

It is often true that the writer who is most reluctant to change may benefit the most from trying a new method. We have noticed several times that stiff, formal executives who need to acquire easy informal styles are determined to continue handwriting their drafts. Yet dictating to a device that resembles a telephone, though difficult for them at first, later produces excellent results, because talking into a telephone encourages informality.

Sometimes companies can be as difficult as individuals about changes. For example, many writers work best by revising triple-spaced copy. Single- or double-spaced copy looks final to them, and they make few changes on it. But one indignant vice president was unable to get copy as ordered because it was inconvenient for the stenographic pool. A little flexibility on the part of both company and writer is often necessary for easier writing.

For some writers, the selection of a physical method is idiosyncratic. One uses longhand; another finds it much too slow. One likes to dictate to a person because it helps achieve a conversational style, but another is distracted by the presence of a secretary. Some writers feel a peculiar union with their typewriters or word processors; something happens as soon as they touch the keys. Others hate the keys and enjoy the convenience of turning a dictating machine on and off as they please. Experimenting for a few weeks may help a writer to discover a congenial method that will take some of the unpleasantness out of writing. If one dictating machine is unsatisfactory, a writer should try others. They come in a variety of shapes and types these days, and some writers are much happier with one kind than with another. Also, many writers find that they like to dictate some writing, such as letters, but prefer longhand for more formal writing. It is always easier for a writer if the physical method is congenial.

Even successful authors are eccentric about their methods of work. Henry Miller noticed that writers and painters often seek uncomfortable positions and miserable circumstances as though discomfort helps them work.[4] Hemingway always wrote standing; others must write at the same time every day or in the same place. An executive in one of our courses was convinced that a special kind of cigar helped him to write, and one medical writer believed that her favorite perfume was necessary. Coffee is essential to many. Even the sight of an empty mug helps. Habits and sense appeals that sound silly may prove useful. We know very little about the relation of habits and senses to the writing process. When some seemingly unteachable poor spellers practice with thick crayons on huge sheets of paper, they generally learn to spell correctly. If muscular actions aid spellers, why should not writers experiment to see what assists them? Habitual actions and experiences encourage some writers to achieve a working mood—sharpening pencils, raising the blinds, enjoying a particular scent, drinking from the coffee mug, or sitting in a special position may start the words flowing.

## ORGANIZING AND OUTLINING AS AIDS TO STARTING

Many of our students who had difficulty starting to write found that formulating a plan or outline is the most helpful of all devices. Some of them were used to outlining long reports but did not plan letters or short memorandums. Jotting down the few points to be covered often brought to mind a good beginning or ending. And writers in the habit of dictating found outlines particularly useful. Their minds were relieved of worry about forgetting a point during dictation. Their struggle to remember to include everything had often interfered with their expression of preliminary thoughts. But once they had planned, they no longer sat staring at a secretary or machine and wondering how to begin. Their first point was right there before them, and it was usually a better beginning than they would have thought of in a rush to dictate without planning. When they did not organize their thoughts before writing, they frequently used a poor order that buried an ideal beginning in the middle of their letter or memorandum. Planning enabled them to place this idea first and to start easily.

Some writers resist planning because they dislike the outline method that they were taught in school—the detailed plan with many subdivisions descending from Roman numerals to small letters in double parentheses. But that kind of plan is seldom necessary for a short work (Chapter 4). Simply jotting down the main ideas, indenting minor ideas, and then numbering them all in the order that one plans to use is much less painful. Yet this simplest of plans can make a work more logical and

15

effective because it helps writers to find and use the best order for the material.

Writing these outlines revealed to our students and to us that they were familiar with few methods of organizing ideas, perhaps only one or two. Their writing suffered from being forced into one of the few plans familiar to them, and the familiar plans did not provide suitable beginnings. The methods of organizing discussed and illustrated in Chapter 3 enable such writers to choose from many plans and therefore to choose better. The thinking involved in the selection of plans prepares writers to begin their papers more easily.

## KINDS OF BEGINNINGS

A knowledge of the many beginnings appropriate in functional prose, such as those discussed and illustrated in Chapter 5, also helps writers get started. Once they know where to find the advantages and disadvantages of a number of beginnings, they are less likely to sit waiting vainly for some ideas to pop into their heads. Who can honestly say, "I just can't think of a way to begin," while looking at dozens of ways illustrated and discussed?

If the readers of this book have trouble starting to write for some reason not mentioned here, we hope that they will note the information and the methods that solved similar problems—self-analysis, experimentation with new methods of writing, knowledge of many ways to organize, planning and outlining, and familiarity with many ways of beginning. If none of these work, writers should start with some other sections of their paper and let their minds develop beginnings at a slower pace. Many of the best beginnings are written last.

## FREEZING DUE TO FALLACIES

Fallacies freeze many writers. Trying to remember and attempting to apply the prescriptions and proscriptions of the fallacies that they regard as sacred can worry writers even before they put the first word on paper and can demoralize them thereafter. In any list of serious common offenders against good writing, fallacies rank high in destructiveness. Harmful advice has been drummed into the ears of engineers, scientists, and managers so insistently and convincingly that when they write or discuss writing they seem to be in some looking-glass world where all the principles of writing are upside-down, grossly distorted, or completely misunderstood. There Miss Mouldypate incants, "Never end with a preposition, never end with a preposition, never, never end with a preposition," as her ball misses a wicket. Mr. Chipson-Shoulder pleads as he hands out straitjackets, "But it is the best way to say it, the only way to say it, my way to say it. And besides, the vice president-elect says it that way." And little Adam Panacea takes his thumb out of his ear long enough to recite

*Please use a small word.*
*Small words are good*
*For you and for me.*

Bits of advice from fallacy land have a strong influence on writing. If cooking were controlled by such misconceptions, indigestion and poisoning would threaten at every meal. Style in the writing of engineers, scientists, and managers has been poisoned by erring precepts that are no more accurate than a word passed around a circle is for the last listener. We label and attack these fallacies in the hope that writers will note the falsity and forget the errors for good.

Unfortunately, some writers have had poor advice impressed upon them so strongly that they cling to the misinformation tenaciously. We have observed otherwise intelligent English majors going from professor to professor, trying to get one of them to accept the fallacy "Never repeat a word." We have offered comfort to weary managers who, editing close to a deadline reports written in primer style, subordinate some material and rewrite the worst sentences only to have scientists or engineers or managers' assistants complain, "But it is a rule that you use short sentences and short words for good style." Ah, yes—"See Jane. See Dick. See Spot. See Spot run." See readers lose time and patience. Also see writers lose their tempers when managers or other supervisors enforce a fallacy and ruin a report. The demon most feared by capable writers is the supervisor who tries to remove all articles, pronouns, and transitions because the supervisor is addicted to telegraphic style, which clouds or buries any meaning the writer managed to express clearly. The college student who will not repeat a word, the writers who submit reports in primer style, and the managers who know only one route to brevity— telegraphic style—are all insistent that they have been taught these fallacies in elementary or high school and that, therefore, the fallacies are sacred. Few of these writers and editors remember much else from that peerless teaching, but they unfailingly remember every fallacy. Fortunately, freeing them from these incorrect rules quickly helps them to write and edit better.

Few people can concentrate on applying a dozen or more incorrect rules without feeling so constrained that they hate to write. When they are forced to write, everything—diction, sentences, paragraphs—becomes awkward and unnatural, and every revision is made slowly and painfully. The most useful help writers can give themselves before they write is escape from fallacies so that by the time they do write they will have eliminated the anxiety and strain caused by erroneous rules.

The common fallacies that caused the most problems for our students are discussed in the appropriate chapters and listed in Appendix A, "Fallacies to Forget." Sometimes it takes little more than avoiding these erroneous teachings to turn dull, awkward writers into interesting ones.

## NOTES

1. Nicolas Segur, *Conversations with Anatole France*, trans. J. Lewis May (London: John Lane, The Bodley Head, 1926) 22.
2. *Writers at Work: The Paris Review Interviews*, Second Series (New York: Viking, 1963) 249, 267.
3. *Writers at Work* 223.
4. *Writers at Work* 171.

# CHAPTER 2
# The Flow Method

*The desire to write grows with writing.*
                                        *Erasmus*

Men and women in the professions who work properly to improve their writing can improve it with ease and pleasure. When the level of instruction is appropriate to their intelligence and training, learning to write better offers them the interest, challenge, and satisfaction of any opportunity to use reason and imagination in solving problems. But poor instruction or instruction at too low a level bores, confuses, and discourages professionals and thus makes their writing worse. Good writers are not born; they are made and unmade.

When professionals ask us, "Can I improve my writing by myself? What book do you recommend?" we always answer frankly that a competent editor or instructor experienced in teaching well-educated people is the best help that they can get. And many of them have proved this by learning to write better under the guidance of good teachers.

But others have been less lucky: they could not find competent instructors, they were not free to take courses, or for some other reason they had to study by themselves. When they tried to teach themselves, they met discouragement and failure. Some erred by trying to learn all the rules in a handbook of writing and became hypochondriacal about their writing. Others attempted to learn by copying the writing of their supervisors only to find that succeeding supervisors disapproved of their imitative styles and the styles proved difficult to unlearn. And some never wrote a sentence without stopping to change it according to some dimly remembered precept of a teacher or supervisor. They became tense and confused and soon hated the thought of writing.

Many of the engineers, scientists, and business administrators with whom we consulted were using methods of writing and of improving writing that were cumbersome, wasteful, and even painful. We devised for them methods of studying and of writing that would avoid unnecessary labor, take advantage of what is known of the psychology of writing and

19

of writers, and enable each of them to progress individually. These writers experimented with us to help us evolve methods that assist most of the writers and editors who try them. We believe that so many are helped because they are eager and conscientious. They know that those who improve their writing advance their professions and advance in their professions.

The order of writing and studying that we suggest replaces some common misconceptions, obstacles, and emotional problems with a knowledge of sound principles, the practice of efficient methods, and the satisfaction of accomplishment. At each step we indicate the sections of this book that will help writers to determine how much study and practice they need. By avoiding details of little use and concentrating on common errors and weaknesses, we try to help writers to employ their time economically. By exploring some of the causes of common errors and weaknesses, we try to help writers to work with understanding and, therefore, with more enduring success. And assuming that writers in the professions are intelligent people, we point out their opportunities to use their own judgment, we discuss the philosophy behind our recommendations, and we consider, at least occasionally, the future of language and style.

Before planning, which is the first step in the flow method, writers should have general knowledge of the type of paper they plan to write. This advice is not restricted to beginners, because some experienced employees in business, industry, government, science, and technology may have written only one type of paper for as long as twenty years. A transfer, a new supervisor, a new job, changing times may suddenly require them to write other types. Faced with this change, some of them become more nervous than beginners and more reluctant to reveal their ignorance by asking questions. Almost by tradition their first resort is the files of other employees, particularly the files of their supervisors. Here they may find no suitable examples, poor examples, average examples, and good examples; unfortunately, they do not have the knowledge necessary to judge quality. For both beginners and experienced writers we therefore recommend reading the advice in Chapters 14 and 15 on whatever type of writing they are planning.

Writers should also attempt to estimate and understand their readers. A glance at the entries under *readers* in the index gives some idea of the influence of readers on writing. Even before authors take the first step in writing, choosing a plan, they should know who their readers will be.

## STEP ONE—STUDY YOUR READERS, CHOOSE A PLAN, AND OUTLINE

In choosing a plan, writers must estimate closely the readers' interest, their reading skills, and their knowledge of the subject. Some well-meaning advisers press a fallacy on writers: "Put yourself in the reader's shoes." (For other fallacies, see below and Appendix A.)

When one chemist persisted in sending unnecessary details of his research to a vice president who was not interested, a consultant advised, "Before you write you ought to consider how much of this information Mr. Jones wants and needs. After all, he is a busy man, and his desk is piled with reading. Surely you don't think that he wants all these details."

"But I did consider it," the chemist protested. "My writing teacher told me to put myself in the reader's shoes, and I did. I said to myself, 'OK, Joe, you are now in Jones's place; you are a vice president. How much of this material do you want?' And I wanted it all."

A better admonition than "Put yourself in the reader's shoes" is "Forget yourself and become your reader." When Joe was told this, he said, "Oh, well, if you mean I am Jones without any scientific training and interested only in results and using them for company decisions, why then I have to cut out these details. But that's not putting myself in the reader's shoes."

Once he was set on the right path, Joe managed easily to understand Jones's needs, but some people never see any point of view but their own. Concentration on self is often unintentionally comic, like the remark of an actor, "I've had a rough time with *The Hostage*; I get thrown about a bit. One time I broke the leading lady's ankle."

The ability to know and understand the reader's point of view is essential in writing functional prose. It is essential to the selection and rejection of material, to the choice of an organization (Chapter 3), to the style of paragraphs and sentences (Chapters 12 and 11), and to the choice of words (Chapter 10). It aids every decision that a writer makes during revision. It influences every facet of a paper. If readers are varied, a good writer weighs and considers and then adjusts the paper to their variety. This is a more difficult and a far more useful process than mentally promoting himself to vice president and deciding what he as vice president would like.

Writers' successes in assuming another point of view depend on their flexibility and knowledge of the world. Going to the theater, reading about the lives of others, and perceptively observing people—these provide a basic understanding for writing. This is one major similarity between the good writer of expository prose and the good executive: both use their understanding of people, and their understanding of people is good. The relationship is so close that we consider good young writers in our classes to be potentially good administrators, and if they are offered and take the chance to become administrators, they do not disappoint us. It is also true that we have observed poor writers become better writers after holding administrative positions that forced them to work with people and thus to learn more about them. A good writer is a good practicing psychologist.

Effective style also demands that writers choose the best plan for their type of paper, their material, and their skills. But writers cannot choose well if they lack knowledge of suitable organizations. And they do not choose at all if they seize the first order that comes to mind, the only order that they know, or the order in which ideas pop into their heads by association. Writers who are not familiar with a number of plans should read Chapter 3.

Having chosen their plan, writers should test it by outlining the paper. Chapter 4 helps them find the method of outlining that best suits them and the writing in question. It is designed to aid those who do not understand how to outline, those who think that they should write in the order in which ideas occur to them, and those who dislike and avoid outlining. The chapter discusses the general principles of outlining and some special uses of the outline in the professions.

We have observed many writers dictate a work, struggle to revise it, and then in desperation finally plan and rewrite it. Organizing thoughts before writing is pleasant and profitable, but organizing after writing is irritating and inefficient. Planning is not the second, third, or fourth step; it should be the first.

## STEP TWO—WRITE THE FIRST DRAFT

Once writers have an outline, they are ready to write. For best results they should prepare a rough draft. It is essential that while writing the rough draft authors protect themselves against interruptions—their own interruptions (stopping their writing to rewrite or revise) and outside interruptions. Writing and revising are different activities that are best performed separately. A writer who does not delay to polish a sentence, stop to dawdle over the selection of a word, or pause to answer the question of a colleague writers faster, enjoys writing more, and produces better results.

Many writers tell us that the rougher their first drafts are, the more they are able to accomplish at one sitting; and the more freely they let the words flow, the more the writing seems theirs. Authors frequently find that the flow assumes a rhythm of its own as, helped by their outlines, they move briskly along. Experience teaches them to stop only when the flow stops. Thus they accomplish in a half day what used to take weeks. Although they know that the draft needs editing, they are happy with what they have accomplished and often pleasantly weary. They have enjoyed writing. Other writers tell us that thinking about the report while they are completing their research and then selecting an organization and writing an outline is like priming a pump. Previously they had had to coax out every word. After selecting an organization and outlining, they feel as though the floodgates have opened. Sentences rush forth, and the flow grows stronger as they write. The secret is never to stop the flow by switching to correcting or rewriting or even reading. All that experienced flow writers dare allow themselves is a brief glance now and then at the outline, and many of them do not need that because to avoid even that minor distraction they have examined the outline meticulously before writing. When the flow stops, these writers have written more at one sitting than ever before. After three or four such experiences they feel confident that the flow will always be there for them. Never again will they approach a blank page feeling "How dry I am." They need only let the words flow.

When the first draft must be the final draft, it can contribute little to the improvement of writing, for improvement requires careful revision. After writers revise a number of papers, their first drafts will be better, but still their best writing will be that which they had time to revise carefully. Much of the unpleasantness associated with writing is due to postponing writing too long and then trying to accomplish the impossible—a polished paper in the first draft.

## STEP THREE—COOL

After completing a rough draft, writers are likely to feel a warm glow of contentment, and they should bask in their well-deserved reward. This is not the time to be critical of their brainchild, for they are too close to creation to view the result objectively, too likely to confuse what is in their minds with what is on the paper. At this time writers may not notice that words or even whole lines are missing, they may overlook glaring errors, and their most muddled passages may seem beautifully clear. Revision must wait until the authors are cool and objective.

How long that takes depends on the individual. Trained revisers, like editors, may be able to correct a draft as soon as it is typed. But most authors need more time to become objective, four to eight hours being satisfactory for many. Some who use word processors, for example, revise on the printouts, not the screen. Writers should experiment to find the amount of time that must elapse before they can view their work as another reader views it. Sooner or later every author experiences the icy shock reflected in a thought like "How did I ever write that?" or "Now what does this mean?" or "I'm glad I didn't send that out." At this moment of truth writers are ready for Step Four. After they have cooled, writers can face self-criticism without the pain experienced while writing and revising at the same time. Now the writing is no longer the writer, and self-criticism is less discouraging.

## STEP FOUR—REVISE

Here is the real work—revision. Poor writers are inclined to scorn it and to assume incorrectly that good writers do not need to revise. Writers of informational prose believe that articles, stories, and poems pour fourth in their complete and final forms. Until authors are disabused of this error, they are at best halfhearted about revising and at worst unwilling to change a word that they have written.

Stubborn refusal to change is a major block to improvement. "Won't it become awkward or too polished if I revise it so much?" asks a chemist who was advised to make subjects and predicates agree and to see that all pronouns had unmistakable antecedents. "But style doesn't sound easy and natural if a writer hacks over it that much," says an engineer when asked to supply transitions between sentences. And a geologist wails, "Why doesn't my writing just flow onto the page in its final form the way writing flows for creative writers?"

A dangerous fallacy supports these mistaken beliefs. Believers in this fallacy greet every suggestion for change with, "But that's the way it came to me." Behind the protest clearly lies the belief that the "way it came" to the writer is sacred and immutable. In fact, the writing of many beginners seems, at least in the authors' eyes, to be carved in granite. For fallacies besides the sanctity of the first draft, see Appendix A.

Belief in the sacred permanence of the first draft makes the writing of that draft slow and difficult. If for a good reason, like extreme awkwardness or lack of sense, a passage needs complete rewriting, the writers who believe that expression is sacred in the form that it first came to them try to change the passage as little as possible, rather than to improve it as much as possible. They do this even when common sense demands that they throw the sentences away and start again. Such writers also object to removing from a paragraph a sentence on another subject. And if anyone suggests cutting a section from a report, they bleed. The belief that every idea is expressed best as it first comes to a writer makes revision painful and unsuccessful. Writers work better if they are convinced that they are blessed when their first expression of an idea is satisfactory, that they are lucky when a first revision succeeds, and that such good fortune in writing is rare. Successful revision demands a measure of ruthlessness. If writers try to revise and cut their work as calmly and thoroughly as they would the work of someone else, they will, after a time, revise less painfully. And their readers will comprehend less painfully.

Successful authors not only revise, they revise assiduously and interminably. This is proved by their own remarks; the comments of their biographers; and the corrections on their typescripts, galleys, and page proofs. Usually the more easy and natural their work seems, the more it has been revised. Writers of less experience are reluctant to believe this, for the misconception that natural writing is easy has captured their minds.

Yet Ernest Hemingway told George Plimpton that he rewrote every day whatever he had written before and also went over the whole when it was completed, then corrected and rewrote the typescript, and finally revised the proofs. "You're grateful for these different chances," he mentioned. Asked how much he rewrote, he said, "It depends. I rewrote the ending to *Farewell to Arms*, the last page of it, thirty-nine times before I was satisfied."[1] And Aldous Huxley answered a similar question asked by George Wicks and Ray Frazer, ". . . I write everything many times over. All my thoughts are second thoughts. And I correct each page a great deal, or rewrite it several times as I go along."[2] Frederick Lewis Allen told Robert van Gelder that he writes two or three drafts.[3] Allan Nevins complained that he "rewrites so much that his typing bills eat up large amounts of the profits of his books." He writes four or five drafts of all his work and revises "to cut down and to brighten the phrasing."[4] H. G. Wells wrote first drafts "full of gaps." When they were typed, he made changes and additions between the lines and in the margins. Then the second drafts were typed and he revised them. He said that he repeated the process "four, five, six or seven times."[5] John Gunther's corrected typescripts were interlined with handwriting and even pasted in places; van Gelder, who saw them, thought that there were more words written in than there were on the original page. Gunther also revised this revised typescript and made many changes in proof.[6]

Every good writer on business and professional subjects that we have met has spoken of the time and labor that revision takes. "I write just as soon as my experimental procedure is established," a chemist told me. "And by the time my results are complete, I have revised and rewritten sections of the paper ten and twelve times and feel that I do not have so much more to do for them." A company vice president said, "I try to draft an answer to every important letter the day that I receive it so that I see the letter in typescript the next morning. Then I feel fresh and can catch an error or change a stupid approach." "We write my speeches over and over," admitted a government official, "and I read each new version aloud several times and change it as I read." "Sometimes I seem to change a word every time I see it," said a medical writer, "except the medical terminology, of course. And I sweat over how much of that to use."

Writers seem eager to know whether others revise as much as they do. Even Somerset Maugham described feelingly "the heavy cost of naturalness":

> I think no one in France now writes more admirably than Colette, and such is the ease of her expression that you cannot bring yourself to believe that she takes any trouble over it. I am told that there are pianists who have a natural technique so that they can play in a manner that most executants can achieve only as the result of unremitting toil, and I am willing to believe that there are writers who are equally fortunate. Among them I was much inclined to place Colette. I asked her. I was exceedingly surprised to hear that she wrote everything over and over again. She told me that she would often spend a whole morning working upon a single page. But it does not matter how one gets the effect of ease. For my part, if I get it at all, it is only by strenuous effort. Nature seldom provides me with the word, the turn of phrase, that is appropriate without being farfetched or commonplace.[7]

Writers of functional prose who feel that they should be able to pour out clear, readable first drafts are usually relieved to learn that authors whom they admire are not able to do this but must revise many times.

We have never met a writer of informational prose who wrote an easy and natural style in the first draft—or even a clear style. But we have met some writers who protested that they did not have time for revision and many writers who did not know how to revise.

Those who say that they do not have time for revision have usually not tried very hard to find time. They have brushed revision aside as a luxury not suitable to a writer of functional prose and have failed to consider the importance of their readers' time. High-salaried readers mulling over the meaning of confused writing and finding it necessary to query what they read are much more expensive than one writer's revision time.

The problem of time is particularly difficult when writers first try to improve their papers. Then they face the most faults and the least time, and they lack experience in revising. As writers begin to see how long it takes them to revise efficiently, they also become aware of the good results of purposeful revision; therefore they then plan their work to allow adequate time for the steps of writing. By starting to write earlier they can provide an adequate interval before revision and can leave more time for editing. A welcome, and sometimes unexpected, result is the increased thoughtfulness of their papers, which benefit from those excellent second thoughts that would otherwise arrive too late.

Purposeful revision by a writer who has cooled to an objective frame of mind is economical. The real waste of time is an interruption of one's writing of the first draft to change it or an attempt to revise a paper that one has just written. Later revision is more efficient revision.

Revision, unlike writing the first draft, permits interruption and working for short periods, like five or ten minutes. Distractions that dam the flow during the writing of a first draft are tolerable during revision because they do no damage so serious as interrupting the creative flow of expression. Writers should seize available bits of time, such as fifteen minutes when a meeting starts late, ten minutes when they wait for a telephone call, a half hour now and then in the laboratory while they are waiting for results of experiments. If they have a pencil and a few pages of the manuscript that is being edited always with them, they will find many opportunities for editing during otherwise lost time.

### First Revision—Weighing Content

During the first reading of a paper writers should have two questions in mind:

1. Does this paper contain all the material that my readers need?
2. How much material can I remove without interfering with my readers' understanding and needs?

Good writers consider these questions first while planning the paper, then again after outlining it, and finally during this revision of the first draft.

Writers who think only of themselves may include material to impress readers with the author's knowledge, ability, or diligence. They may even write an entirely unnecessary paper just to produce another publication. To discard their own writing ruthlessly is the first lesson that they must learn. Good writers fill their own wastebaskets rather than their readers'.

Other writers may include a wealth of information in their reports but not the information that their readers need. Many a specialist sends to management useless technical information but omits the information necessary for the decisions management must make. More careful consideration of readers' needs should help such an author to avoid this mistake. Writers should ask some of the following questions: Why are people reading this paper? What training and background do they have for understanding the subject? How well do they read? What kind of thinkers are they? How much of my report will they read?

Writers should strive to identify their readers correctly. Countless writers in business or industry have explained, "Well, I'm writing this for the boss and the files, and my boss knows all about this, more than I do; so I don't have to worry about clarity." Pity the poor substitute for the supervisor if one has to function while the supervisor is absent. In one instance a large fire in a plant was followed by investigations conducted by five government committees, each interviewing managers separately. Temporary supervisors were moved from other divisions of the company to the site of the fire while the regular supervisors were working with the investigators. The temporary supervisors never saw the bosses who knew "all about this." They had to function with the reports in the files, which confused them. Only occasionally could an emergency supervisor find a report writer to shake loose some meaning. It was a very costly period for the company, and many employees had much to explain and to change at the end of it. Readers of files need full information clearly expressed, not some incomplete reporting in the jargon of one particular laboratory or plant or office.

Truth to tell, most writers of reports also must depend on the clarity and completeness of the reports when the writers are called on to discuss them after the work is no longer fresh in their minds. One of our students arrived one morning to find on his desk a request for clarification from his general manager, who could not understand the first paragraph on page 10 of a report distributed the preceding year. When the writer pulled the report from the file, he found that he could not understand his own paragraph. He read the preceding and following pages without finding any enlightenment and then sat worrying what to do.

Before he had an answer, his telephone rang, and an irate general manager informed him that when he requested information at the start of the working day, he expected to receive it before lunch. The writer apologized, mumbled excuses, and finally lapsed into silence. "Well," said the general manager, "if you don't know what it means now, chances are you didn't know what you meant when you were writing about that polymer, which has just become of great interest to the Government and to us. You may be a beginner here, but before you leave this afternoon everyone in this division will know your name, and it will go down in the history of your department. Now sign up for the first report writing course you can find." The report writer explained all this to the writing instructor and added, "At least I wasn't fired."

He might just as well have been because after a year and a half, tired of his identification as the writer who didn't know what his own writing meant, he left the company, even though by then he had improved. When writing, he should have included himself among the readers of his papers. Then perhaps he would have been able to discuss his statements intelligently when upper management needed desperately to understand them.

## Second Revision—Increasing Clarity

During the second reading writers should strive for clarity. The word *strive* is used advisedly, for clarity is not easy to attain. If their other papers were not clear, the writers should search their current work for the faults that clouded the meaning in those papers. They should rephrase any ambiguous expressions even though they think that their readers will know what they meant to say. A reader should never be given the opportunity to think, "Well, I know what you mean to say because I know what you ought to be saying, but you haven't said it." As soon as a reader must supply what a writer intended to say, the writer has failed.

Almost every chapter in this book contributes to the achievement of clarity, but if writers address the wrong readers or are so muddled that they cannot understand their own writing, more than the study of the chapters is necessary. Sometimes it helps to test writing on friends who are similar to the readers of the document. We recommend that a few pages that have been completely revised be tested first. The friends should be urged to ask about anything that they do not understand. A careful study of their responses can aid writers to achieve clarity when they revise. They might consider the following questions: Is more or less background useful? Are the words understood by the reader? Does the reader need a simpler approach? These and other questions may be answered by friendly readers, but not if writers act as though they will bawl or brawl if criticized unfavorably.

## Third Revision—Meeting Standards of Correctness

In the third reading and in as many subsequent readings as are necessary, writers should correct their papers. The errors most common in business, government, industry, science, and technology are described and corrected in Part 2 and in Appendix B. Writers should read these sections, mark the mistakes that they make, and study the correction of the mistakes. Nearly all writers can benefit from meticulously examining ten pages of their writing for the errors described. At the end of this exercise they will know their common errors, the ones to search for in the third revision.

All writers should pay particular attention to the advice about coherence between sentences and between paragraphs (Chapter 12). They should observe carefully the examination of coherence on pages 364–366 with a view to scrutinizing their own writing sentence by sentence and supplying connections where there are fewer than two. This important step in revision supplies ease and grace to writing and guides the progress of readers so that they will continue reading. It also alerts writers to any sentences or paragraphs that are out of place.

Experienced authors should check ten pages of their writing annually to make sure that they have not acquired or reacquired errors. The poor usage corrected in Part 2 appears in their daily reading, and it is a short step from reading errors to writing them.

Chapters 6 through 8 sometimes illustrate more than one way to correct an error, and writers should practice the methods illustrated; otherwise, correcting the error in only one way will make their writing monotonous.

## Fourth Revision—Achieving Brevity

The fourth revision should be a strenuous attempt to reduce the number of words. After writers have studied how to avoid the seven kinds of wordiness discussed in Chapter 13, they will find revising for brevity profitable. Beginners have reduced long papers by one quarter of their length. Even experienced writers find that removing unnecessary words often leads to deleting whole sentences, and thus writing becomes keener and livelier.

After authors have revised several papers for brevity, they will find fewer superfluous words in their first drafts. Achieving brevity, at first a giant step, will later be quick and easy. But because wordiness is contagious and the disease is rife, writers must always be vigilant.

## Fifth Revision—Improving Style

The fifth and final revision, an attempt to develop better style, is advanced work. A few writers will be satisfied with the clear, brief style achieved by the end of the fourth revision and will go no further. Others will not have the time to apply the principles of effective style to everything that they write but will wish to apply them to their most important writings. Many writers, however, will be driven by pride in themselves and their work to revise thoroughly all papers that might affect their reputations.

A decision as to how much polishing of a given work is necessary or desirable is a matter of individual judgment. The decision will be a wiser one if writers consider their readers rather than themselves. Directions for laboratory work, for example, do not justify much polishing. Once they are correct, clear, and brief, they are ready. But safety advice is another matter. Because of its importance to readers, a writer should revise and rewrite safety advice beyond the point of clarity, correctness, and brevity to achieve effective emphasis. Writers should not polish just the papers that interest them, for those are likely to be written better anyway. Writers should revise for easy reading any papers addressed to many readers. To save the time of many readers is a professional courtesy. It is also good business.

Writers should polish important documents, because they affect reputations, company style, and individual success. Chapter 9 is a general introduction to style; it prepares writers for a study of style and diction, sentences, and paragraphs.

For writers who never seem able to find the exact words that they need, Chapter 10 examines the principles of word selection. Because the study of words is interesting and the challenge of finding effective diction attracts many intelligent people, it is not necessary to urge writers to be thorough in their improvement. But many writers are tempted to select words by guess or by instinct, and they should know that they can choose words more efficiently if they understand the principles of selecting words.

Chapter 11 helps writers to improve their sentence styles. Concentration on one length, one grammatical type, or one rhetorical type is discouraged by an introduction to kinds of sentences and their uses. This chapter stresses effective subordination in the sentence, the correct use of parallel construction, and the application of other principles of sentence emphasis.

Writers should also revise to achieve good paragraphs. Chapter 12 gives advice on constructing paragraphs, checking transitions between paragraphs and between sentences within a paragraph, using coherence as a device of style, and achieving proper emphasis in a paragraph.


**THE FUTURE**

Those who have improved their writing by using the instructions in this book may be asked, sometimes even before they have completed their studying, to supervise or edit the writing of others. The section entitled "Epilogue: The Editor and Supervisor and the Future Editor and Supervisor" helps them to avoid some common errors of those who supervise writing. Necessarily brief, this section is not a textbook on supervising or editing, but just advice for a supervisor or editor who wants to begin in the right way.

Many writers receive the advice of supervisors or editors. The Epilogue of this book helps them to benefit from that assistance and to avoid some common errors of writers under supervision.

## THE FINAL CHAPTER

The Epilogue is the end of this book, but there is no end to improving writing. If writers are alert, everything they read can help them. The articles, books, letters, memorandums, and reports that they read not only convey information but also illustrate techniques of writing. Aware writers notice errors and weaknesses to avoid and note telling devices and successful methods to borrow. They may plan, for example, to avoid tangled sentence structure that made it necessary for them to reread or letter openings that annoyed them or the unnecessary material that angered them by wasting their time. But a gracious tone in a letter, subheads that made their reading easy, an interesting beginning that persuaded them to read further—these are now more than reading pleasures; they are reminders of techniques that the writers have studied; they are inspiration.

The leisure reading of writers can contribute even more to their progress. While they enjoy an essay, a story, or an article, they absorb, without conscious effort, a knowledge of paragraph structure, an acquaintance with sentence patterns and rhythms, and a feeling for well-chosen words. And these are not so pleasantly and quickly acquired in any other way.

Anyone can learn much from the best writers. But many writers choose as models the poor and average writers of their professions and neglect excellent authors who are more helpful and more interesting. Intelligent writers familiar with the basic principles of writing learn more by reading the best books and critically observing the techniques of the best authors than they learn in any other way.

## NOTES

1. *Writers at Work: The Paris Review Interviews*, Second Series (New York: Viking, 1963) 222.
2. *Writers at Work* 197.
3. Robert van Gelder, *Writers and Writing* (New York: Scribner's, 1946) 25
4. Van Gelder 83.
5. Van Gelder 130.
6. Van Gelder 229.
7. Somerset Maugham, "The Summing Up," *Mr. Maugham Himself* (Garden City, NY: Doubleday, 1954) 558.

31

# CHAPTER 3
# Effective Organizing

*Order and simplification are the first steps*
*toward the mastering of a subject—the*
*actual enemy is the unknown.*

*Mann*

Chaos is so frightening and repugnant to human beings that they associate it with their most terrible experiences. Job speaks of death not only as a land of darkness but as a land "without any order." Many equate the word *disorder* with *sickness* and *disordered* with *deranged*. Perhaps that accounts for the extreme irritation and dislike that readers show when writing lacks plan. No other defect arouses so much anger and aversion.

Although writers generally try to avoid arousing antipathy in readers, sometimes writers in business, government, and industry invite it by neglecting to plan. Supervisors notice with surprise that professionals trained in logical thinking write disorganized reports, and they wonder why. One reason is lack of practice. Some of these professionals are used to following blindly an organization required by a school, department, or professor; by industry or government; or by their profession. No one criticizes their organizing as long as they place material under the required main divisions. This makes them careless of everything except the outline of the whole, gives them no acquaintance with variety in organization, provides no experience in creating and developing plans for papers, and, worst of all, leads them to consider themselves logical thinkers who have no need to plan writing. When such writers must depart from familiar organizations, they are at a loss. Other writers fail to plan long papers because they are deceived by their success with short papers that require little organization, and therefore they do not realize that long works require careful and detailed planning.

Whatever the reason, those who write illogically are seldom familiar with methods of organizing papers. Even when they attempt to plan, their chances of success are limited by their lack of practice and by their ignorance of the advantages and disadvantages of various organizations. Writers can choose well only when they have a reasonable number of choices and know something about each one.

Planning a work is like planning a journey. On some journeys travelers may start without plans or goals because they intend to wander as the fancy strikes them. But those are not business journeys. Successful business travelers plan their journeys—choose their transportation, their connections, their stops. They know what they expect to accomplish at each place, whom they have appointments to see, what they expect to say, and what influence they wish to exert. They send letters announcing their coming and reinforcing their visits. They know when and how they will end their travels. They are no idle wanderers aimless and unfettered. Good writers of the prose of information choose their method of development, their connections, their stops. They know what each paragraph should accomplish and how that accomplishment is related to their goals. They send warnings as they approach new points and remind their readers of what they have said whenever the readers need reminding. These authors are not aimless, moving from topic to topic as the fancy of the moment dictates and caring about only their own convenience.

The first step in their successful planning is a consideration of organizations to eliminate those unsuited to the material. This selection is usually simple. For example, a writer rejects chronological order if there is no time sequence in the subject, geographical divisions if the topic has none, and climactic order if the main ideas have no suitable variations in importance.

Second, thoughtful writers consider their readers and do their best to select the organization that will help readers the most. This choice may involve more deliberation than does the selection of a plan suitable to the material, but it is not necessarily more difficult.

If readers know less about the subject than the writers do, the writers should make certain that any necessary explanations are placed where readers need them. If readers would find Section II of the paper easier to understand after reading the descriptions and definitions in Section IV, good writers move this explanatory material or even the whole of Section IV to a position before Section II. This is a much easier shift in an outline than in a completed paper. As they examine their outlines, these writers are likely to find other ways to help readers who may find the material difficult: comparing the unfamiliar with the familiar, supplying illustrations and examples, explaining abstruse concepts in more detail or in other words. Different adjustments may be desirable for other readers. For fellow specialists good writers carefully remove unnecessary explanations and use technical language to achieve brevity. If busy executives are likely to skim over only a page or two, a wise writer places the most important points first. When a long report contains material for many departments of a company and each department reads only what concerns it, writers may choose to organize their reports by departments—manufacturing, packaging, marketing, advertising, etc. Then they may supply such helps as informative subheads, a detailed table of contents, or an index to guide readers to the sections desired.

If there are two groups of readers—for example, some with technical training and some without—writers must try to satisfy both. For the nontechnical readers they provide summaries of the main points of their reports and detailed tables of contents that will enable readers to find any parts easily. But writers may organize the main body of reports to please their technical readers. If it is impossible to suit one presentation to two groups of readers (a review for college freshmen of recent developments in the writers' field, for example, and the same topic for a meeting of industrial chemists), the writers may have to prepare two reports, each with its own material, organization, and style. It is better to face this necessity at once than to struggle unsuccessfully to prepare one paper for both groups.

When they have just one reader, good writers are particular about the preferences of that reader. For a supervisor of sales who usually thinks about sales by areas of the United States, they plan a report by geographical sections to save the reader the labor of extracting this material and reorganizing it. A report on a problem for a reader who usually considers first a precise definition of the problem, second the recommended solution, and third the reasons for the recommendation courteously follows that order. If a supervisor has turned down a suggestion twice, astute organizers begin with new evidence of the value of the suggestion; they do not open with the twice-rejected suggestion. Within the limits imposed by the material, the readers rule. Writers find it easier to please sovereigns if they plan to please them.

## PRINCIPLES OF EMPHASIS BY ORGANIZATION

When planners do not know their readers, they should rely on the principles of emphasis that make writing more effective for all readers. Thus they should ask: Do the first and final topics contain major ideas? If they do not, the writers reshuffle to improve the organization. A planner with several topics of varied importance should consider the advantages of climactic or reverse-climactic order. In a short report topics arranged in the order of increasing importance may hold the readers' interest. But in a long report writers may place the topics in the order of decreasing importance to ensure that readers have the main points even if they do not read the whole report.

Planning also enables good writers to estimate the proportion of space assigned to each idea. When they find that they have devoted too much space to minor ideas, they try to remove some details or to express the unimportant ideas more succinctly. If that is not possible, they may compensate in one of the following ways for the emphasis that length of presentation places on minor ideas:

1. Giving to major ideas the positions of strongest emphasis—the beginning and the end
2. Using strong and vigorous language for the main ideas and less emphatic language for bulky minor ideas
3. Presenting vivid illustrations of major points.

Thus recognition of the poor emphasis in the first plan helps writers to organize better. But writers who do not discover until their final draft that they have given only a little space in an unstressed position to their main ideas seldom have time to reorganize. If they attempt to correct the poor emphasis by labeling their main points *important, major,* or *significant*, they soon discover that proper emphasis demands more than hastily inserted adjectives.

## COMMON ORGANIZATIONS: THEIR ADVANTAGES AND DISADVANTAGES

As they consider plans for whole works, writers may note that one section of a paper should be organized chronologically, another has place relationships suitable for organization by geography, and a third lends itself to arrangement in steps of increasing importance. Thus selecting a plan for the whole work may lead them to assign plans for some of the parts.

Consideration of the readers may also suggest organizations for parts. Planning a memorandum for a supervisor who likes only papers that start with a problem and end with a solution, a writer may find that comparison is better than problem-and-solution for the presentation of a new method. To satisfy the supervisor-reader, the writer begins with a statement of the problem, but in discussing the solving of the problem, the writer compares old and new methods.

Good organization of the parts and the whole is as essential to good style as good bone structure is to a beautiful face. When a plan has a spark of appropriate novelty or freshness, readers are delighted. They may not say so because they may not analyze why they find reading easy or think a report interesting. Sometimes, indeed, critics may praise the style when they mean the organization, as when they say that a model with good facial bone structure has a beautiful face. In his classic address before the French Academy in 1703, Georges Louis Leclerc de Buffon commented on this close relationship of organization and style:

> Style is but the order and the movement that one gives to one's thoughts. If a writer connects his thoughts closely, if he presses them together, the style will be firm, nervous, and concise; if he lets them follow one another leisurely and at the suggestion of the words, however elegant these may be, the style will be diffuse, incoherent, and languid.[1]

36

A good plan helps a writer. Even thinking about plans is salutary, for it makes writers consider their subjects as units. While considering the limits of the subject and the best order for their thoughts, writers begin to evaluate the material, and new facets and relationships catch their attention. To organize thoughts, writers must shape them, compress or extend them, weigh them, relate them to others, and see them in the perspective of the whole. Writers who have planned their work and outlined are usually ready to write. Their preliminary thinking is complete, and their main ideas are expressed. They are quite different from writers who moan that they cannot begin and from writers who without thinking rush to compose and dictate one beginning after another but never touch upon their subject.

Writers of functional prose can begin to plan as soon as they know their readers and the main points of their papers. They seldom have to pull ideas from the air; scientists, technologists, and managers usually know what ideas and information they must convey. Thus they can consider readers and a subject in relation to many plans, like those below. They need never stare at blank pages blankly.

## Order of Time

### Suitable Subjects

Chronological order is often desirable and sometimes necessary in writing for industry. Insurance companies, for example, usually request that accidents be reported in the order in which the events happened. And day-to-day reports on the progress of construction are reports in the order of time. Factory procedures or processes, like distillation, are clear to readers when presented in the order in which the steps occur. Histories of a study, a subject, a company, a society, an industry, a life are usually presented chronologically, as are experimental sections of reports. Detailed instructions, such as those for a machine operator or a laboratory technician, are clearest when written in the exact order in which they are to be carried out. When such instructions are completed, they should be an accurate checklist of the steps performed.

### Advantages

The order of time is used so frequently because it has many advantages. It is easy to keep a chronological account coherent. Time words and phrases are excellent connectives: *four years later, the next day, the following step, after this, then, before, until,* and similar words and phrases denoting the passage of time come readily to a writer's mind, even for the first draft. When such natural connectives mark the order, readers do not become confused about the sequence of events. Moreover, it is easy to record events in the order of occurrence and easy to check for omissions. This makes chronological order desirable whenever an account must be written and checked quickly and whenever every event or the exact order of events is important.

### Disadvantages

Though unsurpassed for instructions, chronological order is not desirable for all subjects. It does not necessarily provide proper emphasis. Chronological order may assign to unimportant material the best positions of emphasis—the beginning and the ending—and bury important subjects in the unemphatic middle.

To correct this weakness, writers often disturb the order of time by beginning and ending with important subjects. Thus the writers of epics begin at a critical point in the middle of the action. A writer of functional prose might open an article on the medical uses of nuclear energy with a description of the treatment of patients today and then shift to the earliest medical use, the first step in a chronological discussion. Or a writer might open with an imaginative picture of the medical uses of nuclear energy in the future, shift to the first medical use, recount the history thereafter chronologically, and conclude with the situation today. This provides a strong beginning and ending and, except for the opening, a chronological sequence.

Chronological order, no matter how well it suits the material, is a poor choice if it buries information that a busy reader in management wants to find quickly and easily. Sometimes this buried material is all that such a reader wants. For instance, the following chronological organization of a progress report buries the accomplishments of the "current period" so thoroughly that a hurried reader might have difficulty finding them:

Background

Summary of earlier progress

Report on current period
    Problems
    Methods
    Accomplishments

Forecast of future work
    Description
    Projected dates

For such a reader a reporter should select a better organization, perhaps order of importance, which is described later in this chapter. If chronological order is essential, the writer should at least help the reader by placing first a brief summary of current work and a reference to the page on which the detailed account of current work begins.

Another serious disadvantage of time order is the tendency to become monotonous. To keep their readers' attention, writers sometimes introduce other organizations when boredom seems imminent. They may set up and knock down an opposing argument, organize by climax, describe by position, or introduce any other plan that they can combine with the chronological. The difficulty is to switch without confusing readers. An experienced writer can move readers from a time sequence to a climactic presentation effectively, but a beginner may have to work hard to achieve a smooth transition.

A minor monotony of chronological order—the tendency to place time words at the beginning of nearly every sentence—is easily corrected by one of two methods. The simpler remedy is to move the time words: *On that day the first patient was lowered into the reactor* may be changed to *The first patient was lowered into the reactor on that day.* The second method is the substitution of other thought connections for the chronological. This often requires some rewriting of the preceding or following sentences. The example above might become *The first patient was therefore lowered into the reactor.* A too sensitive or a self-conscious writer should avoid worrying unnecessarily about the monotony of time connections. Writers of short works need not be concerned about having too many time connections if they vary the positions of the time references and thus avoid calling them to the attention of readers.

## Order of Place

### Suitable Subjects

Organization by place—sometimes called space order, spatial order, or geographical order—is frequent in business writing. Reports may be organized by the divisions of a company; sales and market research reports may be organized by geographical area; details of a description may be organized from top to bottom, near to distant, east to west, left to right, etc. Place may be important to a reader, for example to a sales manager comparing districts. It may clarify a concept, as viewing the outside of apparatus and then the inside does. It may give a prospective purchaser of business property a good picture: a description of the community and of the exterior, then of the interior, and finally of the contents of the buildings. Thus spatial order suits many readers; a drawing or photograph may help them to follow the text more easily.

Drawings and photographs should be combined with words for effective presentation. It would hardly seem necessary to mention that the numbering of parts in drawings and photographs should be related to their use in the text, but many published illustrations seem to have been numbered by artists ignorant of the order of use. The writer then has to jump in the text from *8* to *3* to *5* although the parts might just as easily have been numbered *1, 2, 3* for the convenience of both the reader and the writer. Also, if a writer uses numbers for space concepts, it is wise to avoid other numbers in that passage. Two sets of numbers, even when expertly handled by a writer, may confuse a reader temporarily and necessitate rereading. Handled inexpertly, they may confuse permanently.

### Advantages

Organization by place has many of the advantages of chronological organization. Spatial words and phrases provide connections between sen-

**Figure 1.** Climactic order in storytelling.

tences and paragraphs: *where, wherever, farther left, on top of, directly north, two inches below,* etc. Another advantage shared with chronological order is that writers do not have to spend thought and judgment on what should come next; once they have established spatial order, what comes next is out of their hands, and they can devote themselves to phrasing and sentence structure. This order, moreover, presents a clear, logical picture, thus enabling the reader to find a part of the work easily.

### *Disadvantages*

Spatial order does not, however, stress points of interest or by itself hold a reader's attention. In a short paper a reader may be carried from place to place by the transitions and the logical arrangement, but in a longer paper these devices are likely to become monotonous unless other plans are introduced within sections for emphasis and variety. Many writers employing organization by place have, moreover, the same problem with introductory phrases that writers employing chronological organization face; they should consult page 39 for methods of meeting this problem.

## Order of Increasing Importance

### *Suitable Subjects*

Arranging thoughts and events in the order of increasing importance, an ancient organization for storytelling, is popular in business, scientific, and technological writings today. In analyzing fiction, critics usually diagram this ladderlike arrangement simply (Figure 1).

Writers of functional prose used to regard climactic or nearly climactic order as the best and perhaps the only order for reports on experimental work:

Statement of problem
Description of equipment
Discussion of procedure
Statement of results
Discussion of results
Conclusions and recommendations

Journal articles were traditionally organized this way, with a history of the art or a discussion of the background of the problem preceding either the statement of the problem or the description of the equipment (Figure 2, Paper A):

Background and history
Statement of problem
Description of equipment
Discussion of procedure
Statement of results
Discussion of results
Conclusions and recommendations

Writers also arranged reports on other work in a crescendo of interest (Figure 2, Paper B):

Background
Statement of problem
Analysis of problem
Description of solutions proposed
Advantages and disadvantages of these solutions
Conclusions and recommendations

Critics and teachers have diagrammed the method as a triangle with the apex representing the least important point and the base the most important material (Figure 2). Because a triangle suggests bulk more than it suggests rising interest, many students find a shaded diagram more meaningful. In Figure 3 the light areas are the least important. The gradual darkening indicates the increasing significance of the material, and the darkest section represents the material of greatest interest, placed at the end of the report.

### Advantages
The great advantage of arranging material in a crescendo of interest is that the gradually increasing importance holds a reader's attention. If material has the proper degrees of interest, a reader progresses always

41

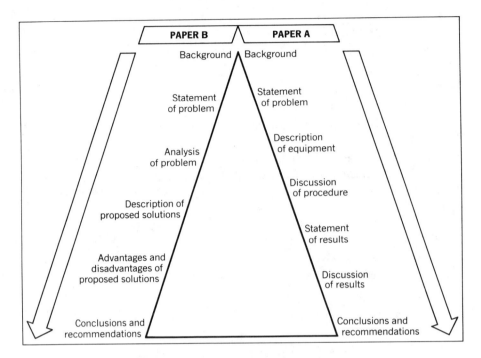

**Figure 2.** Climactic order in two papers.

to more significant material and thus, according to the theory of climax, the material lures the reader to the next page and the next. Another advantage is that this order is traditional in science and technology, and readers find it familiar and safe.

### Disadvantages

Climactic order has disadvantages when topics do not have suitable degrees of interest. In many papers the crescendo is imperfect. The analysis of the problem, for example, may not be less interesting or less important than the description of the solutions; or the description of solutions may be more interesting than the advantages and disadvantages of the solutions. Even the description of equipment may be the most important topic in some experimental papers. And sometimes the importance of a topic may vary with the reader. When interest does not increase gradually as it should but is now small and now great or is even, the principal advantage of climactic order—the holding of a reader's attention—is lost.

Moreover, a paper cannot hold a reader's interest until it has engaged it, and the first material in an informational paper organized by climax is the least important and often the least interesting. A story writer easily overcomes this disadvantage by choosing for the first incident some event that attracts the reader's interest. Although some writers of functional

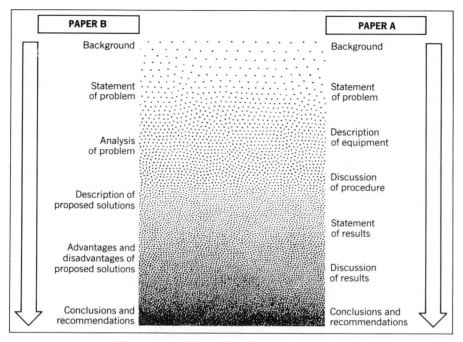

**Figure 3.** Climactic order in two papers.

prose also use this technique effectively, to ask all of them to do this is to ignore the nature of their material, the difference in their freedom of choice, and the waste of their time on what may be an uncongenial and unsuccessful exercise.

A writer using the traditional organization of scientific and technological papers must consider a formidable obstacle in the attitude of the reader. A reader of informational prose is not in the mood of a reader of fiction, who is delighted by suspense, pleased to spend leisure time reading, and entertained by effective dilatory presentation. A reader of informational prose is likely to be busy and hurried and wants to grasp as much as possible in as little time as possible. If the organization of a report places the material the reader wants at the end, the reader thumbs through the pages to find it. And fruitlessly turning pages of tables and charts does not put anyone in good temper. When the reader finally comes upon the conclusions and recommendations, there may be little time left to read them. And the reader who does have the time may no longer have the patience and energy required to understand the full statement of conclusions and recommendations without reading the rest of the report.

One method that most writers can use to make this organization more helpful to the reader is to begin with a summary of main points, with an informative abstract, or with an abstract that stresses the findings (Chapter 12). These beginnings orient the readers and offer clues that help

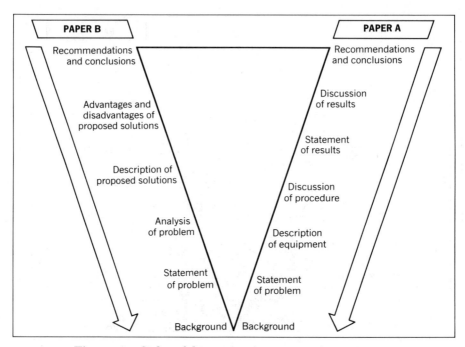

**Figure 4.** Order of decreasing importance in two papers.

them determine the answers to such questions as How much of the report concerns me? Must I read this immediately? or What is this memorandum asking me to do—read for information? act? approve or disapprove action? Readers who have been told why they are reading are often spared the irritation of a second reading and can respond more rapidly.

## Order of Decreasing Importance

### Suitable Subjects

Busy readers complained so much about the inconvenience of climactic organization that writers in government and industry reversed this organization and placed the important material first for such readers (Figures 4 and 5). This order of diminishing interest or decreasing importance obviously suits only subjects with appropriate variations of importance in their divisions.

### Advantages

For hurried readers organization in a diminuendo of importance has the advantage of the front-page stories in their morning newspapers; the most important material is first. When they stop reading, they have covered all the main thoughts that they can grasp in the time expended. Like newspaper readers, they know that the material at the end is least

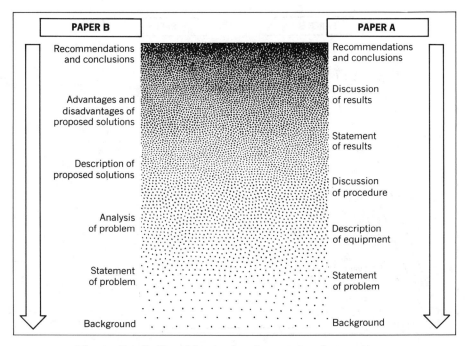

**Figure 5.** Order of decreasing importance in two papers.

necessary to them, and they can in good conscience put it aside for later reading or can forget about it.

A report organized in the order of reverse climax makes a strong initial impression because it begins with the material that interests readers most. Many managers and other busy readers welcome such a presentation.

### Disadvantages

The interest of anyone reading an entire report organized in the order of decreasing importance gradually diminishes. Sometimes writers can stimulate dying interest by adroit placement of material of concern to specialists, but more often they can do little to recapture their readers' close attention.

Moreover, this order may fail to arouse the interest of some readers because it demands that readers be able to understand conclusions and recommendations before they have read the rest of the report. This organization is therefore obviously not suited to all readers and all subjects.

## Order of Emphasis by Position

Because of the disadvantages of organizing by the increasing or decreasing importance of subject matter, writers sought a plan that would

45

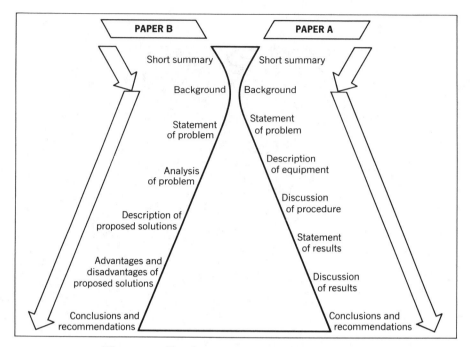

**Figure 6.** Emphatic organization of two papers.

eliminate some undesirable features and retain the advantages. They evolved the organization presented in Figures 6 to 9, which begins with a short summary, or abstract.

Usually the short summary, or abstract, is the beginning of a report, as in Figures 6 and 7. Sometimes it is a separate work, as in Figures 8 and 9. It may be a separate work sent without the report to certain readers, it may be part of a letter of transmittal accompanying the report, or it may be a separate page that accompanies the report but is not part of it. One danger of presenting it as an individual work is that it may soon be detached physically from the report and then the report is organized in the order of increasing importance with a weak beginning instead of the strong one the writer intended. Even a summary that is not physically detached from a report may become separated from it in a reader's mind. An executive may read a letter of transmittal and postpone his study of a report. When he eventually does begin to study the report, he may think, "I've read that," skip the letter of transmittal, and begin his reading of the report with the next topic, the point of lowest interest.

The prefixed summary or abstract should be brief: a sentence or two suffice for a memorandum of a few pages; as much as a page may be necessary for a long report, the limit being one page. This brief epitome may be (1) a summary of the significant ideas, which may be scattered in the report, (2) a summary of the results, conclusions, or recommen-

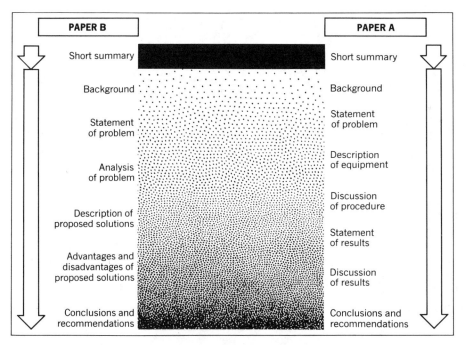

| PAPER B | | PAPER A |
|---|---|---|
| Short summary | | Short summary |
| Background | | Background |
| Statement of problem | | Statement of problem |
| Analysis of problem | | Description of equipment |
| Description of proposed solutions | | Discussion of procedure |
| Advantages and disadvantages of proposed solutions | | Statement of results |
| | | Discussion of results |
| Conclusions and recommendations | | Conclusions and recommendations |

**Figure 7.** Emphatic organization of two papers.

dations, or even (3) a summary of one result or conclusion or recommendation. (Chapter 12 discusses the abstract in more detail.) A writer must consider readers carefully in order to select judiciously the thoughts to be presented to those who read only the summary; the writer must express these thoughts in language intelligible to those readers and must bring all the concepts of the paper within their understanding.

### Advantages

When a writer can do all this, a separate short summary and a report in climactic organization constitute a superior plan. Its popularity is well founded, for it enables busy executives to read a few sentences of a memorandum with confidence that they will find in those sentences the information they need. They can read one page of a report of two hundred pages, know that they have the gist of the matter, and judge on the basis of the summary whether they will read more of the report. Even specialists benefit. Glancing quickly through the summaries of a number of reports guides them to the reports they should read immediately and to those they may put aside until later. If summaries are well written and the readers are well informed on the subjects, they can read ten or fifteen summaries and arrange the reports in the exact sequence in which they should read them. The value of this procedure to a busy specialist should not be overlooked.

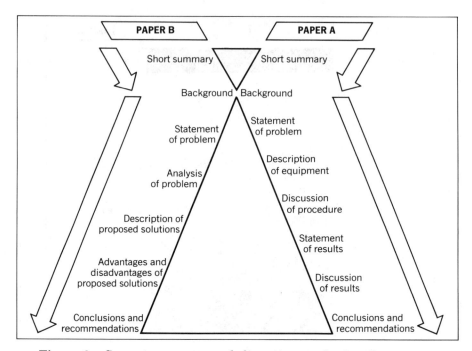

**Figure 8.** Separate summary and climactic organization—two papers.

The organization (Figures 6 to 9) uses the positions of emphasis to great advantage. It opens with a short summary or abstract of the significant thinking of the paper, then presents the material in the order of increasing interest, and ends with a full presentation of the material most important to readers. By thus combining the desirable stress of a beginning and an ending that both contain important material with the effectiveness of climactic presentation, it offers advantages to many kinds of readers.

### Disadvantages

In government and industry the popularity of this organization has created a disadvantage. The initial summaries, especially in letters and memorandums, tend to become monotonous formulas (Chapter 5). Once writers are aware of this danger, they can avoid it by remembering that their readers know from the format of the writing that they are reading a letter or memorandum or report. They do not need to be told so in the first sentence: "This memo summarizes . . ." (Chapter 5). It is usually equally obvious without a statement that they are reading a summary or a recommendation. A writer who starts with a main idea of the report rather than with a statement that this is a report writes more interesting and more varied first sentences because the different ideas of various reports suggest different constructions.

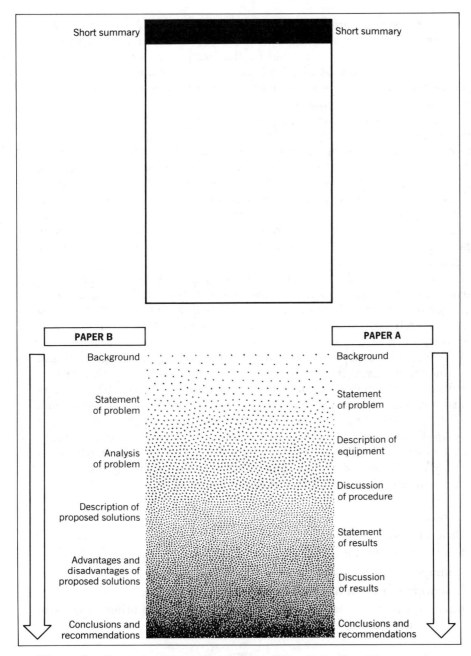

**Figure 9.** Separate summary and climactic organization—two papers.

Supervisors who demand formula beginnings should remember that although readers may not object to them at first, after they have read dozens of them, they are irritated by the routine sameness and finally they do not read them at all. Such beginnings become like the inside address, the salutation, and the complimentary close of a business letter—words that many readers skip.

Some material should not be organized by this plan because the initial presentation of the main point of the paper may antagonize a reader. If the main point is a recommendation that has been considered carefully several times and rejected, a weary reader may become annoyed. "Not that again," he protests and throws the paper aside. If a reader is stubbornly opposed to an idea, only a foolish writer waves a red flag in the opening sentence unless the writer wishes the bull to snort, stamp, and attack. Another beginning—the reason that led the writer to recommend again a much rejected or strongly opposed idea—will prove more successful.

Occasionally an initial summary presents difficulties that are not insurmountable: (1) readers need some slight background or explanatory material to understand the summary or (2) writers are immersed in their final research and cannot take the general view of the material that is necessary for a summary.

If there is no acceptable way to incorporate slight background and explanatory material in the summary, they may precede it. But writers should carefully limit this material to the minimum that their readers require, should rigorously revise it to achieve succinct expression, and should, by one method or another, strongly emphasize the summary when they finally present it. The best way to achieve emphasis is to use strong, vivid expression. If this does not highlight the summary sufficiently, writers may use devices of format. One or more of the following devices will stress summary beginnings that do not stand first:

Larger type

Different type

White space around a centered summary

A single line around the summary to box it

Indention of the main points of the summary

Indention and the use of numbers, dots, dashes, or other devices to mark each main point of the summary

An initial summary is sometimes difficult to compose when a writer's mind is concerned with the details of a long research project. Such a writer has lost the general view of the subject and cannot recapture it because parts of the work dominate his thinking or because he is pressing hard to meet a schedule. Some research groups assign the summary to another person whenever a report writer meets such a difficulty. A person more remote from the current work can view the entire project more objectively and isolate the main points more easily than the writer can at that time. However it is achieved, the initial summary should express these principal points clearly and briefly.

## TWO POPULAR ORGANIZATIONS

For more than twenty years writers in our classes have asked us, sometimes plaintively, to recommend a plan or two that will serve for papers of any length when the writers must rush a report and therefore have little time for careful selection of a method of organizing. Having long deplored panaceas for poor writing, we have been naturally reluctant to guarantee the success of any method for organizing a wide variety of papers. It is true, however, that our students in science and technology and business and industry have found that the two following plans usually serve well enough when a writer is in a hurry. The first is the traditional organization for papers in science and engineering; the second is a plan for writers who wish to stress results.

The first begins with a statement of the problem, presents the methods or analyzes the solutions, and ends with the results and conclusions. In simple expression the plan follows:

PLAN I

1. What was the reason for this research? or What brought this up? or Why am I reporting?
2. What was done? Who did it? When? Where? How?
3. What is new? or What are the findings?
4. How can they be used? or (for theoretical papers) What is the significance or potential significance of the findings?

As Plan I does not begin with material of great interest and does not reach any until Step 3, writers of medium-to-long papers should consider preceding Step 1 with a summary of findings or of other important material.

The second plan begins with the most important results and conclusions, follows with the supporting details, presents any other findings, describes the methods, and ends with the answer to What's next? In simple wording the plan follows:

PLAN II

1. What's new?
2. What are the supporting details?
3. What are the less important findings—if any?
4. What was done? Who did it? When? Where? How?
5. What's next?

51

A writer using either of these plans must be prepared to modify them to accommodate the material. Because the plans are useful for many kinds of reports—memorandums, papers of various lengths, progress reports at stated intervals, and even published papers—omissions and additions may be necessary, but these do not include major changes that destroy the unity or emphasis of the plan. Writers on theory, for example, may usefully stress in the first two steps of Plan II the main significance or potential significance of their work. If the results are negative, they are of interest to those who by reading the report can avoid repeating the work; therefore they should be included.

Plans I and II serve well for reports so brief that only a sentence or two is needed to develop each topic, and they serve equally well for long works that answer each question in detail. The one disadvantage derives from the popularity of these plans with writers: they may use the same plan for the same readers so many times that the monotony destroys the effectiveness and the reading becomes boring, particularly if the same plan is used for every paper or even for every biweekly report.

## SOME PROBLEMS IN ORGANIZATION

### Planning a Straw Man Argument

Stating an opinion, conclusion, decision, or belief and then knocking it down is a technique often used when a writer is combating a popular or entrenched opinion, when a reader may present the opposing argument if the writer does not mention it, or when only an opinion stands in the way of the idea a writer is advancing. This can be an effective method of organizing argument, but writers who choose it must be sure that they will not become contentious. This order requires judgment and taste. One major problem is to set up the opposing argument fairly, but not so firmly that it cannot be knocked down. Only when their arguments are more persuasive than their opponents' should writers choose this order.

A supervisor who has excellent reasons for recommending changes in a shipping department may start with the advantages of the present methods and then prove that changes would retain these advantages, provide other advantages, and eliminate some defects of the present system. This would be tactful and convincing.

But if one is suggesting marketing a drug that has been withdrawn for seemingly good reasons and that has been reconsidered at intervals and rejected, it would be tactless and weak to begin by rehashing the old arguments for rejection and then answering them. A more effective beginning might be some new, startling market information about requests for the drug and about the strong complaints from doctors and hospitals that could not obtain it. If these details are impressive, a reader will continue and may be persuaded by the information and by the answers to familiar arguments.

When writers begin by knocking down an argument, they must also use other plans. For example, they must organize the arguments they expect to combat. Some suitable orders are climactic, reverse climactic, and chronological. The writers may wish to organize their own arguments more effectively than an opponent's. Here, of course, is an opportunity for subtle deception. It is possible to organize an opponent's arguments ineffectively for poor emphasis and to organize one's own arguments to achieve convincing emphasis. Only careful readers experienced in organizing will detect the deception. This trick is common in political oratory and writing because it deceives well and is difficult to combat. Any writer for industry who is honestly concerned with presenting the truth seldom uses this trick. However, writers in government and industry should know that the practice exists, for they are also readers who must beware of such cheating and writers who may have to reply to it.

Some writers of argument organize all their paragraphs by presenting the opposing point of view first and then knocking it down, the harder the better. Their organization becomes a constant seesawing from an opponent's point of view to the arguments against it, and readers grow tired of switching back and forth. If such a paper does not conclude with a strong statement of the writer's best point or a summary of the writer's point of view, the conclusion may stress the opponent's arguments, not the writer's.

To a writer who has selected a pro-and-con organization, transitional expressions come naturally: *but, yet, still, however, on the other hand, on the contrary, nevertheless, notwithstanding, in opposition to this, in contrast with that, at the same time, although that is true, but then, but after all, conceding that point, moreover, further, besides, in addition, also, too, furthermore, again, first, second, third, finally, hence, therefore, consequently, thus, as a result, because of this, in short, in brief, in fact, indeed, to be sure.* There is, therefore, no excuse for monotonous repetition of a writer's favorites.

### Comparisons and Contrasts

Organizing comparisons and contrasts defeats many writers of technical information, as their circuitous and repetitious papers show.

Sometimes the defeat is due to their attempts to express in words material that belongs in tables and graphs. Whenever writers find themselves struggling with expressing the details of comparisons, they should consider using tables, graphs, or other visual aids. Even in a memorandum or in a letter, appending a table or a graph is better than trying to express the information in sentences—better not merely because it lightens the task but because it makes reading and understanding easier. It leaves writers free to comment as they wish on information without detailing all of it in the text. If they think that it will assist readers, they may use representative tables or graphs in the text, the detailed tables or graphs appearing in an appendix. Whatever arrangement they choose, they should try to help readers by expressing in sentences the details that may be understood best in that form and placing in tables, graphs, or other pictorial form the details suited to such presentation.

Another weakness in many comparisons is the use of an organization that encourages and demands repetition. The following outline, the plan of a paper comparing two drugs, shows a poorly organized comparison that drags on interminably.

```
Information about Drug A
    Price
    Manufacturing costs
    Other costs
    Relief
        Remarks of patients
            After one dose
            After four doses
            After eight doses
        Relief measured by objective tests
            Test I
                After one dose
                After four doses
                After eight doses
            Test II
                After one dose
                After four doses
                After eight doses
Information about Drug B
    Price
    Manufacturing costs
    Other costs
    Relief
        Remarks of patients
            After one dose
            After four doses
            After eight doses
        Relief measured by objective tests
            Test I
                After one dose
                After four doses
                After eight doses
            Test II
                After one dose
                After four doses
                After eight doses
Advantages of Drug A
    Price
    Costs
    Patients' testimony after one dose
```

Advantages of Drug B
    Patients' remarks after four and eight doses
    Relief measured by objective tests
Disadvantages of Drug A
    Patients' remarks after four and eight doses
    Relief measured by objective tests
Disadvantages of Drug B
    Price
    Costs
    Patients' remarks after one dose
Comparison of Drugs A and B
    A more economical
    A more immediately efficacious after one dose according to patients' remarks
    B more efficacious after four and eight doses according to patients' remarks
    B more efficacious after one, four, and eight doses according to objective tests
Recommendation: That B be studied further with a view to reducing manufacturing and other costs


A paper such as the one outlined here makes readers feel that they have been riding on a merry-go-round. The comparison is less repetitive and much clearer when organized as a unit:


Comparisons of Drugs A and B
    Prices
    Manufacturing costs
    Other costs
    Patients' remarks
        After one dose
        After four doses
        After eight doses
    Results of objective tests of relief
        After one dose
        After four doses
        After eight doses
Summary of advantages of B (if the paper is long enough or the material complex enough to require a summary)
Recommendation: That B be studied further with a view to reducing manufacturing and other costs


Outlining before writing usually makes excessive repetition obvious, but many writers do not outline. They doggedly explore all the topics and even strive valiantly to make them sound different. Such overlapping discussions should be avoided.

A writer presenting

1. The characteristics, advantages, and disadvantages of fractionating tray $x$
2. The characteristics, advantages, and disadvantages of fractionating tray $y$
3. The characteristics, advantages, and disadvantages of fractionating tray $z$

still requires additional sections to compare the three. Varying the phraseology for these discussions usually is a strain and leads to artificiality, whereas failing to vary leads to monotony. And the paper will tend to become three papers, one on each tray, with the comparative study at the end not strong enough to pull the three together. Unless some readers are known to be interested in only one tray, the writer should organize by some other plan, such as the following:

1. A comparison of the advantages of fractionating trays $x$, $y$, and $z$
2. A comparison of the disadvantages of fractionating trays $x$, $y$, and $z$
3. Recommendation

Whether writers should place advantages or disadvantages first depends on where they want the emphasis to fall. Many writers prefer to place last the points that bolster the recommendation. Thus a writer recommending Computer I rather than Computer II ends with the comparisons favorable to Computer I; a writer comparing two factory sites ends with the material that favors the site that the writer has selected; and a supervisor discussing the promotion of one of two assistants compares the two in the order that places last the strong points of the assistant the supervisor wishes to promote. If this climactic order does not suit their material, writers should choose some other order—for example, chronological or geographical. Antonio Ferri in his article "Hypersonic Flight Testing" gives similarities and then differences in two paragraphs arranged chronologically:

> In the early stages of high-speed aerodynamics, we described what we observed by means of three interchangeable adjectives: hyperacoustic, supersonic, and hypersonic. All three meant but one thing—that the observed phenomenon had occurred at a speed above the speed of sound. They differed only in their etymology; the first is all Greek, the second all Roman, and the last is Greco-Roman.
>
> Today, however, the word hypersonic is used only in conjunction with flight speeds above 5 to 6 times the speed of sound, while supersonic now defines the flight velocity range above the speed of sound but below these values. The word hyperacoustic has become an anachronism.[2]

It is sometimes more effective to present one subject separately and then to compare it with another or with others. If the first subject is more familiar to readers or if the first subject is the only familiar one, then to progress from the known to the less well known or from the known to the unknown is sensible because these plans assist readers. The familiar material may clarify the unfamiliar material and thus prepare readers, as the following example from the same issue shows:

> If you or I had invented the world, chances are we'd have botched the job. For eventually we would get around to inventing a way to hold the solid stuff of our world together. This is too all-encompassing a job to entrust to the selective interactions of chemical bonds. So it is likely that we would cook up some kind of "universal intermolecular force" and let it go at that. But in our haste to keep the world from falling apart, we would no doubt forget to invent some means, at the same time, to keep these universal forces of adhesion in check. That would be disastrous: Before our home-made world was ten seconds old, all its moving parts would lock solid, stuck fast with the universal glue of our own inventiveness.
>
> Now nature has been somewhat less reckless than we would have been, and somewhat more cunning too. She did invent a "universal intermolecular force" to keep the solid matter of her world together, but she made it an exquisitely refined and well-controlled force. Not that it's a weak one; it is, after all, the force that holds the atoms of a steel building together at the places where they are not held together by covalent, metallic or other chemical bonds, and it often accounts for more than 80% of the binding forces in solids.[3]

A common error to avoid in comparisons is varying the order of the topics compared. If, for example, the characteristics to be compared are given for the first subject in the order *age, sex, education, income, marital status*, and *number of children*, then the same order should be used for the other subjects. Changing the order confuses readers and has even confused the writers. Writers should think of the topics as though they were tabulated and should follow the table headings exactly to enable readers to compare as easily as possible.

## Choice of Induction or Deduction

"Shall I use induction or deduction?" asks a writer of informational prose and sometimes wastes time writing a section of a paper both ways to see which is preferable. Whether to move from the particulars to the general or from the general to the particulars, from proof to proposition or from proposition to proof, may usually be decided by considering one's readers. Which order will be clearer, more convincing, more helpful to them? If a generalization will help readers to understand the details, examples, or illustrations, then it should stand before them. If the details, examples, or illustrations will help readers to understand the generalization, they should stand first. Occasionally a writer places a generalization in the middle of a discussion if readers who have become familiar with the preceding particulars can grasp the generalization. This order reserves the position of final stress at the end of a paragraph, section, or paper for a striking example. The example is frequently a more effective ending than a proposition that a reader grasped earlier in the paragraph. If, after readers know the details, a generalization is so simple and obvious that it must come as an insult to their intelligence, it should stand before the particulars, or it should be omitted.

Sometimes a simple proposition may be introduced successfully by a question. Walter C. Michels uses such a question and answer and follows them with particulars of proof in his talk on the occasion of receiving the Oersted Medal of the American Association of Physics Teachers:

> Who is to be blamed for the increasing confusion between science and technology, for the short-sighted policy that seems to be emerging? I submit that the fault lies largely with us, the community of physicists. We have not used the past eighteen years to accomplish the educational job that the relative freedom of that period made possible. A public that does not understand science and the role that science plays in our rapidly moving, technologically dependent civilization cannot be expected to act intelligently in its own self interests. The understanding necessary for a continued national interest in science cannot be brought about by advertising or by propaganda; it can result only from a serious and long term educational program.[4]

### Explanation of the Chosen Organization—Desirable or Undesirable?

The question of whether writers should announce or describe their organization for their readers is answered best by consideration of the readers. If a plan is simple and easy to follow, discussing it may be oversimplification; for example, if readers are familiar with a plan because it is traditional in a profession or because it is used frequently by those reporting to them, no preliminary description is needed. But if a complex plan is new to them, readers may find a description of it helpful. This description may be given in a table of contents or in a descriptive abstract. Meaningful headings and subheadings keep readers apprised of the steps of an organization, and transitional words and phrases make a plan clear. Although readers should not be uncertain about the plan of a work, it is generally not desirable to explain it at length. If an organization is so difficult that readers cannot follow it without explanations, the writer should search for an order that is easier for the readers to understand. Prolonged elucidation of a plan of development stresses the plan at the expense of the ideas, but more subtle guides to a plan clarify it without shifting the emphasis from the ideas. Finding an organization that does not require much clarification and finding methods of directing readers without subtracting emphasis from ideas are both easier before a paper has been written. Outlining, which is discussed in the next chapter, provides a writer with a general view of a work and its parts while the planning is still flexible.

## CONSIDERATIONS IN THE CHOICE OF A PLAN

Some writers who never gave much thought to organizing their ideas may study planning only to feel overwhelmed by their newfound richness of choice. They need not despair. They have fewer choices than they think because many influences narrow the selection: the material, the readers, the writer's goal, and the writer's ability and personality.

The preceding pages of this chapter warn several times that a writer's material must be compatible with the chosen plan. A writer does not succeed by seizing any organization of ideas and trying to impose it on unsuitable material; for example, a writer who tries to impose chronological order on material that does not have the necessary time relationships is heading toward frustration and failure. Such a writer may, however, organize one section chronologically if it has sufficient chronological relationships—the experimental procedure, for instance—even though the paper itself requires another plan for the whole. A writer should never choose a plan that requires stretching or squeezing the ideas. One cannot make out of a half yard of fabric a garment that requires four yards, nor can one choose a garment requiring two yards of fabric and squeeze in five extra yards. But when the plan is appropriate, the ideas fall easily into place and fit nicely.

Readers are another strong influence on the choice of plan. The writer must consider their education, background, training, familiarity with the subject, and interests. An intricate and subtle plan, for example, is not suitable for poor and inexperienced readers. A plan that places highly technical subject matter first should be altered or abandoned if the readers addressed lack the technical knowledge necessary to understand the ideas without any explanation and background. But for a different kind of reader, a manager with a technical background, for example, who wants to grasp the gist of a report immediately, a summary of technical material in technical terms may be a good beginning.

Writers' goals also influence the selection of organizational methods. Writers who wish to persuade should choose plans that stress by position the most effective arguments or evidence. But writers who wish to inform might better concentrate on organizations that help readers to understand, such as the following:

moving from the known to the unknown
moving from the simple to the complex
dividing into simple segments

59

Thus the writers' goals influence the selection of plans for the whole and for the parts.

The ability and personality of a writer may also limit the choices. A timid writer with little experience in organizing papers may hesitate to use an unfamiliar, difficult plan for a long and complicated paper even though the plan suits the material well. Instead, such a writer may fall back on a familiar organization. But a braver soul with more planning experience is likely to choose the new plan as a challenge and work hard to solve any problems. Each chooses according to personality, and each may succeed.

In spite of these limitations, writers still have a useful variety of choices if they are informed about possible plans. Imagination and creativity help writers to see the potentials of different organizations in relation to their material and readers, as well as to their goals and personal preferences. Common sense and logic help them to outline their papers easily according to the chosen plans, as the next chapter shows.

## NOTES

1. Georges Louis Leclerc de Buffon, "Discourse on Style," trans. Rollo W. Brown, *The Writer's Art* (Cambridge: Harvard UP, 1932) 279.
2. Antonio Ferri, "Hypersonic Flight Testing," *International Science and Technology* (April 1964): 64.
3. Louis H. Sharpe, Harold Schonhorn, and Charles J. Lynch, "Adhesives," *International Science and Technology* (April 1964): 26.
4. "Men with Ideas," *International Science and Technology* (April 1964): 98.

# CHAPTER 4
# Easy Outlining

*If a man can group his ideas, he is a good writer.*

Stevenson

A writer can move and change parts of an outline easily. But once a paper has been written, the parts are closely woven into a whole. And a writer is naturally reluctant to destroy this whole. Thinking of the labor involved in reassembling a completed paper discourages change. But an outline encourages a writer to improve the organization of a paper. Because an outline prevents thoughts from prematurely hardening, shifting the parts of an outline is easy. It is like shaking a kaleidoscope: another design emerges, new possibilities reveal themselves, fresh relationships appear. We do not know any way for writers working by themselves to shake up their thinking and shape their papers as successfully as by outlining. Outlining encourages forethought instead of afterthought.

## MOVING MATERIAL IN A COMPLETED TEXT

For writers forethought is less painful and demanding than afterthought. Moving a section later by hand or machine is the least of the difficulties if the writing is to be coherent. Figure 10 is used for discussing the problems. Once a writer has moved Paragraph ii from Page 3 to the position after Paragraph x on Page 7, the work has just begun. The writer must now return to Page 3 to connect smoothly Paragraph i and Paragraph iii. Usually paragraphs are connected by more than one transitional method (Chapter 12). Then, on Page 7, Paragraph ii must be connected with Paragraph x, now preceding it, and with Paragraph xi, which now follows Paragraph ii. In addition to providing coherence between these paragraphs, the writer must read Page 7, and possibly Page 8, to be sure of having observed the following points:

1. The thought should flow logically from Paragraph x to Paragraph ii to Paragraph xi.

**Figure 10.** Two typewritten pages of a report.

2. The language of x and xi must be harmonious with the language of ii.
3. Paragraph ii should be compatible in thought with the sections before x and after xi.
4. If Paragraph ii contains any material that a reader needs to understand the material in Paragraph x or in the paragraphs preceding it, that material must be supplied where it is needed.

A clear writer checks these points whenever it is necessary to move material in a written document. No wonder writers grit their teeth at the thought of the checking, connecting, and harmonizing necessary when a section is moved in a completed part of a paper. One of the most welcome contributions of outlining is that it calls attention to sections that are out of place and enables writers to move them by simply crossing out and writing in or by renumbering or both. Then the harmonizing of style, the logical flow of thought, and the coherence are provided in the normal course of writing and editing. And everyone is happy except possibly the dentist whose bank account misses the results of all that grinding of teeth.

Most writers outline by listing their main ideas on sheets of paper (one sheet for a short work) and then numbering and renumbering as they experiment with various organizations. When they have settled upon a plan, they supply divisions of main ideas and then divisions of divisions; later they select a plan for each set of subdivisions. Some writers find an even more flexible method helpful. They jot each main idea on a card and then lay out the cards and move them about to try various plans. Laying out this game of solitaire is most helpful to those who have difficulty finding and accepting any order but that in which their ideas first come to them. Sometimes accidental placing of the cards suggests relationships that writers had not noticed.

In whatever form it is written, an outline is a trial, a pattern to try on and refit and redesign as necessary. It is skeleton with movable parts. If writers have placed a finger where a toe should be, they can change this easily; if a bone is missing, they can supply one; if the skeleton has two heads, they can discard one. But once the bones are fleshed, it is difficult to detect skeletal errors, let alone to correct them. Few writers have x-ray eyes and osteopathic techniques. But all writers can improve their outlines.

## GENERAL ADVANTAGES OF OUTLINING

### Unity

Working on an outline improves the thinking of writers. Outlining forces them to define their thesis clearly, and once a thesis has been clearly stated, departures from it become obvious. Writers who cannot see that a section of their finished report destroys the unity would have spotted such a digression in an outline and could have removed it easily.

### Coherence

Errors in coherence are also more obvious in an outline than in a completed report. Writers who do not plan before writing may struggle long and hard to connect sections of their writing because they do not know what is wrong until someone points out that they are trying to connect two parts that do not belong together.

But an outline turns a searchlight on errors. A safety chairman, for example, placed the four following recommendations in the order in which they had occurred to him while he was dictating a memorandum:

1. That black lines be painted at eye level on all glass doors
2. That supervisors enforce regulations about safety goggles
3. That temporary black tape be placed on glass doors until the black lines can be painted
4. That caution signs be placed on the temporary scaffolding at the entrance to Building A

An accident to a chemist who had tried to walk through a glass door and had nearly succeeded prompted point 1. That accident reminded the writer of an accident to a laboratory assistant who had neglected to wear safety goggles. Then he thought of the time that it would take for the black lines to be painted, and he decided to order temporary tape. This reminded him of the temporary scaffolding and the need for warning signs.

When he wrote in that order, he could not connect the topics. But as soon as he had jotted an outline, the safety chairman telephoned the maintenance department and ordered 3 and 4. Then he sent that department a memorandum of confirmation and an order for 1. He also saw immediately that regulations about safety goggles belonged in a separate memorandum to the supervisors of a few laboratories. Thus he wrote less, secured prompter service, and saved the reading time of supervisors. Yet when writers are busiest, they tend to ignore planning and to rush immediately to a final draft of what came to mind. Because writing by free association wastes time, planning is most necessary when a writer is most hurried.

## Emphasis

An outline is thus useful for improving unity and coherence, and it is also useful for more advanced work—the achievement of good emphasis. An outline gives a general view of the whole paper and a specific view of the parts that writers need when they consider emphasis. Outlining shows which subjects are in the positions of importance, which subjects will receive stress because their presentation will be lengthy, which organization will emphasize important ideas, and which parts of the paper need attention because the chosen plan stresses them too much or too little.

Few writers can see all this without outlining. And even fewer have the time and energy and patience to move parts of a completed paper, compensate for lengthy development of minor ideas, and rephrase sections to improve emphasis. Such strenuous revision is not always possible without a complete rewriting. Moreover, many writers of functional prose who rewrite to strengthen unity and coherence are reluctant to rewrite to improve emphasis. Therefore the only way for them to achieve better emphasis is to outline their papers before writing them. Then they can experiment with arrangements of topics to find the best one for their material, their objectives, and their readers. Also they can experiment imaginatively with that plan, adapt it as they desire, and thus attain effective emphasis.

## SPECIAL ADVANTAGES OF OUTLINING

### For Busy Writers

Outlining offers special advantages to those who are interrupted frequently—sometimes for long periods. After a long interruption writers might spend an hour or more reading preceding pages and finding their way back into their work. Consulting an outline enables them to do this in much less time; sometimes the bird's-eye view offered by an outline helps them to orient themselves in a few minutes. For those who can work only at intervals, an outline is essential. They may write an outline piecemeal by filling in missing sections as they think of them or as they complete their research. The outline will help to keep them informed of progress in their research, for omissions will be obvious. And when, as they are writing their paper, annoying interruptions drive ideas from writers' minds, their outline often helps them to recall the thoughts or at least to find their place quickly and begin rephrasing.

The outline also reinforces memory at other times. It reminds planners what the next topic is, whether they have already covered a point, and how important one topic is in relation to other topics and to the whole work. And outlining helps to avoid repetition, digression, omission, and poor emphasis. A completed outline is a boon to writers when they have been doing other work and are suddenly asked a question about the subject of a report. Many writers have told us that when their minds went blank and they did not have time to search their report for an answer, their outline supplied it quickly. Those who can refer to an outline do not seem confused or ill-informed about their work. And the usefulness of the phrasing in an outline has won the blessing of many trying to write on one of their bad days. When words just will not come, it is a boon to find phrases and sentences ready in an outline.

### For Co-Writers

Another advantage of the outline is its usefulness when several people are writing one report. The writers can see the positions assigned to their work; can establish relationships between the sections, especially those immediately preceding and following; and can think more easily of parts in relation to the whole. When one person writes a report and others approve it, the altering of an outline until it is approved prevents much tiresome rewriting. Thus outlining helps to eliminate some of the irritations and dissatisfactions of the writing by groups or committees that is necessary in government and industry.

### For Progress Reports

Outlining is also useful to writers of progress reports, for it keeps them aware of goals, warns them against attractive but insignificant digressions, and keeps them alert to the relative importance of ideas. Reading outlines of preceding reports and then outlining the current one can help writers to avoid losing sight of the main objectives of a long study and to refrain from overemphasizing minor details of recent research.

### For Abstracts, Summaries, Tables of Contents

An outline is most useful in the preparation of an abstract or summary, especially if one is required before a paper is completed. Writers of abstracts or summaries who have no written plan of the whole work may emphasize minor points at the expense of major, include material that will not appear in the final paper, or even omit main points. But writers with an outline have a much better general view of their subject and probably have ready in their outline some of the wording for the abstract or summary.

An outline may also provide the wording for the descriptive table of contents and for the subject headings of a paper, for tables of contents and subject headings are often little more than expanded or contracted outlines.

### For Short Works and Dictation

Letters, memorandums, and two- or three-page reports benefit from outlining. Many people, especially those who dictate easily and fluently, fail to plan these works. Some who always outline lengthy works rush without a plan to dictate short ones. As they dictate, they are expressing one idea while they are trying to remember the three ideas that should come next. They fill in with unnecessary words while they grope for the next thought, express themselves with halting ineptness while they try to remember what they just finished saying, and in their frustration even become angry with their secretary or their reader because the dictation is not progressing well.

Of course it is not. When they combine phrasing the dictation of the moment with developing or recalling the next thought, their minds are likely to rebel and perform only one of the tasks, and that one indifferently. Those who dictate without a plan usually break most of the rules of coherence and emphasis and tend to be turgid and verbose. Good and even adequate impromptu speakers are tempted to dictate impromptu. But what is acceptable to an audience, which has the advantage of a speaker's personality, gesture, voice, and phrasing as well as the atmosphere of the occasion, is a far different matter on paper. Sensible writers realize later that the works they dictated without a plan are their poorest. Unfortunately they sometimes mail or distribute them immediately and regret their costly haste later.

Those who doubt that this applies to them—and some successful impromptu speakers stubbornly resist the truth—should try dictating from informal outlines for a week and should compare the results with the results of impromptu dictation. Reading one's own hasty dictation is a useful revelation. It makes strong men and women shudder, especially if they have complained about the letters and memorandums of other writers, and leads even stubborn doubters to the realization that they dictate best from an outline.

Writers in our courses who dictated or wrote hastily and carelessly and scorned all kinds of planning were amazed at the improvement that resulted from a few moments spent outlining. Some writers had been using one letter plan for ten years or more, regardless of the contents and of the readers. They found much to improve when they spent a few moments thinking; they often developed better organizations for some types of letters and some kinds of readers. Thus many letters benefited from a few minutes spent planning one letter. These writers were surprised too at how their outlining suggested a new idea, located an annoying digression, and pinpointed unconscious, unnecessary repetition in even the shortest work. No wonder that the secretaries commented that the bosses had become better and happier writers. Every secretary knows that dictation that is not outlined is marked by omissions, awkward insertions, time-consuming backtracking, and separation of related ideas. Much incoherence in brief works is due not to inability or failure to recognize related ideas but to lack of planning.

One supervisor, planning a review to give to an assistant, wrote four points:

1. Lack of aggressiveness
   Two examples when leadership might have prevented errors
2. Need for courses in new methods
   Two suggestions
3. Good team work
   Getting along well
   Doing his share of routine work
4. A new assignment—his teaching new workers

The supervisor had placed last her main point—her wish to have her assistant spend most of the next four months teaching other workers.

As she looked over the points, she decided that unfavorable point 1 would make a poor beginning. If she started with 4 and related her first and second points to it as necessary improvements, then 3 would follow as a concluding reason for expecting success in the teaching. She also considered starting with good teamwork as one reason for selecting her assistant to teach new workers. Then 1 and 2 would become improvements for the assistant to make in preparation for the teaching. As the supervisor considered other organizations, the main point of her writing dominated, and her other thoughts related themselves to that main point. Her final memorandum, which had more focus than if she had discussed points 1, 2, 3, 4 in that order, influenced the assistant as the writer wished.

Outlining a short paper also helps in other ways. Some executives learn to study their jottings, divide them into two memorandums, and then direct them to different readers. Their readers respond better when they do not have to wade through what does not concern them to find what does. One executive examined an outline of a letter, telephoned to discuss three possibilities, and wrote a confirmation of the action selected. That saved the energy and time involved in a long letter and a long reply. The most common improvement is the removal of unnecessary material. Apparently writers tend to expand a short work when they first think of the contents, and if they do not examine and plan, they dictate much more material than a reader needs. Outlining helps to keep short works to proper length.

## KINDS OF OUTLINES

The outline form that writers should use to achieve these good results is determined by their own preference. Whatever form helps them the most is the best form for them. Many writers avoid formal outlines except when they write outline reports. Having to struggle with carefully numbered parallel topics distracts them from their ideas and irritates them. They evolve satisfactory systems of their own. Many of them indent to mark the relative importance of topics and do not number topics until they have reached a final decision about order. Others use cards for the outline of a short report, but for a long report they number the cards and have their secretaries type an outline from them. Some outliners use colored pencils, strips of paper, or blackboards. Others find word processors useful for outlines because the vexing problem of coherence does not arise in the outline. Whatever suits writers is best for the outlines they prepare for their own use.

But for outline reports submitted to readers, standard forms are necessary because readers can understand them more easily than they can understand a writer's idiosyncratic outlining. The principal types are the sentence outline, the topic outline, and the paragraph outline.

The example of a sentence outline (1, below) illustrates closed (or close) punctuation, which may be used in any of the outline forms; the example of a topic outline (2, below) illustrates open punctuation, which is also suitable for any outline form; and the example of an outline that combines forms (4, below) uses periods after sentences but no punctuation after topics that are not sentences. Writing for a conservative reader, most experienced authors use closed punctuation; for a modern reader, open or a combination (4).

## 1.  Sentence Outline

  I. In a sentence outline, each division is one complete sentence.

 II. The use of complete sentences has advantages.
- A. An outliner is forced to formulate and analyze thoughts thoroughly.
- B. Information is conveyed fully and clearly in a form familiar to a reader.
  - 1. A reader does not have to expand incomplete sentences to understand the meaning.
  - 2. A reader is not distracted by the choppiness of the topic outline.
- C. The thought relationships are expressed more exactly than in a paragraph outline.
- D. Should the outliner wish to expand the outline into a paper, many sentences are ready for use.

III. The use of complete sentences has disadvantages.
- A. Repetition of sentence patterns is likely, as in topics III.B.2.a and b of this outline.
- B. When subdividing is thorough, the use of sentences for minor subdivisions may be awkward, inappropriate, and cumbersome.
  - 1. Subdivisions in many outline reports, particularly scientific and technical reports, may be numerous.
  - 2. Many subdivisions lend themselves better to expression as phrases or single words.
    - a. This is true of lists.
    - b. This is true of details.

    (1) Subdivisions after the fourth are frequently unsuited to expression in sentences.

        (a) Sections a and b of this outline would be clear as single words.

        (b) Scientific and technological information like a reading of temperature or pressure belongs in phrases.

    (2) Expressing details in sentences gives them too much emphasis.

        (a) This is illustrated in topics (a) and (b) of the preceding section.

        (b) Indented material easily extends to so many lines that the bulkiness emphasizes minor information.

        (c) A sentence is the most important grammatical form in an outline.

## 2. Topic Outline

  I. Divisions in topic outlines being clauses or phrases
 II. Advantages of topic outline
    A. Clarity of thought relationships
    B. Minor grammatical forms for minor topics
III. Disadvantages of topic outline
    A. Ambiguity more possible than in sentence outline
    B. Writer not forced to formulate ideas fully
    C. Form not familiar to all readers
    D. Need to achieve parallelism
        1. Writing parallel constructions (Chapter 11) more difficult than writing sentences
        2. Under one topic, placing divisions of equal rank in parallel form
            a. Topics A, B, C, D, etc., under III
            b. Topics 1 and 2 under D

## 3. Paragraph Outline

    I. In a paragraph outline each division is a summary of a paragraph.
   II. The paragraph outline, which may be difficult to prepare for long papers, is particularly useful for short papers.
 III. It exhibits paragraph structure but not necessarily thought relationships.
 IV. It provides sentences that are useful when a writer expands an outline into a paper.
   V. It is not useful as a report form.
 VI. The thought is often less complete in a paragraph outline than in a sentence or topic outline.

### 4. An Outline That Combines Forms

Many writers combine the sentence and the topic outline to obtain some of the advantages of each. They express major topics in sentences and minor topics in clauses, phrases, or single words. This requires attention to parallel construction if an outline is written for anyone besides the writer. The following partial outline combines forms:

I. Some writers use sentences, phrases, clauses, and single words in an outline.
   A. Sentences for main topics following
      1. Roman numerals
      2. Roman numerals and capital letters
      3. Roman numerals, capital letters, and Arabic numerals
   B. Phrases, clauses, and single words for minor topics
      1. Under such topics equal divisions to be parallel in expression
         a. Nouns and their modifiers, as in topics 1, 2, and 3 under A in this outline
         b. Single words, as in topics (1), (2), (3), and (4)
            (1) Nouns
            (2) Verbs
            (3) Adjectives
            (4) Adverbs
II. Combining grammatical forms offers some of the advantages of the sentence outline and of the topic outline.

## USES OF OUTLINE FORMS

Outline forms are usually employed as follows: sentence outlines or sentence outlines with minor subdivisions in phrases or single words—for outlines to be submitted to readers instead of complete reports, for outlines to be used only by the writer, and for outlines to be used by several writers in planning a paper; topic outlines—for a writer's own use and for the use of several writers in planning a paper; paragraph outlines or paragraph outlines with some subdivisions in clauses, phrases, or single words—for a writer's own use in planning short papers.

The paragraph outline is a very useful but neglected form. Many writers would benefit from outlining by paragraphs when they plan a letter:

1. Thank him for hospitality at convention
   Dinner and theater
   Flowers for Helen
2. Request samples of polymers discussed with him at convention
3. Ask him to serve on Hospitality Committee next year
   Four national conventions
   Three regional meetings
4. Send copy of article on heat transfer he requested

70

Looking this over, an outliner may decide to interchange 2 and 3. The two items on hospitality may be easily connected, and the two professional courtesies (of sending an article and requesting samples) belong together. According to the amount of development given the ideas, the division of the material into paragraphs may be one paragraph for convention hospitality and one for the professional courtesies, either paragraph standing first; or the division may be four paragraphs in one of the following orders:

I

Thank him for convention hospitality
Invite him to serve on hospitality committee
Mention that requested article is in mail
Request samples

II

Tell him I'm sending article
Request samples
Invite him to serve on hospitality committee
Thank him for hospitality

III

Thank him for hospitality
Tell him I'm sending article
Request samples
Invite him to serve on hospitality committee

Order I provides a pleasant beginning suited to a reader who is pleased by social grace, as the correspondent seems to be; it separates the thoughts on hospitality and on technical matters. Order II also separates these but starts with the technical article, a good beginning if the request was urgent. This plan ends with gracious thanks. Order III groups at the end the two requests, a necessary precaution if the reader is absentminded enough to overlook one request if the two are separated. Thus consideration of the reader determines the choice of plan.

Of major importance for busy executives is the simplicity of such outlining, the ease with which one can change the order by renumbering or inserting arrows, and the help to an executive who may be interrupted while dictating or writing. The paragraph outline of a letter, memorandum, or short report is almost too obviously useful. This outline resembles the papers that one searches the office for while they are lying on top of the pile in the middle of one's desk. It is easy to overlook.

However, once writers use it, they are less likely to overlook it again. In fact, one executive became so enthusiastic about the value of the paragraph outline in improving his writing and so eager to recommend it repeatedly that his harassed colleagues wished he had not been enlightened.

# CHAPTER 5

# Dozens of Beginnings

*Well begun is half done.*
>                              *Horace*

First words are often unfortunately last words. Those who receive reports and articles may never go beyond beginnings because they are not captives. Like hurried newspaper readers who do not turn to the page where a story is continued, readers in industry and government often turn to another report or article. Americans are used to thumbing the pages of newspapers, magazines, and books to decide on the basis of beginnings what they wish to read. And many influential readers in administration must estimate from first words whether to read further or to discard; otherwise they cannot keep their heads above the flood of paper. Not surprisingly, then, they judge writers by their opening sentences. Although good writing and poor writing in the middle of a paper may escape notice, opening sentences, spotlighted as they are, may influence readers to continue or to stop reading and may be the basis of their judgment of a writer's ability.

Writers who wish to improve the beginnings of their papers should study the following kinds and choose the best for their subjects, their readers, and their writing skills. Imaginative, experienced writers may use any of these beginnings in their papers. Less experienced writers should consider the advantages and disadvantages of each kind before experimenting. Writers can also benefit from examining critically the opening sentences in their reading to see how and when various beginnings are used. Studying the best writing in their company will acquaint them with opening sentences that are appropriate and acceptable there. But they should not confine their study to the writing of their colleagues. Books, magazines, and newspapers furnish varied examples of good openings.

As evidence of how many types of beginnings readers see daily, the first edition of *Effective Writing* cited examples from just one Sunday edition of *The New York Times*. More than twenty-five types had appeared, and more were easy to find in the magazines and journals to which the readers of *Effective Writing* were likely to be exposed. This second edition illustrates from a wider variety of publications. Teachers have told us that they and their students found examples of beginnings most helpful, and writers teaching themselves always request many diverse examples. The number of examples of beginnings frequently used has therefore been increased in the hope that a variety of beginnings and examples of the most common types of openings will help writers to choose well for each paper and to improve the beginnings not only of papers but also of all sections of papers, including individual paragraphs.

Studying and imitating beginnings can be interesting and productive. Study and practice prevent the stultifying waste of time of the writer who laments, "I just don't know how to start. I've been trying for three days now, but I can't think of a good beginning." Once writers become acquainted with ways to introduce the subject, they can spend their time choosing wisely from appropriate beginnings. One executive who writes many kinds of memorandums, reports, and letters keeps a list of beginnings handy on a card on top of her desk to consult whenever she hesitates over an introduction. Another executive who writes most of his business day has memorized the standard beginnings but keeps a file of new kinds and of original or unusual adaptations of familiar beginnings that he finds in his reading, and he soon uses them in his writing. Just being alert to good introductions has helped many beginners to start papers effectively. Writers who fashion a successful opening sentence are likely to find it easy to continue. Like Gibbon writing *The Decline and Fall of the Roman Empire*, they discover that once they have the first sentence, the rest follows.

## BACKGROUND

Background, or "history of the art," used to be the traditional opening section of articles in scholarly journals of science and technology and therefore became a common beginning in academic, industrial, and business reports on science and technology.

Writers began with the first mention of their subject in a journal and then discussed chronologically other articles, books, and convention papers on that subject. In historical surveys this beginning served well enough, but as an introduction to new research it often created unfortunate effects. Some writers gave the impression that, like sophomoric researchers, they were reluctant to discard anything they found in their own preliminary research and were using the section on history of the art as a catchall. Other writers seemed timid about beginning and, like reluctant bathers, immersed first one toe and then another until they had exhausted all their trivial findings. A background section seemed to many readers to be material to skip, particularly when it was offered to hurried readers in business and industry who wanted only the minimum necessary to understand the text and wanted that conveniently placed with the material it explained.

Today historical or explanatory introductory background appears mainly in technical papers addressed to readers with expertise in the subject of the paper and is much briefer than the earlier pages that surveyed research on the subject chronologically. For example, "Superheavy Elements: An Early Solar System Upper Limit for Elements 107 to 110" opens with a single paragraph of background:

> Superheavy elements (SHE's) are elements with atomic numbers greater than 106 which may have been produced in stars by the same processes that produced thorium, uranium, and plutonium. Myers and Swiatecki suggested that a region of relative stability, which would permit some of these elements to have appreciable half-lives, existed near atomic number $Z = 114$ and neutron number $N = 184$. More recent calculations, in which alpha and beta decay were considered in addition to spontaneous fission, suggest that the most stable nuclide is $^{294}110$, with a half-life ($t_{1/2}$) of $10^5$ to $10^9$ years. Numerous attempts have been made to synthesize SHE's in the laboratory and to search for them in nature. The possibility of an SHE with $t_{1/2} \approx 10^9$ years has led many investigators to search for extant SHE's; claims have been made for their discovery in halos in mica, in meteorites, and in hot springs, but these claims have been criticized. Other investigations are discussed in several reviews; there is currently no strong evidence for extant SHE's. Anders and co-workers claimed evidence for an extinct SHE with $Z = 115$ (or 114 or 113) in the Allende meteorite, based on xenon isotopic abundances that appear to be due to fission of an unknown transuranic nuclide. This result is difficult to understand in terms of calculated half-lives for these nuclides, all of which are less than 1 year, based on the more recent calculations. However, the calculated half-lives are uncertain by many factors of 10.[1]

This paragraph is preceded, as is customary in many journals, by an abstract that tells readers the gist of the article:

> *Abstract.* The abundance of samarium-152 in the Santa Clara iron meteorite is found to be $108 \times 10^7$ atoms per gram. This quantity, if attributed to fission of a superheavy element with atomic number 107 to 109, limits the amount of superheavy elements in the early solar system to $1.7 \times 10^{-5}$ times the abundance of uranium-238. For element 110, the limit is $3.4 \times 10^{-5}$.[1]

In the same issue of *Science*, Garry A. Rechnitz opens with the publications appearing in the field at the time of his article, then briefly mentions the new research directions stimulated and their relation to the scope of his article:

> One of the most rapidly expanding research areas relating to analytical measurements is the development of potentiometric membrane electrodes

with selectivity for ions, dissolved gases, and biological materials. Activity in this field is so intense that new publications are appearing at a rate approaching 500 per year. Entirely aside from the appealing practical possibilities for such potentiometric membrane electrodes, it appears that new research directions have been directly stimulated by the timely infusion of concepts from physics and biology. Some of the consequences of the latter are examined in this article.[2]

The abstract preceding a paper on stem vasculature in the *American Journal of Botany* summarizes the article, and the opening paragraph of background gives past and future information about a series of papers. The background is, then, the relationship of this article to the other papers of the writers on the subject:

> This is the sixth paper in a series devoted to an exploration of stem vascular organization in the Araceae. Initially we have presented preliminary observations outlining the approach and demonstrating the technical methods used and the kind of results obtained (French and Tomlinson, 1980). We then proceeded to describe vascular patterns in the subfamilies Pothoideae (French and Tomlinson, 1981a). Monsteroideae (French and Tomlinson. 1981b), Calloideae and Lasioideae (French and Tomlinson, 1981c) and Philodendroideae (French and Tomlinson, 1981d). In this paper we complete the description of the three remaining subfamilies. A final paper will summarize our results and conclusions.[3]

Authors of textbooks are partial to background as a beginning. Some textbook treatments in the past were long, discursive, and unnecessarily detailed. Today the best textbook writers limit their openings to strictly relevant material. Stephen J. Pyne begins the preface of *Introduction to Woodland Fire: Fire Management in the United States* with the most strictly relevant material briefly summarized:

> Within our solar system the Earth, and perhaps the Earth alone, is a fire planet. Only on Earth are combined the essential components of combustion. With lightning, it has a source of ignition; with atmospheric oxygen, an oxidizing agent; and with organic matter, a fuel. Jupiter has lightning, Mars has traces of free oxygen, and the moons of the outer planets have atmospheres rich in flammable hydrocarbons. But only the Earth contains all the essential constituents, the processes needed to mix them, and a suitable environment for their interaction. To complement its ignition source, the Earth also has a suppressant: water. The Earth can start fire, sustain fire, and suppress fire. The things that make Earth unique among the planets have made it hospitable to fire. And fire, in return, has had much to do with shaping the history of the planet.[4]

To establish background for an article on an artist's plans for a crater in the Arizona desert, Fred Hapgood briefly describes the geological formation of the crater:

Millions of years ago a vent tore open fifty miles beneath what is now central Arizona. Through this lesion the pressures of the deep interior took slow aim at the roof of that world, the floor of our own, and commenced a bombardment of molten rock that is continuing still. Over time this barrage, the remains of which are known as the San Francisco Volcanic Field, walked north and east, just nicking the edge of the Painted Desert, near what are now Colorado, Utah, and New Mexico. Right at the point where field and sands touch, a small, exceptionally well formed cinder cone can be seen, standing off from the other volcanoes, deeper into the desert. From a certain angle the cone has the shape of a bow wave, as though something alive were running under the sand, out toward the midst of the coral and orange landscape. This is Roden Crater.[5]

Historical background provides a succinct introduction to a section entitled "Stellar Collapse and Supernova Explosions":

No treatment of compact objects would be complete without some discussion of supernovae. As we have already described in Sections 9.1 and 10.1, it came as no surprise to some when the Crab and Vela pulsars were discovered in supernova remnants. Ever since Baade and Zwicky (1934) had shown that the gravitational energy released by the collapse of an evolved stellar core to a neutron star would be more than adequate to power observed supernova outbursts, many researchers regarded collapse and supernova explosions to be events intimately associated with neutron star (and, possibly, black hole) formation.[6]

"Cash Flow—It's Not the Bottom Line" begins with a paragraph stating the background of a contention:

A growing number of securities analysts, financial writers, and accounting policymakers contend that financial statements providing information of a company's cash flows yield a better measure of operating performance than do the company's income statement and balance sheet. According to recent surveys, corporate and government officials have accepted this view; they rated cash flow data the most important piece of information contained in published financial statements. The trend toward wider acceptance of this yardstick has been building since the early 1970s. Accelerating the trend have been several developments—including new financial reporting rules on such issues as foreign currency translation, equity earnings, interperiod income tax allocation, and lease and interest cost capitalization—that put greater distance between a company's net income and its cash flow; the adoption of "liberal" accounting practices by some companies; and record inflation levels.[7]

In "The Adrenal Chromaffin Cell" Stephen W. Carmichael and Hans Winkler also present as introduction the background necessary for understanding their work:

77

Under conditions of fear or stress a surge of the hormone adrenaline mobilizes the body for peak physical response. Flooding the bloodstream at up to 300 times the normal concentration, the adrenaline interacts with receptors on cells in various organs, increasing the heart rate and blood pressure and prompting the release from the liver of extra sugar to fuel muscular work. Taken together these reactions constitute a "fight or flight" response that prepares one to combat an enemy or flee from danger. They are the end result of a secretory event in the adrenal medulla: the inner part of the two adrenal glands just above the kidneys. There specialized cells known as chromaffin cells manufacture, store and secrete a complex mixture of hormones, the most important of which is adrenaline.

Chromaffin cells are of interest not only as the root of the fight-or-flight reaction but also because they offer insights into the workings of other secretory cells, notably neurons, or nerve cells.[8]

C. Arden Miller writes of the past in shocking contrast to the present and future as background that prepares the readers for surprising new facts and controversial ideas about infant mortality in the United States. The introduction is background history, it is true, but this particular background has the vigor of argument.

In 1979 the U.S. Public Health Service established nine infant deaths per 1,000 live births as a goal the nation should be able to meet by 1990. Early last year a Government official testified in Congress that the goal could be met if previous favorable trends continued. In December of that year, reviewing provisional data for 1983 and the first nine months of 1984, the same official acknowledged that the trends were less favorable and that the goal would not be reached by 1990.

The infant mortality rate had been declining steadily for many years, dropping from 47 in 1940 to 13.1 in 1979. The decline continued at an average of 4.6 percent per year until 1983, reaching a rate of 10.9 infant deaths per 1,000 live births. Then, according to the provisional data, the speed of decline diminished markedly, to 2.7 percent (10.6 deaths) in 1984. Although the Public Health Service asserted in May that the goal of an infant mortality rate of nine will still be reached by 1990, the likelihood that it will not be is strong.

What happened? As is usual in complex social issues, a number of possible contributing factors can be cited, and it would be difficult to prove that any one of them by itself tipped the scales. Still, one definite change that took place not long before the rate of decline flattened out was the reduction by the Reagan Administration in the funding of several programs for children, mothers of young children and pregnant women. I and many other observers think these cutbacks have contributed significantly to the change of trend in the infant mortality rate by weakening national policies (which had never been more than marginal) for the care and protection of pregnant women. Senior officials of the Department of Health and Human Services deny the connection. They point instead to such factors as the high rate of teen-age pregnancy, the use of tobacco, alcohol and drugs by many pregnant women and the complex racial mixture of the U.S. population;

they also cite the possibility that high-technology medicine is merely post-poning the death of some infants who earlier would have appeared in the statistics relating to naturally aborted pregnancies. The Administration has declined a proposal to study the effect of the cutbacks.[9]

Journalists are masters at compressing necessary background infor-mation. They provide good examples for writers looking for short begin-nings for memorandums or brief reports:

> After 50 years of allowing banks to operate in relative secrecy, federal regulators now think some more sunshine may be a good thing.[10]

Reporting on a conference, Peter J. Mouginis-Mark begins with the details and main points of background necessary to prepare readers for the summaries of papers that follow:

> The two Venus sessions at the 15th Lunar & Planetary Science Confer-ence last March in Houston demonstrated the maturation of geological stud-ies of Earth's sister planet and a great increase in the amount of high-resolution imaging data available for analysis. Planetary geologists have long had a strong interest in Venus because its size, density and distance from the Sun make it the planet most like Earth. While detailed exami-nation of Pioneer Venus spacecraft data continues to yield insights into Venusian global and regional properties, new radar images of the planet have introduced a new period of studies.[11]

Walter Pincus, *Washington Post* staff writer, begins a presentation of new information about the "40 Years of the Bomb" with brief background information:

> Since the first and only times it was used 40 years ago, the atomic bomb has been a source of controversy, fear and myth—and a great deal of mis-information. Much of the history of "the bomb" has been obscured or mistold.
>
> Debate continues, for example, over whether the United States should have invented the bomb or should have dropped it on Japan. Yet, as in much of today's discussion over strategic weapons, all the facts surrounding the development and dropping of those first nuclear weapons have not been understood.[12]

In many of these beginnings the treatment of background helps readers to follow and understand a text and to feel comfortable reading in a new field or in one that is only somewhat familiar. Some writers want back-ground beginnings to do more than this: they want opening words that capture the attention of readers, hold readers by the interest inherent in the subject, and convey immediately the first step in their thinking on the subject. In attempts to reach these goals such writers turned to other beginnings, many of them illustrated in the following sections: summary of main points, scope of paper, questions, wit and humor, specific details.

Before very long, however, background as a beginning demonstrated strong staying power. Openings composed of background and one or even two other topics soon became popular.

Background combined with contrast of the past and present is used by many writers. "Ventricular Tachycardia," for example, moves from background to a summary of the present:

> Ventricular tachycardia (VT) has been the subject of intense investigation during recent years, in both the clinical laboratory and the basic research laboratory. The prevalence of these arrhythmias in our population (especially in the setting of ischemic heart disease) and their often dramatic clinical consequences have provided a strong impetus toward understanding the mechanisms of VT and developing more effective methods of therapy. Despite the intense research effort directed at furthering our understanding, our knowledge of the mechanisms causing the arrhythmia is still incomplete, methods of treatment are imperfect, and even the nomenclature by which we categorize VT is often inconsistent and confusing.[13]

Jeremy Shapiro combines background with cause and effect to introduce Paul S. Bender's *Resource Management*:

> Hardly anyone would disagree that business has entered a period characterized by radical changes and great uncertainties. Inflation, scarce resources, rapidly fluctuating markets, and worldwide competition are just a few of the symptoms, or causes, of the unstable business environment. At the same time, there is an air of bewilderment in many companies, large and small, about new informational and communications technologies— what they are, how they can be used to reduce planning uncertainties (and increase profits), and where they are headed.[14]

In *Science*, background combined with definition leads to the scope of the article:

> Solar chemistry may be defined as that area of photochemistry in which the excitation energies fall within the spectrum of solar radiation at the surface of the earth. The potential importance of solar chemistry in the generation of energy-rich molecules from cheap, energy-poor raw materials has been discussed by many authors, and the role transition metal complexes could play in such energy storage schemes has received special attention in recent years.[15]

Similarly, in the *New England Journal of Medicine* the background required for understanding "Hepatitis B Vaccine in Patients Receiving Hemodialysis" precedes a statement of the scope of the report:

> Hepatitis B viral infection has been a major complication in patients receiving long-term hemodialysis, since this form of therapy was introduced in the 1960s. Unlike healthy adults, patients undergoing dialysis who are

80

infected with hepatitis B virus usually have mild, asymptomatic infections. However, a high proportion become chronic carriers of the virus, and most remain highly infectious, as indicated by a high prevalence of the hepatitis B e antigen. Control of infection in such patients is an ongoing problem, and the new hepatitis B vaccines, which have proved to be highly immunogenic and efficacious in healthy adults, may provide a solution.[16]

"Antiemetic Efficacy of Dexamethasone," also in the *New England Journal of Medicine*, summarizes the present state of some side effects of chemotherapy and follows with a description of the scope of the study:

> Nausea and vomiting are frequent and serious complications of cancer chemotherapy. The management of such side effects continues to be a major problem in cytotoxic drug therapy. The commonly used "standard" antiemetics, such as prochlorperazine (Compazine), have been demonstrated to be only moderately effective in controlling these symptoms.
>
> Because of the limited effectiveness of the standard antiemetics in controlling nausea and vomiting induced by cancer chemotherapy, newer agents, such as cannabinoid drugs, metoclopramide, and more recently, glucocorticoids, have been employed with some success. Although the mechanism of steroid activity as an antiemetic is unknown, it has been suggested, though not demonstrated experimentally, that inhibition of prostaglandin synthesis by corticosteroids may play an important part. Both methylprednisolone and dexamethasone, used either alone or in combination with other antiemetics, have shown promise in several uncontrolled trials and, as a result, are now widely used in clinical practice.[17]

In the *Quarterly Review of Economics and Business* Walter Adams and James W. Brock give detailed background and then move easily to the use of this history in the present:

> Almost from the beginning, the antimonopoly strictures of the Sherman Act have been bedeviled by a persistent philosophical paradox. The policy objective of the law was clear: to promote competition by prohibiting conduct which eventuated in monopoly. But how was the law to encourage competition if at the same time it punished the competitor whose success was crowned with monopoly? What distinction, if any, was to be drawn between monopoly resulting from superior performance ("good trusts") and monopoly based on predatory conduct ("bad trusts")?
>
> In a landmark decision, *U.S. vs. Alcoa*, Judge Learned Hand articulated the dilemma as follows. . . .
>
> The apostles of the New Economic Darwinism have seized upon this dilemma as a basis for their attack on the antimonopoly statutes. In their view, monopoly is simply the result of successful competition—proof of the monopolist's superior performance. Sound public policy, they argue, should not penalize but encourage such conduct.[18]

Stuart Diamond, writing in *The New York Times*, gives the background of his article in two examples and a short summary of recent plastics development:

> An ice skating rink in Port Washington, L.I., has no ice. There is, instead, a quarter acre of white plastic, which provides slightly more friction. "It's great for training and it's cheaper—no refrigeration," said Bryan Trottier, the hockey star of the New York Islanders and owner of the plastic skating rink.
>
> A footbridge in Charlottesville, Va., has no steel. The structural supports are fiberglass-reinforced plastic. So are the beams of various small buildings around the country. The lighter plastic is easier to install, does not corrode and is as strong as steel. "It is high strength and the beam size is increasing," said A. Keith Liskey, vice president of the Morrison Molded Fiber Glass Company of Bristol, Va.
>
> There have been some dramatic leaps in the development of plastics in the past few years. Technological innovations have made the material much stronger, more durable, heat-resistant and able to mimic the properties of ice, steel, copper, aluminum, wood and bones. The benefit is sometimes cost, and often increased function. Between 1977 and 1983, industry sales more than doubled, to $15.6 billion a year.[19]

An article in *The Washington Post* illustrates the journalist's ability to compress. A summary of historical questions about the MX missile is followed by a one-sentence statement of the problem:

> The giant MX missile, which promises to be the most accurate and reliable land-based intercontinental ballistic missile in the nuclear arsenal of either superpower, has been dogged for a decade by questions about its military purpose and how to deploy it.
>
> The problem is that the MX is designed to threaten Soviet ICBMs without having the capability of carrying out the threat.[20]

*The New York Times* offers details of the height of the boating season as a contrast for a warning:

> The height of the boating season has arrived in the New York metropolitan area. On Saturdays and Sundays in July and August, local waterways are jammed with vessels carrying friends and food. But despite the good times that boating offers, a sober side cannot be ignored: Boating is dangerous if safety measures are not observed.[21]

Robert Giroux in the book review section of the same paper begins with mention of the 421st birthday of Shakespeare, skims over the historical background before 1857, and then contrasts the history of Delia Bacon's opinions to reach a climactic conclusion at the end of the second paragraph:

> It is interesting to recall, as those who celebrated Shakespeare's 421st

82

birthday last Tuesday probably did, that it was an American who originally launched the public attack against his authorship of the plays. Before her book appeared in 1857, no published writer of substance doubted Shakespeare had written Shakespeare's plays. Starting with Ben Jonson and down the years through Dryden, Pope, Samuel Johnson, Goethe, Lessing, Hazlitt, Coleridge and many others, no serious critic questioned that the man from Stratford was the greatest poet and playwright in the English language.

But in pre-Civil War New England, a frail and pale-faced woman in a black dress, taking to the lecture platform in Boston, Cambridge, New York and Brooklyn, changed all that. Her biographer, Vivian C. Hopkins, notes in "Prodigal Puritan" (Harvard University Press, 1959) that her eloquence and the intensity of her slender figure with her large eyes, dark hair and very white skin mesmerized audiences. She won the friendship, as well as the crucial support, of three important men of letters—Emerson, Carlyle and Hawthorne—though she did not necessarily convert them to her ideas. Her followers called her a "seer" and a "prophetess." She described her vision in these words: "God has at last given me the utterance that I have all my life lacked." She felt she had divine assurance that "this great secret, in which the welfare of mankind is concerned, will not perish with me." Her secret? Francis Bacon was the author of Shakespeare's plays. Her name was Delia Bacon.[22]

Although beginning with background may suggest the heavy state-of-the-art openings in journals of science and technology, openings like the one in the article on Delia Bacon are not unusual in journalism. Indeed, many journalists choose to begin with background history or information or both. Gregory Sandow in the records section of the *Saturday Review* uses both in a lively way:

> In the early days of records, a complete opera recording was news. "A complete *Traviata* on 38 discs!" the opera buff cried, and then hired a truck to get the records home. Later, in the first years of LPs, we began to get two or even three versions of all the standard works. We called them by their singers' names: the Callas *Norma*, the Tebaldi *Boheme*. Now, we are in the "Age of the Conductor." Nobody talks about composers much, or even singers. We don't get records of Mozart, Wagner, or Bizet. Instead, we get *Cosi fan tutte* (Angel DSCX-3940) with Ricardo Muti the big attraction, *Tristan und Isolde* (Philips 6769 091) with the spotlight on Leonard Bernstein, and what's emphatically billed as Herbert von Karajan's *Carmen* (Deutsche Grammophon 2741 025).[23]

L. Erik Calonius also contrasts past and present, ending with a quotation about the present:

> Time was when American ambassadors such as John Adams and James Monroe literally made U.S. policy. Special Envoy Monroe doubled the size of the U.S. with the Louisiana Purchase, and President Jefferson initially was shocked; he doubted the constitutionality of the transaction.

Things have changed. Electronic communications and air travel have drawn the world into a smaller sphere. And the American presidency has become a much more imperial institution. The role of ambassadors has shrunken.

"I was in the Oval Office many times when the president would reach for the telephone and phone up the prime minister of the United Kingdom or the chancellor of Germany," says former National Security Adviser Zbigniew Brzezinski, "and because it occurred so frequently, the chances were great that we probably wouldn't even bother to tell our ambassador that the conversation took place."[24]

Discussion of the background of a broad topic from the 1970s to the present leads to a statement of the limited part of the topic to be discussed in "Hazardous Substances in Western Developed Countries":

A characteristic of the 1970s in many Western developed industrialized countries, including the United States, was the expansion of government intervention and regulation in the areas of occupational and environmental health and of consumer protection. This expansion was an outcome of a demand by the majority of their populations for better protection of the health and safety of workers and consumers and of the environment. This demand has remained undiminished in the 1980s. Although some governments (including that of the United States) have recently tried to weaken their regulations, these attempts have not been popular and have not responded to a general mandate. On the contrary, poll after poll has shown that protection of workers, consumers, and the environment still has widespread support among Western populations, even when the questionable assumption is made that this protection slows down economic growth.

Industry has for the most part resisted these interventions and regulations. One form of resistance has been to move entire industrial processes from countries with regulations to countries with fewer or no regulations. The extent and nature of this movement are unknown. In spite of occasional news flashes about specific examples of exportation of hazardous substances, there does not seem to be much popular awareness of the issue, and there has been no systematic study of the public-health problem.[25]

The background that helps readers to understand written works is also often combined with other types of beginnings. F. Peter Woodford describes in delightful specifics the poor writing in "journals—even the journals with the highest standards—" and thus supplies an introduction to the question "Does it matter?" and to his answer:

In the linked worlds of experimental science, scientific editing, and science communication many scientists are considering just how serious an effect the bad writing in our journals will have on the future of science.

All are agreed that the articles in our journals—even the journals with the highest standards—are, by and large, poorly written. Some of the worst are produced by the kind of author who consciously pretends to a "scientific scholarly" style. He takes what should be lively, inspiring, and beautiful and, in an attempt to make it seem dignified, chokes it to death with stately abstract nouns; next, in the name of scientific impartiality, he fits it with a complete set of passive constructions to drain away any remaining life's blood or excitement; then he embalms the remains in molasses of poly-syllable, wraps the corpse in an impenetrable veil of vogue words, and buries the stiff old mummy with much pomp and circumstance in the most dis-tinguished journal that will take it. Considered either as a piece of scholarly work or as a vehicle of communication, the product is appalling. The ques-tion is, Does it matter?[26]

In *Scientific American* Carlo M. Croce and George Klein use the same method in their opening paragraph and then discuss different "mecha-nisms by which an oncogene may be activated," background information helpful to readers:

> Every human cell contains oncogenes, genes that have the potential to cause cancer. These genes apparently carry out normal functions until a malignant change takes place. What is it that changes the oncogene from a normal part of the cell's genetic machinery into a source of cancerous, or neoplastic, transformation?[27]

Comparatively few articles use combinations with background as the second or third topic, but in *Technology Review* an article begins with comparison followed by background:

> During World War I aviators took to the air to observe troop movements on the Western Front. Soon aerial bombardment began, and dogfights raged in the sky as each side strove for "command of the air." The arms race in space has begun to develop along similar lines. But this time the threat of escalation affects the world's entire population. Satellite dogfighting could lead to nuclear war.
> The United States and the Soviet Union first saw the potential of satellite reconnaissance in the fifties. Since then numerous satellites have been used to provide information on industrial activities, military maneuvers, and missile tests. Reconnaissance satellites monitor arms-control agreements and stand ready to furnish prompt warning of a nuclear or conventional attack. Satellites help ships and submarines navigate more accurately and could help guide missiles, increasing their precision. Worldwide commu-nications, military and civilian, depend critically on satellites.
> Inevitably, the Russians and the Americans have been tempted to seek ways to destroy each other's satellites.[28]

## CAUSES AND EFFECTS OR RESULTS

Causes and effects or results are useful beginnings for reports in busi-ness or industry because they satisfy a reader's curiosity as to why the company is involved in the research and what the research has produced.

Readers of journals of science and technology also favor these beginnings because the first words they read make them feel that they are getting to the heart of the matter. Such beginnings, too, quickly explain the writer's concern with the subject, as the next two examples show.

Jonathan P. Hicks begins a news story with a result and four causes:

> Corporate profits in the second quarter of the year appear to have been weak, a result of slow economic growth, the continued strength of the dollar, the widening trade deficit and a trend toward doing more business with foreign suppliers, according to economists.[29]

Barry Sussman opens a news story with three possible causes:

> While various members of Congress are promising major tax reform plans next year, many citizens are reforming their own taxes, and have been for years: They cheat, and they don't think anything is wrong with it.
>
> There are many explanations for income-tax cheating. One is that there is a bit of avarice in most of us. Another is the relative ease with which the revenue man can be dodged. Still another is the knowledge that wealthy people may, quite legally, avoid paying any income tax, making others feel they are suckers if they pay their full share.[30]

The author of "Computers: Our Aging Workhorses" hypothesizes a cause and its result:

> The Navy supply system's aging general-purpose computers are overburdened and increasingly less capable of coping with the burgeoning demands made on them. Often, the fleet tends to overlook the critical importance of its own supply system. But if that system should suddenly break down because its computers no longer can handle the work load, every line officer on board a ship will soon take note of the effects. The flow of ammunition, spare parts—any type of material support—will grind to a halt, and the Navy's operational capability will be paralyzed.[31]

"The Folly of Stock Market Timing" opens with a result as a preliminary to examining possible causes:

> Many readers of this magazine, either as senior executives or as members of governing boards or both, have been responsible at one time or another for overseeing on a part-time basis the management of money for the benefit of others—in pension funds, endowment funds, trust funds, and charitable foundations. While no definitive published survey on the long-term investment performance of such funds exists (in contrast to the performance of the money managers, on whom there is a plethora of data), my observation of a large number of trusteed funds over the years suggests that many and probably most have underperformed on a long-term basis the market as measured by, say, the *Standard & Poor's* "500."[32]

In the *Journal of the Optical Society of America*, "Surface Enhancement of Coherent Anti-Stokes Raman Scattering by Colloidal Spheres" starts with an effect and its cause, part of the background of the study:

> Inelastic light-scattering signals from a molecule located on either side of the surface of a small particle may be quite different from such signals when the molecule is in an extended medium. When embedded within the particle, the molecule is stimulated by the local fields at its particular location, and there is included in the emission at the shifted frequency induced fields necessary to satisfy the boundary conditions. These effects have been elucidated for spheres, concentric spheres, circular cylinders, and spheroids, and there have been experimental studies with polymer latex spheres impregnated with fluorescent dye.[33]

The beginning that presents causes and effects or results is likely to be one that also supplies a bit of background, as shown in *Technology Review*:

> By World Bank estimates, at least 800 million persons in developing countries have a diet so limited they do not have the energy for routine physical activities. The World Health Organization (WHO) estimates that another 300 million children are retarded in growth and mental development and at increased risk of disease and death as a result of malnutrition. In some of the poorer countries in Africa, Latin America, and Asia, 70 to 80 percent of the children are growing up with their genetic potential for growth and development impaired and their health compromised. Even in countries that have made some progress in reducing hunger, such as India, Indonesia, the Philippines, Colombia, and Brazil, the majority of preschool children has lower-than-normal growth curves or weight for age. China has all but eliminated overt clinical malnutrition among its children, but chronic undernutrition still greatly reduces growth and development among some segments of its population. Even in the United States, surveys by the Department of Health and Human Services show a lower-than-normal growth rate among children of underprivileged and minority groups. Malnutrition obviously is a problem that cannot be solved by economic development alone.[34]

Sometimes cause or effect or even both may be hypothetical:

> The late Quaternary sediment cover on the slope and upper rise off New Jersey is uneven in thickness, disrupted stratigraphically, and highly variable in age and deposition rate. It is our premise that these attributes are largely the result of geologically recent downslope transports.[35]

> The Defense Department's plans to build a new generation of nuclear weapons in the 1980's will require a major increase in the production of bomb-grade plutonium. So great is the demand that, even with defense reactors running at full capacity, some analysts have predicted that short-

ages will appear by the end of the decade. Consequently, officials in the Department of Energy (DOE) have been eyeing a source of plutonium that has previously been politically and technologically off limits: the spent fuel rods from commercial nuclear reactors.[36]

## COMPARISON AND CONTRAST

Beginning with a comparison seems to be as natural for writers as beginning with historical background. The poet may open with a simile:

*Oh, my luve is like a red, red rose*
*That's newly sprung in June*

and continue with another:

*Oh, my luve is like the melodie,*
*That's sweetly played in tune.*

Writers of informational prose also use figures of speech (Chapter 10), but not usually for emotional impact. They compare to clarify ideas, and for that purpose they move from the familiar to the unfamiliar, from the simple to the complex, from the old to the new.

"Robots—The New Breed," Chapter 3 of *Technology Edge: Opportunities for America in World Competition*, opens with a contrast of the present and the future:

> Robots in one Japanese factory now manufacture replicas of the motors and gears that are their own internal organs. Sometime in the next decade, machines that are still more advanced will achieve the beginning of "robotic life." They will complete the assembly of the first robot wholly constructed and assembled by others identical to it.[37]

Gerard K. O'Neill also opens his "Overview" with a contrast:

> In the traditional "smokestack industries," making automobiles, steel, commodity chemicals and similar products, the international battles for market share began long ago. The United States has already lost some of those battles, and it may lose more. None of those industries offers opportunities for the United States to recoup its economic fortunes in competition with Japan and Western Europe.
>
> But newer, higher-level technologies have opened market opportunities that the United States—or any nation—could still exploit for major economic growth. For true success, however, the three criteria for growth that I set out in the introduction to this book must be met: better service, better energy efficiency and less damage to the environment. Even that is not enough. Americans need to know that a new industry can grow, and that the United States can share in its economic growth.[38]

Lewis Thomas begins with contrast the chapter "Rheumatoid Arthritis and Mycoplasmas":

> The greatest difficulty in trying to reason your way scientifically through the problems of human disease is that there are so few solid facts to reason with. It is not a science like physics or even biology, where the data have been accumulating in great mounds and the problem is to sort through them and make the connections on which theory can be based. For most of this century—by far the most productive of technology in the history of medicine—clues have been found through analogies to known disease states in animals, sometimes only vaguely resembling the human disease in question.
>
> In rheumatoid arthritis, the only comparable diseases that occur spontaneously in animals are the infections caused by mycoplasmas. In several species of domestic animals, most persuasively in swine, the joint lesions caused by mycoplasmas are indistinguishable in their microscopic details from those of rheumatoid arthritis in man.[39]

Contrasting the past and present has been chosen by many writers for the first paragraphs of their articles. "Peddler Patrol" contrasts yesterday with a week ago:

> As frantic last-minute shoppers struggled from store to store along Fifth Avenue at noontime yesterday, the only sidewalk peddler in sight from 42nd Street to 59th Street was Junior Grandison, selling children's books and Webster's Dictionaries on the southwest corner of 45th Street.
>
> It was a far cry from the scene a week ago, when hundreds of peddlers of scarves, trinkets, watches, jewelry and wind-up toys jammed the sidewalks and caused some of the most powerful merchants on Fifth Avenue to complain publicly to Mayor Koch. The merchants called the spread of illegal peddlers along the elegant avenue "a disaster."
>
> Since Thursday, the Police Department's newly reinforced Holiday Enforcement Patrol has arrested 173 peddlers and evidently scared off scores of others.[40]

Carl Rain contrasts past and present quality control and thus clarifies a new concept by comparing it with a familiar one:

> A few decades ago, quality control in industry was a simple matter: You just eyed a truck axle, machine bearing, or whatever for obvious flaws. Or you bent, tugged, squeezed, or heated a sample till it cracked or crumbled. But today these veteran techniques are being joined by a group of sophisticated new inspection tools known as nondestructive testing (NDT). Unlike visual methods, nondestructive techniques—which use X-rays, ultrahigh-frequency sound waves, or lasers in tandem with powerful microcomputers—allow inspectors to poke *inside or beneath* the surface in order to spot potential trouble. And unlike conventional stress testing, NDT leaves a part intact.[41]

A newspaper article finds a humorous point of similarity between two familiar whipping boys:

> If there is one similarity between the Immigration and Naturalization Service (INS) and the Internal Revenue Service (IRS), it is their mutual inability to publish adequate forms in a timely manner.[42]

And comparison opens "Disappearance of Stabilized Chromatic Gratings":

> The visual contrast sensitivity function for isoluminance chromatic gratings behaves differently from the luminous contrast sensitivity function measured under comparable conditions. The two sensitivity curves cross each other, with the chromatic sensitivity being greater at low spatial frequencies and the luminous sensitivity being greater at high spatial frequencies. (This is analogous to the relation between luminous and chromatic flicker sensitivity curves.) These results are believed to reflect the relatively coarse spatial organization of the opponent-color pathways, compared with that of the pathways that transmit chromatic information.[43]

Comparison is effective too at the beginning of Frank Graham Jr.'s review of Edward O. Wilson's *Biophilia*:

> We speak of "bookish instincts" when the ability to read seems to flow out of thin air into lucky children. But sometimes we forget that before there were books our ancestors read the secrets of plants and the movements of birds and beasts.
>
> Both inclinations are surely rooted in the human psyche, though until recently there were few penetrating explanations for these phenomena aside from allusions to natural curiosity or the influence of the home. More modern studies of reading trace the paths by which the brain has developed over the millennia to process information in terms of words and symbols.
>
> Less attention is given to what seems to be our inborn fascination for other living things and the way they behave. This fascination is especially strong in children, though in many it begins to wane during the steady growth toward maturity that Henry James likened to a "reluctant march into the enemy country, the country of the general lost freshness."[44]

Thus comparison and contrast clarify and vivify informational prose. Writers presenting dull material or those who consider their own writing dull should try enlivening it with these devices, particularly in the opening paragraphs, where dullness is most discouraging to readers. General readers of science, technology, and other specialized subjects can be helped and interested by contrast.

The following Knight Ridder News Service story on biotechnology opens with an interest-arousing comment in a light vein, follows with supporting details, then presents the sharp contrast of what scientists are doing, and offers a point of view. Thus several types of beginnings are combined to arouse and sustain interest, but the sharp contrast dominates:

> The United States is awash in milk. Last year, America's dairy farmers sent 137.7 billion pounds of milk to market, while milk drinkers and cheese eaters consumed only 129.1 billion pounds. Government stores of cheese and butter, made from the excess of this and past years, amount to the equivalent of 12 billion pounds of milk.
>
> The surplus has become such a headache that the government this year took the drastic step of killing dairy cows in hopes of reducing the glut. Under the dairy herd buy-out program, the Department of Agriculture expects to spend $1.8 billion to buy and slaughter about 1.5 million dairy cows and calves, about 8 percent of the nation's herd.
>
> Meanwhile, in the laboratories of at least four big chemical companies, scientists are striving to perfect and bring to market a miracle substance made possible through the new technology of gene-splicing: a protein that can be administered to cows to increase their production of milk.
>
> **Impact on Milk**
>
> Bovine somatotropin, which is expected to be available for commercial use in about four years, increased milk production between 16 percent and 40 percent in tests on cows at Cornell University.
>
> Its development now, when overproduction of milk is already a costly national problem, is a dramatic illustration of how the impending biotechnology revolution in agriculture could sow harm as well as good across U.S. farmlands, some sociologists say.[45]

## DEFINITION

*Definition* has many meanings when it is examined in the beginnings of articles and books: the essential nature of something expressed in words; the meaning of a word, phrase, sign, or symbol; an explanation that describes or identifies to make definite and clear.

Several attempts at definition are discussed clearly and briefly by Melvin Levine in his article "Learning: Abilities and Disabilities" in *The Harvard Medical School Health Letter*:

> "Learning disability" is the term currently used to describe a handicap that interferes with someone's ability to store, process, or produce information. Such disabilities affect both children and adults. The impairment can be quite subtle and may go undetected throughout life. But learning disabilities create a gap between a person's true capacity and his day-to-day productivity and performance.[46]

Definition is common at the beginnings of books or chapters. "Wave Modification," Chapter 3 of *Ocean Wave Energy Conversion*, defines three terms in the first paragraph:

Water waves, like sound and light waves, can be redirected by physical objects. The ways in which the waves are redirected fall into three catagories: refraction, reflection, and diffraction. *Refraction* is the turning of the wave front by a change in the water depth. *Reflection* is the reversal of the direction of wave movement due to the wave striking a barrier. *Diffraction* is the dispersion of wave energy into quiet waters in the lee of a partial barrier. These phenomena are illustrated in the following sections. Our interest in wave modification stems from recent studies into using these phenomena in wave energy conversion.[47]

A chapter on trademarks in *Patent and Trademark Tactics and Practice* starts with a brief definition:

A *trademark* is any word, name, symbol, configuration, device, or any combination thereof one adopts and uses to identify and distinguish his goods or services from those of others. In essence, a trademark is usually a brand name:[48]

And Chapter 14 of *Comparative Endocrinology* defines terms by giving the meanings they have in that chapter:

The term *pineal organ* is used here to mean any epiphyseal structure derived from the roof of the brain diencephalon. The term *pineal gland* is restricted here to those cases where a secretory function has been demonstrated or where indirect information (such as cytological appearance) indicates a secretory process. In cases where there is more than one structure developed in the epithalamic region (e.g., an epiphyseal structure and a parapineal device), the term *pineal complex* is appropriate.[49]

Combination of definition with other beginnings is common. In *Solar Engineering of Thermal Processes* Chapter 6, "Theory of Flat Plate Collectors," opens with a definition of a solar collector and then clarifies the meaning by contrasting a solar collector with more conventional heat exchangers:

A solar collector is a special kind of heat exchanger that transforms solar radiant energy into heat. A solar collector differs in several respects from more conventional heat exchangers. The latter usually accomplish a fluid-to-fluid exchange with high heat transfer rates and with radiation as an unimportant factor. In the solar collector, energy transfer is from a distant source of radiant energy to a fluid.[50]

"The Dynamics of Rigid Bodies," Chapter 4 of *Dynamics*, combines definition and comparison:

A macroscopic object that cannot be deformed by the forces that act on it is called a *rigid body*. As with "particles" or "nonviscous fluids," a rigid body is a mathematical idealization, which is never realized exactly. Nevertheless, the dynamics of a great many commonplace objects, in many diverse situations, is well modeled by a rigid-body description.[51]

The opening chapter of *Black Holes, White Dwarfs, and Neutron Stars: The Physics of Compact Objects* begins with the origin and then contrasts the three species of compact objects with normal stars in two ways. This combination clarifies the meaning and leads naturally to further analysis of compact objects:

> A book on compact objects logically begins where a book on normal stellar evolution leaves off. Compact objects—white dwarfs, neutron stars, and black holes—are "born" when normal stars "die," that is, when most of their nuclear fuel has been consumed.
>
> All three species of compact object differ from normal stars in two fundamental ways. First, since they do not burn nuclear fuel, they cannot support themselves against gravitational collapse by generating thermal pressure. Instead, white dwarfs are supported by the pressure of degenerate electrons, while neutron stars are supported largely by the pressure of degenerate neutrons. Black holes, on the other hand, are completely collapsed stars—that is, stars that could not find *any* means to hold back the inward pull of gravity and therefore collapsed to singularities. With the exception of the spontaneously radiating "mini" black holes with masses $M$ less than $10^{15}$ g and radii smaller than a fermi, all three compact objects are essentially static over the lifetime of the Universe. They represent the final stage of stellar evolution.
>
> The second characteristic distinguishing compact objects from normal stars is their exceedingly small size. Relative to normal stars of comparable mass, compact objects have much smaller radii and hence, much stronger surface gravitational fields. This fact is dramatically illustrated in Table 1.1 and Figure 1.1.
>
> Because of the enormous density range spanned by compact objects, their analysis requires a deep physical understanding of the structure of matter and the nature of interparticle forces over a vast range of parameter space. All four fundamental interactions (the strong and weak nuclear forces, electromagnetism, and gravitation) play a role in compact objects. Particularly noteworthy are the large surface potentials encountered in compact objects, which imply that general relativity is important in determining their structure.[52]

The same book introduces Chapter 17 with a definition combined with a brief presentation of characteristics that explains why supermassive stars are included in a discussion of compact objects:

> Supermassive stars are hypothetical equilibrium configurations with masses in the range $10^3$–$10^8 M_\odot$. They can be quite compact, as we shall see, with surface potentials a small but nonnegligible fraction of $c^2$. Following quasistatic evolution, they can undergo catastrophic gravitational collapse. Consequently, supermassive stars are possible progenitors of su-

permassive black holes. Both objects—supermassive stars and supermassive black holes—are frequently invoked to explain the energy source responsible for the violent activity observed in quasars and active galactic nuclei. For these reasons, it is appropriate to discuss the properties of supermassive stars in a book on compact objects. As we shall see, the gross equilibrium and stability properties of supermassive stars can be readily understood by utilizing some mathematical machinery we have already developed for white dwarfs and neutron stars (e.g., the energy variational principle).[53]

Expanded definitions also appear in the opening of Chapter 11 of *Patent and Trademark Tactics and Practice* and of an article in *Scientific American*. The expanded definition of ownership rights, for example, discusses the contents of the written document such as the right to patent the invention:

> An *assignment* is a transfer of ownership rights with regard to an invention. It usually takes the form of a written document that includes the right to patent the invention, along with all rights in all patent applications and patents emanating from the invention. An assignment is, in essence, a sale of an invention.[54]

A longer definition of a gyroscope, which begins the article "Optical Gyroscopes," includes a general description of the function of the instrument that leads into a discussion of how it works:

> The word gyroscope brings to mind a picture of the string-powered novelty capable of balancing itself on the tip of a pencil. Such a toy is actually a primitive representative of a family of instruments whose essence is a framework containing a rapidly spinning wheel. The angular momentum of the wheel causes it to resist change in its orientation even as the framework is rotated. A gyroscope can therefore reveal the extent of the rotation; in this way it can also provide directional information for navigation. Indeed, nearly all vehicles more sophisticated than an automobile (and some that are much less sophisticated) depend on gyroscopes to keep them on course. Gyroscopes are at the heart of the inertial-guidance systems of any jet airliner or oceangoing ship, for example. The reason is that the instruments require no external stimulus in order to function. A gyroscope can, for instance, detect motion even within the confines of an isolating box that prevents a voyager from observing the stars and deprives him or her of a magnetic field for a compass.[55]

## DETAILS

"A hyphen omitted from an equation caused the failure of a United States rocket."

"*Ain't* ain't excluded from *Webster's Third*."

94

"A new method will reduce the cost of producing fuel gasoline by two cents on the dollar."

Specific details such as these arouse interest, titillate curiosity, and please readers. Because of this inherent interest, details are good beginnings, as journalists know and demonstrate in their leads. Three examples, below, from an issue of *The Washington Post National Weekly Edition* demonstrate this.

A list of the unusual names of villages on an Indian reservation leads the reader gently into an article on the bleak conditions on the reservation:

> St. Francis, Upper Cut Meat, Parmelee, He Dog, Milk's Camp, Antelope, Two Strike: These are the villages of despair. The roads are dusty, the air dry. Many houses are vacant, their insides charred and their outsides worn by the harsh winds that whip the prairie. Others, occupied by as many as 25 people, lack electricity and running water. Stoves do not work. Woolen blankets hang where walls once stood. Children are bathed in large metal buckets. Dogs sniff the litter strewn in the streets. Some people live in shacks and old cars.
>
> There is virtually no industry here; the grazing lands are leased by white ranchers, and seven of every 10 Sioux are jobless. Only three of 85 employable residents are working in Upper Cut Meat: a secretary, a policeman and a local councilman. In many cases, the drinking problems are so severe that adults have relinquished their roles as parents, forcing young children to take responsibility for entire families.
>
> Rosebud, situated between the mineral-rich Black Hills and the wide Missouri River, is not typical of America's 265 Indian reservations: Its 10,000 residents are poorer than most, its 20 villages and hamlets more isolated. But it is an example of what can happen when essential institutions do not exist or are allowed to deteriorate.[56]

An article on nuclear arms begins with a general statement of some findings and then gives the source and introduces a long presentation of other details of the survey:

> In the four decades since the dawn of the nuclear age, Americans have shifted dramatically from a belief that nuclear arms serve the cause of peace to an overwhelmingly opposite view—one that is fraught with misinformation and confusion but that shows solid consensus on certain questions.
>
> These are among the findings of a 91-page report, "Voter Options on Nuclear Arms Policy," issued by the Public Agenda Foundation, a nonpartisan research organization headed by pollster Daniel Yankelovich.
>
> In the early 1950s, the report noted, Americans by 2 to 1 said that nuclear arms reduced the danger of war. But while nuclear experts have since grown less anxious about the subject, the level of concern among ordinary Americans has increased "massively and intensively," Yankelovich said.
>
> The report showed, for example, that half of those under 30 believe that "all-out nuclear war is likely to occur within the next 10 years."

Another huge gap between public understanding and the way the experts perceive things is reflected in findings on a key policy question. "Virtually all Americans—81 percent—mistakenly believe it is our policy to use nuclear weapons if, and only if, the Soviets attack the United States first with nuclear weapons," the survey found.

"So there is almost universal misunderstanding of what NATO policy really is," Yankelovich said, noting that official policy is to reserve the option for first use of nuclear weapons.

Among findings that show significant consensus among the public:

- Eighty-nine percent believe there can be no winner in an all-out nuclear war—that both the United States and the Soviet Union would be destroyed.
- Seventy-six percent reject the claim that it is a "wild exaggeration" that all life would be destroyed in a nuclear war.
- Ninety percent believe the Soviets secretly built up their own military strength during detente.[57]

The writer varies the presentation of details by quoting opinions, indenting lists with each item marked by a symbol, and summarizing and correcting mistaken public opinion.

A brief statement of point of view precedes and follows specific details in a commentary on retirement:

The dream of retirement—of a dignified period of ease after a lifetime of work—is so deeply ingrained in the American psyche that any suggestion that the dream is outmoded and economically impractical invariably invites protest.

Yet, in the near future we will have to radically alter our traditional views of a "working life." The majority of us probably will find it desirable and perhaps necessary to delay our retirements and work many years longer than most people do now.

Widespread doubts are developing about the long-term solvency of Social Security, and of some government and private pension systems. Pressure is growing to scale back pension benefits of future federal government retirees—and some state and local retirees as well. Labor unions, fighting to save jobs, have had to "give back" future pension benefits won earlier.

Many Americans already sense correctly that they may not have adequate income to support themselves in retirement and will need to stay on the job longer, or learn skills useable in other jobs. Before the end of the century many more people will want or need to continue working to 70, or even 75 and beyond.

The issue is whether we can change the policies of government, private employers and labor unions to respond to these needs. Instead of pretending that a conflict between the promise and the reality of retirement does not exist, the federal government, with the help of private industry, labor unions and the U.S. education system, should be looking for ways to help older people voluntarily remain in the work force. Only retraining, the abolition of mandatory retirement laws, provision of enough jobs and the removal of disincentives to working longer will enable the vast majority of future older workers to have economic security once they finally do retire.[58]

In a less somber vein a sportswriter for *The Miami Herald* describes with humorous details the wind problem at a golf tournament:

> It causes the stateliest of pines to wail in groaning protest. It turns the clubhouse flags into angry, snapping sheets of canvas.
>
> With a single gust, it will knock the drives of the PGA Tour's strongest long men straight to the ground. It will nuzzle an innocuous putt and send the ball careening down a green as slick as the hood of a Buick.
>
> And it seems never to let up. It will change direction—at a moment's notice—but it never stops. Its full-throated rustle is as constant here as the happy, old-world whistle of the occasional freights on the Long Island Railroad tracks that border the golf course.
>
> It's a wind that will gust to 40 miles per hour, a cruel punch line to a golf course already demonic enough. And it will make and break the 86th U.S. Open that begins this morning on the 6,912-yard, par-70 Shinnecock Hills Golf Course on the rolling, scrubby reaches of eastern Long Island.[59]

Combining details with other introductory methods is common in articles for the general reader and those for specialists from other fields. It is in articles written for fellow specialists that one finds beginnings composed only or mainly of details. This use of details is directed less to arousing and holding interest and more to brevity and directness in presentation. A report in *Science* plunges immediately into details:

> Patients with type III hyperlipoproteinemia (HLP) often develop premature coronary artery and peripheral vascular disease and may have palmar, tubo-eruptive, and tendinous xanthomas.[60]

A list of details after a brief introductory sentence begins an article on mass spectrometry:

> "Needle in a haystack" analytical problems have become important in a variety of areas. Specific compounds which must be recognized in the presence of many others include drugs or disease-indicative compounds in biological fluids, pollutants in environmental samples, chemicals used in the classification of plant and animal species (chemotaxonomy), natural insect attractants (pheromones), flavor- and odor-producing compounds, petroleum and synthetic fuels, process control compounds, chemical warfare agents, and contraband agricultural products, drugs, and explosives.[61]

Details by their very nature are frequently used with some other form of beginning, such as background, definition, or summary. Robert Rainer briefly identifies the land his book review is concerned with, and then he offers rich detail to illustrate its environmental importance:

97

One hundred kilometers off the coast of British Columbia, nestled within the Queen Charlotte Islands, there lies a smaller archipelago so renowned for its peculiar natural wealth that it has been dubbed the "Canadian Galápagos." Indisputably, the 138 islands collectively called South Moresby are biologically rich. Bald eagles nest in densities unrivaled elsewhere in Canada, and peregrine falcons make their home here in the greatest concentrations found on Earth. The world's largest black bears, a million seabirds, and the largest sea lion population on the West Coast find shelter in this wilderness only recently become famous.[62]

In *The Atlantic* "No Parking" begins with a simple summary sentence and then presents amusing details appropriate to a humorous treatment:

Restrained by the Constitution from forbidding car ownership outright, the city fathers of New York have tried to discourage driving in Manhattan by every bureaucratic maneuver known to humanity. Heavy fines are levied for minor traffic infractions, and the streets are covered with signs that say DON'T EVEN *THINK* OF PARKING HERE. Tow trucks prowl relentlessly, ever ready to drag automobiles to a lot in the only part of Manhattan that is farther from the rest of the borough than New Jersey. Manhattan's Bureau of Traffic Operations is rumored to have a special squad of battered station wagons whose plainclothes drivers park by Braille, reducing bumpers to tinfoil wherever they go. Legend has it that automobile storage on Park Avenue is so expensive that socialites find it cheaper to pay armed guards to sit in the driver's seat overnight. Garages charge as much per hour as psychiatrists do. "Gee," out-of-town friends say to me, "you'd have to be crazy to own a car here."[63]

In a delightful satire, "Metropolitan Math," Richard F. Shepard spoofs excessive emphasis on and use of details in a full page of details preceded by a one-sentence summary:

It is a fact that all the facts about New York have never been told, are not even known. It is also a fact that almost all facts about New York are subject to change, which is why it is a good idea to get some of them off to the public before newer facts intervene.

An appropriate place to absorb facts is on one of Central Park's 7,674 park benches, where the entire population of Scarsdale, N.Y., could comfortably sit three to a bench while feeding the park's 6,900 squirrels. Or you might ride in one of the 68 horse-drawn carriages that roll through the park while gazing at some of the 2 million trees—100 different types, from American elm to ginkgo—that cover the 26,168 acres of city parkland.

Another 600,000 trees line the city streets, alongside 63,000 parking meters and 1 million parking signs, which apparently were ignored by the recipients of the 10,404,025 parking tickets issued last year.

There are 6,401.9 miles of streets in the city and 150,000 street name signs. The male chauvinism that once prowled Manhattan streets is reflected in the 80 streets named for males, including Hubert, Murray, Howard and St. Nicholas, as compared with the 22 streets named for females, among them Catherine, Cornelia, Ann and Hester.

The author has organized and commented on the details cleverly in the three-column article, varied the style effectively, and stressed wit and humor all the way to the comic ending:

> But, factually speaking, are New Yorkers happy? Last year, 36,270 couples were married at the City Clerk's office, setting a pace of 139 weddings a day, or 20 an hour. In the same year, 2,525 people dialed 732-8400, the number at the Department of Consumer Affairs that offers taped information on where to get help for various situations; 320 callers wanted information about "Grounds for Divorce."
> New Yorkers know how to drown their sorrows, though—there are 7,855 full-time bartenders in town to help them. And they remain optimistic. There are, after all, only 908 undertakers in a city of 7,073,500 residents.[64]

A skilled professional writer with a sense of humor can keep readers awake for three columns of details, but a less skillful writer should beware. To use small points effectively, a writer needs an eye for and understanding of necessary, unnecessary, dull, interesting, meaningful, and insignificant particulars. Resisting the temptation to use every detail that results from research prevents the overwhelmingly burdensome beginnings of scholarly articles that open with every point the writer has been able to find. Many readers call such passages deadly. Although readers may not die, they will certainly slumber under the weight of such inconsequentiality.

Carefully selected necessary details that are lively and interest-catching, however, can be very effective. This requires that writers vary expression and sentence style; emphasize important details; and subordinate or discard dull, insignificant, unnecessary items. If this task demands more work than writers are willing or able to spend on their beginnings, they should not choose to begin with details.

## EXAMPLES AND ILLUSTRATIONS

> It is hardly a secret that residents of a certain city on Chesapeake Bay call their home town "BALL-uh-mer." Or that to Philadelphians, their home state is "Penn-suh-VAY-nyuh."
> Bostonians, of course, still pahk their kahs in HAH-vehd Yahd.
> And some older New Yorkers still call it "TOY-tee TOYD Street."[65]

These examples are the beginning of an article on urban accents by William K. Stevens in *The New York Times*. Geraldine Fabrikant also uses an example where summary or point of view might be dull:

Last fall when the Comcast Corporation had succeeded in reaching its goal of building a $100 million cash position, the company's chief financial officer, Julian Brodsky, sent a memo to the staff. He pointed out that there was a charge for obtaining telephone numbers from information and suggested that employees use the telephone directory when possible.

Comcast's management is hardly spendthrift. Indeed, the little known cable company that last Tuesday made a $2.1 billion bid for Storer Communications Inc. is considered among the best-run companies in the industry.[66]

And Fred Hiatt uses two contrasting illustrations of the issuance of reports of military analysis for an introduction, develops the information about the Grenada reports, and then uses it to lead to point of view. The Grenada material is itself convincing, but opening with a contrasting example, the Falklands war, increases the effectiveness:

Eight months after Britain and Argentina ended their 1982 Falklands war, the U.S. Navy published a military analysis of tactics, equipment and lessons to be learned from the conflict in the south Atlantic.

A year after the American operation in Grenada, the Pentagon has yet to release a comparable critique.

This silence is not because our military knows more about other nations' wars than about its own; numerous analyses of the Grenada invasion have circulated within the Pentagon. Rather, the Defense Department has publicly declared Grenada an unqualified success, and it does not want that judgment sullied by tales of Army radios that could not talk to Air Force radios or U.S. soldiers mistakenly killed and injured by U.S. Navy bombers.

So the Grenada reports, unlike the Falklands analysis, have been stamped secret—even though an airing of the problems might lead to improvements that could, in turn, prove crucial if the United States in its next campaign runs into something more than 700 troops and armed construction workers.

The decision to withhold the Grenada reports reflects a growing tendency within the Pentagon to release only information that will make the Pentagon look good. Embarrassing information is kept secret while self-serving facts are aggressively put forward, blurring the line between public information and public relations. Secrecy allows the Pentagon to dodge accountability for everything from cost overruns to war in Central America.[67]

In *The Atlantic* "Defining Rembrandt" opens with a striking, familiar example and later in the article adds still others and mentions "250 or more":

Last November West Berlin's Staatliche Museen let it be known that *The Man With the Golden Helmet* was not, as long supposed, a work by Rembrandt Hermanszoon van Rijn. The painting had been one of the most popular in the oeuvre attributed to the artist. It appears on coasters, posters, and playing cards. In phosphorescent hues it has been replicated on velvet.

The revelation took the public by surprise and received considerable attention in the press. As it happens, however, connoisseurs of Dutch art had suspected the painting's misattribution for quite some time. *The Man With the Golden Helmet* was but the latest casualty in a century-old battle to "purify" the corpus of Rembrandt's work.[68]

Gale Warner begins an article on low-level radioactive waste with a seemingly innocuous example and then appends the stinger to propel the reader into an examination of an increasing problem:

> Thirty-five miles north of New York City, Union Carbide Corporation operates a chemical reprocessing plant and a small nuclear reactor. The facility produces some materials used for medical purposes and others used in making nuclear weapons. It also produces a liquid that is considered high-level radioactive waste under one section of the Nuclear Regulatory Commission's regulations. But when the waste is put into 55-gallon barrels with only slightly contaminated paper and metal, the average concentration of radioactivity drops, and the resultant mixture fits within the commission's definition of low-level radioactive waste. The barrels then make the 500-mile journey to the low-level waste dump near Barnwell, S.C., where they are placed in one of 32 trenches and covered with dirt.
>
> While the Barnwell dump is considered a shining star in the waste-management business, significant levels of tritium, a suspected carcinogen, have been detected 200 feet southwest of its trenches.
>
> This story is played out hundreds of times a year. The names and places change, but the problems of low-level radioactive waste stay the same. No one knows how to define it, who should be responsible for it, or how and where it should be discarded.[69]

Examples are effective persuaders, most useful at the beginning of writing meant to implant a point of view or change a point of view. Rosalind Resnick uses an example for this purpose in "There's No Place Like Home: Workers Like the Comfort but Employers Are Resisting":

> Beverly Price faced a dilemma: She wanted to stay home with her new baby, born two months prematurely, but she didn't want to abandon her project at work.
>
> Fortunately for the 33-year-old computer programmer and supervisor, her employer—Financial Data Planning, an actuarial software firm in Miami—was understanding. The company agreed to let Price work on her computer terminal at home and report to the office only one day a week.
>
> The arrangement proved a success for both Price and FDP. During the five-and-a-half months she worked at home, Price enjoyed the flexibility of juggling her work and her feedings. FDP got the benefit of Price's increased productivity and continued management of the software project.
>
> Several weeks ago, Price returned to work full time at FDP with expanded responsibilities.[70]

"Aging and Exercise" starts with a generalization followed by specific illustrations:

No research meeting the tough statistical standards of medical science has yet shown that exercise prolongs life expectancy (at least in people, though some laboratory rats do live longer when they exercise). But there are many aspects of "aging" that people find unpleasant and would like to avoid, even if they wind up living the same number of years. These include the tendancy to huff and puff after minor exertion, diminished muscle strength, a general stiffening of the joints, and loss of bone (osteoporosis). Osteoporosis is especially a problem for older women who suffer from fractures of spine, hips, and wrists. Less easily measured, but equally troublesome, are irritability and depression, which often accompany aging.[71]

The opening of "Synthesizing Chemicals by Computer" usefully combines point of view and examples:

Synthetic materials dominate our modern world. Plastics and other artificial products have replaced natural substances in thousands of applications, from automobile parts to clothing. Even the natural materials that remain, including metal, wood, paper, and cloth, are often coated or colored with the products of laboratory chemistry. Most of us owe our health, and many of us our lives, to synthetic drugs that did not exist until recent decades. And virtually all the colors that we see around us were unknown a century ago; we live in a far brighter world today than did our great-grandparents.[72]

In *Applied Optics* Pochi Yeh uses lists of examples to present background briefly:

The photorefractive effect in electrooptic crystals has been widely studied for many applications, including real-time holography, optical data storage, and phase-conjugate wave-front generation. Recently, increasing attention is focused on utilization of the nonreciprocal energy transfer in two-wave mixing. These new applications include image amplification, vibrational analysis, and self-oscillation.[73]

Stuart L. Shapiro and Saul A. Teukolsky single out an important example for the beginning of Chapter 10 and in the following paragraph list others:

In 1967 a group of Cambridge astronomers headed by Anthony Hewish detected astronomical objects emitting periodic pulses of radio waves. This discovery had a profound impact on subsequent astrophysical research. Its significance was highlighted by the award of the Nobel prize to Hewish in 1974.

The existence of stable equilibrium stars more dense than white dwarfs had been predicted by a number of theoreticians, including Baade and Zwicky (1934) and Oppenheimer and Volkoff (1939) (cf. Section 9.1). Baade and Zwicky (1934), Colgate and White (1966), and others suggested that such objects could be produced in supernova explosions.[74]

Opening with examples is particularly effective in short articles that have a light tone:

Tamarac, a West Broward city of 32,000 with a median age of 63 and golf-cart paths crossing its busiest highway, wants to be one of America's Great Cities.[75]

The second product of civilization—following fire but before flow-through tea bags—was trash.

Slightly gnawed animal bones, stalks and stems and frayed animal skins all ended up on a pile outside the old cave.

So much for the good old days, when nobody cared where the junk ended up—and refuse piles turned into archeological treasures. These days, we are much more particular about where our garbage goes.

Mow a lawn. Eat a melon. Read a newspaper. All of it gets dumped into a garbage can. More often than not, the can has a garbage bag inside.[76]

An example at the beginning offers an opportunity to writers who like to begin with narration, including dialogue, because the story beginning attracts many readers. David Wise opens "Campus Recruiting and the C.I.A." with such an example:

Kevin Ward was sitting in a snack bar at the Johns Hopkins School of Advanced International Studies in Bologna, Italy, when he was approached by a stranger.

Ward, then 20 years old, was nearing the end of his junior year abroad. "I was having a pastry and a cappuccino," he recalled, "when this guy came in. He was in his early 30's, dark hair, neatly dressed, you know, the man in a suit. He said, 'You're Kevin Ward; do you mind if I speak to you?' He handed me some literature and said, 'Have you ever thought of a career in the C.I.A.?'"[77]

In the same issue of *The New York Times Magazine* even advertising writers are on the examples bandwagon, an advertisement for the World Trade Center beginning, "FERRY RIDES & FOGHORNS, TANKERS & TUGS, BOUILLABAISSE & BRIDGES & THE SKYLINE & THE STATUE."[78] The business section also finds this type of introduction useful:

Chef salads and shrimp salads have recently joined the McDonald's hamburger empire. There are sun-drenched atriums in some McDonald's, and in a few hospital cafeterias, nurses are lining up to buy Big Macs. Although the hamburger still holds center stage at the world's biggest fast-food chain, it is surrounded increasingly on the menu by chicken, lettuce, seafood and biscuits. If McDonald's latest experiment pans out, there will even be chocolate chip cookies baked daily at each restaurant.[79]

103

Deborah Rankin chooses an example for the opening of "When Insurers Won't Pay Medical Bills":

> When Mary, a 32-year-old teacher, was hospitalized with breast cancer, she thought her only worry was whether her treatment would be successful. She was wrong. After the California woman was discharged and faced some $40,000 in medical bills, her insurance company rescinded, or canceled, her health policy and refused to pay a penny.[80]

In "Executive Fun and Games" the opening example starts with dialogue, a popular combination:

> "No, I don't play golf," the senior executive admitted. Then he added, with a narrow-eyed stare, "But I'll never talk to you again if you print that. I can't let it be known that I don't play. When I'm invited, I always have that emergency meeting that keeps me tied up."
>
> It's like that in the executive suite. If you don't go in for one of the popular out-of-board room pastimes, you'd better keep it to yourself. The world of business and the world of sport enjoy an inescapable symbiosis. When they are not trapped in a whirl of conferences or making phone calls from the corporate jet, many top executives are likely to be off blasting ducks out of the sky or swatting a golf ball 300 yards down the fairway, praying that it misses that marsh on the left.[81]

"The Choice of Technology" begins with one example of a manager's analysis of the results of modernizing by buying an electric furnace:

> The manager of a steel mill deliberates the purchase of an electric furnace that will modernize the mill's operations. The new furnace will reduce the costs of labor, but it will also consume prodigious amounts of electric power. The manager assumes there will be no change either in the costs of scrap and other raw materials or in the price of the steel product. Given such constraints a cost comparison shows that the high price of the new furnace and its auxiliary equipment would reduce rather than increase the overall rate of return on the mill's total capital investment. The plans for the furnace are scrubbed.
>
> If the manager's analysis had taken account of technological changes in other sectors of the economy, the decision to buy the furnace might have been different. Imagine that as new electric furnaces are introduced by the steel industry, public utilities adopt more efficient methods of generating electric power. At the same time, imagine that the automobile industry shifts to new models of cars that increase the demand for high-strength steel. Such changes could readily lead to shifts in the unit cost of electricity and in the price of high-strength steel. The investment in the electric furnace might well have paid off.[82]

Many articles begin with examples that are anecdotal. In the *New York University Magazine* Lisa H. Towle introduces an article on microsurgery with a brief story about a small child in need of an unusual operation. For a publication designed as a promotion piece for the university, the attention-getting beginning is especially appropriate.

> After seven hours of microsurgery, completed during the wee hours of last Christmas Eve, Vladimir Ramos will eventually run and play with other toddlers. Learning to walk will be harder for Vladimir than for most children, however—he has only one leg. Saving the limb was a medical miracle accomplished by a four-physician team from New York University's Institute of Reconstructive Plastic Surgery.
>
> Vladimir has a mop of brownish-blond curls, large long-lashed blue eyes, and the wide, infectious smile of a small boy who welcomes without question. The child holds out his pudgy arms and demands to be picked up, then pulls himself to a standing position in his crib to emphasize his request. Under red jogging pants, his right leg is in a miniature cast, but he seems oblivious to his circumstances. He is intent on only two things: freeing the pen from his doctor's lab-coat pocket, and freeing himself from the confines of his bed at NYU Medical Center, where his resiliency and spirit have made him immensely popular.
>
> Last year, during the afternoon rush hour, Vladimir's mother took him into her arms and jumped in front of an oncoming subway car. he was rushed initially to Columbia Presbyterian Hospital, where emergency-room doctors realized he needed the powerful operating microscopes and expertise available at Bellevue. Within two hours of the tragic episode, the child was there, prepped for surgery. Vladimir's left leg was too damaged to save, as was his right foot. Working against the clock—the doctors only had eight hours to reattach the damaged limbs—the surgical team decided to perform a cross-leg transplant, attaching the boy's left foot to his right leg. It was only the fifth time such a procedure had ever been performed, and the first time it had been done on so young a patient.[83]

Introductory examples appear frequently in *Psychology Today* and are often written at some length and neatly woven into the articles of which they are intrinsic parts. Scott M. Fishman and David V. Sheehan wrote one such introduction:

> Susan, a 25-year-old legal secretary, was about to leave her office one night when she was suddenly overwhelmed by anxiety she had never experienced before—an intense panicky sensation that something dreadful and frightening was going to happen to her. She became flushed and found breathing difficult, almost as though she were choking.
>
> She struggled to maintain her composure, but within seconds she felt dizzy and lightheaded. Waves of fear coursed through Susan. The sound of her heart beating fast and strong and the sensation of blood rushing through her body at great pressure made her think that she might be dying. Her legs were rubbery, but she managed to make it outside for some fresh air. Gradually the feelings subsided. Relieved but still shaky, she made her way home.
>
> Susan had suffered a panic attack. For many people around the world—

4 million to 10 million in the United States alone—such attacks strike with little warning and for no apparent reason. Most of the victims are women, usually in their childbearing years. What they experience is unlike ordinary anxiety, the nervousness most people experience before giving a speech or being interviewed for a job. As Susan described the terror of her attack, "It could not be worse if I were hanging by my fingertips from the wing of a plane in flight. The feeling of impending doom was just as real and frightening."

In the months that followed, the attacks grew more frequent. Occasionally there would be only a few symptoms. Other times the fierce anxiety would strike with multiple symptoms and terrifying force. Lightheadedness; dizziness; rubbery legs; difficulty breathing; a racing, palpitating heart; choking and tingling sensations; changes in mental perception—any or all might be involved.

As the attacks continued, Susan feared that one might occur while she was driving her car, so she gave up driving. She began to avoid situations and places in which attacks occurred. Gradually the phobia grew so numerous that she could no longer work at her job and was terrified even to spend time with others. Finally, the paralyzing fear confined her to her home.[84]

A number of examples open "The Language of Persuasion" in the same issue:

> "I had all the facts and figures ready before I made my suggestions to my boss." (Manager)
>
> "I kept insisting that we do it my way. She finally caved in." (Husband)
>
> "I think it's about time that you stop thinking these negative things about yourself." (Psychotherapist)
>
> "Send out more horses, skirr the country round. Hang those that talk of fear. Give me mine armour." (Macbeth, Act 5)

These diverse statements—rational, insistent, emotional—have one thing in common. They all show people trying to persuade others, a skill we all treasure. Books about power and influence are read by young executives eager for promotion, by politicians anxious to sway their constituents, by lonely people looking to win and hold a mate and by harried parents trying to make their children see the light.[85]

William L. Allman starts an article on a game "well known among game theorists and political scientists," Prisoner's Dilemma, by narrating an interesting example:

> Captain J. R. Wilton, an officer in the British Army, was having tea with his fellow soldiers in the mud near Armentières, France. It was August 1915, and World War I had become a trench-lined struggle for barren stretches of countryside. Wilton's teatime was suddenly disrupted when an artillery shell arced into the camp and exploded. The British soldiers

quickly got into their trenches, readying their weapons and swearing at the Germans.

Then from across no-man's-land, writes Wilton in his diary, a German soldier appeared above his trenches. "We are very sorry about that," the soldier shouted. "We hope no one was hurt. It is not our fault, it is that damned Prussian artillery."

Enemy soldiers might seem like the last people on Earth who would cooperate with each other, but they did. Despite exhortations to fight and threats of reprisals from their commanders if they didn't, peace often broke out among the German and English infantry. Sometimes there were truces arranged through formal agreements, but many times the soldiers simply stopped shooting at each other, or at least shot where it would do no harm. According to an account by one German soldier, for example, the English battalion across the way would fire a round of artillery at the same spot every evening at seven, "so regularly you could set your watch by it. There were even some inquisitive fellows who crawled out a little before seven to watch it burst."

That such cooperation would occur among their soldiers may have been surprising to the World War I high command, but not to Robert Axelrod.[86]

## LURES FOR READERS

"Look at me," some openings seem to cry, and having captured the attention of readers hold on to it long enough to lead them to read further.

My mother almost never drank," says Nancy Wexler, president of the Hereditary Diseases Foundation, "yet one day, as she was crossing the street, a policeman said, 'Aren't you ashamed to be drunk so early in the morning!'"

Those words terrified her mother, for her father and her three older brothers had died of a frightening hereditary ailment called Huntington's disease, Wexler says, which made them lose their balance, twitch uncontrollably, and eventually lose their minds. Mrs. Wexler thought she had been spared: She was 53, past the age when symptoms of the disease usually begin. None of her family or friends had noticed the slight weave in her walk, for it had come about very gradually. But those startling and brutal words from the policeman sent her rushing to consult with her doctor.

When the doctor confirmed that she was doomed to become totally incapacitated and die within 10 to 20 years, her husband, a Los Angeles psychoanalyst, called their two daughters home from graduate school. He had to explain not only that their mother had Huntington's but also that they, too, stood a 50-50 chance of getting the disease.[87]

Thus Maya Pines catches readers at the start and leads them into her article in *Science 84*. Wayne S. Wooden arouses interest with an unusual, frightening beginning in *Psychology Today:*

America is burning, and America's children are in many instances igniting the torches. According to 1980 government statistics, the United

States has the highest rate of arson in the world. Arson now appears to be one of the nation's fastest growing crimes; the number of cases quadrupled during the 1970s and property loss increased to almost $1 billion a year. In addition, more than 6,000 civilians and 100 firefighters die in fires each year in the United States.

As a researcher with a special interest in juvenile delinquency, I became interested in arson because of another distressing statistic: Children are responsible for two out of every five cases of arson, a share that also appears to be increasing rapidly. And unlike other juvenile crimes, arson is disproportionately a white, middle-class activity. What motivates these kids to set fires? What could possibly drive them to set fire to their world?[88]

"The Last Headline," reprinted from *The New Yorker* in *Audubon*, takes a fresh, imaginative approach to reporting a conference of more than a hundred scientists to present a two-year study of the results of a nuclear holocaust:

It's a strange phenomenon of journalistic life today that the greatest potential "story" of our time—the self-extermination of mankind in a nuclear holocaust—is one that, by its very nature, can never be written. "NUCLEAR WAR ERUPTS—WORLD ENDS" is a headline that we'll never see. Nor, of course, will any columnists ruminate over this event, or any "analytical" pieces reveal to us how it happened and what it all meant; nor, for that matter, will any witness—any Solzhenitsyn—tell us years later how the billions died, or why. No monument will be raised to them, no speech given to honor their sacrifice, no poem or song written in their memory. We must shift all that from the future—a future that, for us, will never be—into the present. The time to file our reports, to write our headlines, to raiss our monuments, to write our poems and songs—and, most important, to act to prevent this inconceivable fate—is now, while we are still around.[89]

Reviewing books on the nuclear age, Scott Haas uses details and questions about this subject to attract readers:

The United States and the Soviet Union have stockpiles of more than 37,000 nuclear warheads, and the waste products of commercial nuclear plants can be reprocessed to make plutonium for even more nuclear weapons. Who are the men and women responsible for perpetuating the nuclear age we live in? How do they think and feel? And what are some remedies for the nuclear dangers we face?[90]

Oddities also invite attention, as Michael Harwood's description of a catbird's mimicry shows:

In spring and summer the gray catbird whistles its good-humored improvisations loudly and tirelessly, caroling even in moonlit midnights. Like many a country musician, it learns phrases from other singers and weaves them in and out of its own musical inventions, but some catbirds are much more talented at singing and mimicry than others. At least a few can imitate not only another bird's song but also behavior at the same time. One catbird was seen flying up to a midair singing position, as a bobolink would, and then descending while singing the bobolink's song; another, flying across a river, impersonated a kingfisher in voice and manner of flight.[91]

In *National Geographic* Esmond Bradley Martin opens his article with an interesting bit of news:

> Sometime in the late 1970s half the white rhinoceros population of Uganda suddenly disappeared—a single rhino, probably shot by a gang of poachers. At the time, nobody realized that it was one of only two left in the country.[92]

And so does a brief report in *The Miami Herald*:

> Four teams of scientists soon will make an unusual trip to Antarctica to study a mysterious and alarming "hole" in part of the atmosphere that protects the Earth from harmful solar radiation, the National Science Foundation said Sunday.
> Ozone above a broad area of Antarctica has been disappearing every spring since the mid-1970s, and researchers do not know whether this is an early warning sign of future changes in ozone over the entire planet or whether it is just a local phenomenon.[93]

A touch of mystery entices the readers of *Audubon*:

> "It was a dark age," says Alan Johnson, a forty-three-year-old ornithologist whose home for a good half of his life has been on France's Mediterranean Coast, in the watery reaches of La Camargue—a partly salty, partly reedy basin at the mouth of the Rhône River which serves as a stopover on one of the world's great avian flyways.
> "Throughout history the flamingo had lived here, and then suddenly, for seven years, the birds stopped breeding or disappeared entirely. Something had gone radically wrong."[94]

Novelty also catches the attention of readers and leads them into the article:

> Most of the ranchers around here had themselves a pretty good laugh last summer when Perry Bushong forked over perfectly good money for 10 scruffy donkeys, taught them to make friends with his Rambouillet sheep and Angora goats and sent them out to work as the Hill Country's first guard donkeys.
> Now the same ranchers are desperately calling Eddie Tom, the local horse trader over in Leakey, begging him to find them anything four-legged that has big ears and brays like an asthmatic foghorn.[95]

Local relationships also catch interest, a fact well known to the old vaudevillians who set every joke in the city they were performing in that day. What better beginning for an editorial in *Florida Naturalist* than urgent news about Florida's official state animal?

> The Florida Panther is Florida's official state animal, and our state's most critically endangered mammal. In 1983 the state legislature took the panther's endangerment seriously and created a technical advisory committee to recommend steps to secure the panther's survival.
>
> Gov. Bob Graham appointed a group of highly qualified experts to the Technical Advisory Committee. The panel studied the available information and has now made its first recommendations to the Florida Game and Fresh Water Fish Commission.
>
> A central part of these findings shows that the 30 panthers remaining in the only known breeding population are suffering from an inadequate food supply. The situation may be so serious that a high proportion of the panther offspring in this small breeding nucleus may not survive due to disease, parasites, and other indirect results of malnutrition.[96]

Carrying local interest a step further, a writer can show how a broader subject concerns the reader. Timothy Ferris does this with new theories of physics in "Physics' Newest Frontier":

> We talk of "the universe" as if it were far away, but of course it is right here, too. Its history concerns not only the remote stars and galaxies, but our world as well—down to the atoms that make up these words, this page, you and me.[97]

In "The Psychic Toll of the Nuclear Age" Robert Jay Lifton places the bomb right at the reader's doorstep:

> When I first heard that the United States had dropped an atomic bomb on a Japanese city, I reacted in a way I am not proud of. I remember telling a friend that if a single American life had been saved, the bombing was the right thing to do. I was 19 years old, a medical student, and I wanted what other Americans wanted—a quick and victorious end to the war.
>
> I find it painful to recall that sentiment now, that terrible wartime acquiescence to mass killing. I do so only because it suggests the special difficulty we have all had in coming to terms with the new image thrust upon us in 1945 by the two American atomic bombings—the image of our extermination as a species as a result of our own technology.[98]

Some subjects for short articles, memorandums, notes, and letters have only one interest-arousing characteristic: newness. Journalists use this quality effectively in brief news stories:

> Minoxidil had been marketed by Upjohn Co. of Kalamazoo, Mich., as a

treatment for high blood pressure. Some users found it not only lowered blood pressure, it raised hair.[99]

Sometimes the new is the old:

> A tractor stuck in the mud on the shore of the Sea of Galilee has led to the discovery of a 2,000-year-old boat.[100]

The appeal of the new combines well with summary:

> The long-awaited autobiography of Ansel Adams will be published in October by New York Graphic Society Books/Little, Brown. This narrative of Adams' life ranges from his childhood on the dunes of San Francisco to his active ninth decade, when he was interviewed in *Playboy*, grappled with President Reagan on environmental issues, and saw "Ansel Adams Day" proclaimed by the California legislature.[101]

Such a combination appeals to readers in business and industry whose first question is always, "What's new?"

> An experimental, nontoxic spray that neutralizes the poison in poison ivy, oak and sumac is being recommended for widespread use among U.S. foresters this summer and might be available to the public by the beginning of next year's season, perhaps earlier.[102]

Sometimes a beginning contains odd, unexpected, eccentric news:

> The U.S. Seventh Cavalry is returning to the scene of Custer's Last Stand.
> More than a century after Lt. Col. George Armstrong Custer led his troop of more than 200 horse soldiers into a 90-minute bloodbath that no trooper survived, the U.S. Army has agreed—though somewhat reluctantly—to bury some of them with full military honors.[103]

## ORIGIN

The use of origin to begin written works started with authors recounting their first flicker of interest in a subject or describing the first mention of the subject by others. This beginning had a small following that found it useful for introducing books. The preface of *Black Holes, White Dwarfs, and Neutron Stars* begins with the origin of the writers' association with the subject and follows with the origin of the subject itself:

> This textbook is the outgrowth of a course on the physics of compact objects, which we have taught at Cornell University since 1975. As a class, compact stars consist of white dwarfs, neutron stars, and black holes. As

111

the endpoint states of normal stellar evolution, they represent fundamental constituents of the physical Universe.

This book, like the course itself, is a product of the burst of scientific activity commencing in the 1960s which centered on compact objects. During this period, pulsars and binary X-ray sources were discovered in our Galaxy. These discoveries proved to be milestones in the development of the field. They furnished definitive proof of the existence of neutron stars, which had previously existed only in the minds of a few theorists. They made plausible the possibility of black holes and even pointed to a few promising candidates in the night sky. More important, perhaps, these discoveries triggered new theoretical studies and observational programs designed to explore the physical nature of compact stars. A whole generation of experimental and theoretical physicists and astronomers has been trained to participate in this exciting, ongoing investigation.[104]

The foreword of *Ocean Wave Energy Conversion* treats the beginnings of recent interest in the subject:

> The topic of wave energy conversion came to public attention in a paper by Steven Salter published in *Nature* in 1974. At about the same time Sir Christopher Cockerell's experiments on a wave contouring raft were reported in a paper in *New Scientist*. That this renaissance of interest in wave energy should take place in the United Kingdom is particularly appropriate because these islands are bordered on their western coasts by some of the world's roughest seas. For the same reason, in 1975 the United Kingdom's Department of Energy started an extensive program of research into various methods of extracting energy from sea waves. This program is still continuing, and a total of some 15 million pounds sterling (30 million U.S. dollars) has already been committed to the work.
>
> As the development of ocean wave energy has continued, other countries have started their own programs. In Japan, also an island surrounded by rough seas, there has long been interest in the topic. As long ago as the 1940s, Commander Masuda invented and built a wave powered air turbine that actually generated a small amount of electricity from the waves in the Sea of Japan. The Norwegians Budal and Falnes also developed a wave energy conversion buoy and expect to have a prototype in operation by 1982/ 1983. Work on a number of different systems has also been done in Sweden, France, and the United States.[105]

In the fifth anniversary issue of *Science 84*, "20 Discoveries That Changed Our Lives"[106] has articles that might be expected to consider origins; still it is noteworthy how many open with that topic.

Bernard Dixon begins "Of Different Bloods" with the discovery of human blood groups:

> The modest and unprepossessing son of a journalist, Karl Landsteiner received his medical degree from the University of Vienna in 1891. But for him medicine was only a stepping-stone toward chemistry. In 1900 he discovered human blood groups, an achievement that turned the once rare and

dangerous procedure of blood transfusion into a daily, lifesaving routine. Safe transfusion, however, was only one consequence of Landsteiner's work, which revealed for the first time specific differences between the cells of one individual and another.[106]

Ian Hacking combines analysis of present use with origin:

Our world is inundated with statistics. Every medical fear or triumph is charted by a complex analysis of chances. Think of cancer, heart disease, AIDS: The less we know, the more we hear of probabilities. This daily barrage is not a matter of mere counting but of inference and decision in the face of uncertainty. No committee changes our schools or our prisons without studies on the effects of busing or early parole. Money markets, drunken driving, family life, high energy physics, and deviant human cells are all subject to tests of significance and data analysis.

This all began in 1900 when Karl Pearson published his chi-square test of goodness of fit, a formula for measuring how well a theoretical hypothesis fits some observations. The basic idea is simple enough. Suppose that you think a die will fall equally often on each of its six faces. You roll it 600 times. It seems to come up six all too frequently. Could this simply be chance? How well does the hypothesis—that the die is fair—fit the data? The result that would best fit your theory would be that each face came up just 100 times in 600. In practice the ratios are almost always different, even with a fair die, because even for many throws there will always be the factor of chance. How different should they be to make you suspect a poor fit between your theory and your 600 observations? Pearson's chi-square test gives one measure of how well theory and data correspond.

The chi-square test can be used for hypotheses and data where observations naturally fall into discrete categories that statisticians call cells. If, for example, you are testing to find whether a certain treatment for cholera is worthless, then the patients divide among four cells: treated and recovered, treated and died, untreated and recovered, and untreated and died. If the treatment is worthless, you expect no difference in recovery rate between treated and untreated patients. But chance and uncontrollable variables dictate that there will almost always be some difference. Pearson's test takes this into account, telling you how well your hypothesis—that the treatment is worthless—fits your observations.[106]

Leonard S. Reich combines comparison with the ubiquitous origin:

The burgeoning electronic technologies of the late 20th century are built on an earlier technology at least as impressive to its generation as silicon chips and microcomputers are to us. The triode vacuum tube, developed by American inventor Lee de Forest in 1906, transformed the fledgling technology of radio and launched the modern electronic age.

It started with Thomas Edison. In 1883, working to improve his new incandescent lamp, he noted that an electric current would flow in the vacuum between the lamp's carbon filament and a positively charged metal

113

plate. He could see no major commercial potential in the phenomenon, dubbed the Edison effect, and moved on.[106]

William L. Brown describes selection before 1900 and after, thus combining contrast with origin:

> For thousands of years Indians of the Western Hemisphere grew corn, varieties pollinated by the wind and bred largely by chance. Despite their lack of scientific insight, they transformed a wild grass from Mexico into one of the world's most productive plants. Sixteenth- and 17th-century farmers continued the practice of corn improvement. Distinct varieties were developed by selecting the best ears at the time of harvest and using seed from those ears to produce the next year's crop. This kind of selection continued until about 1900 and resulted in scores of high-yielding, randomly pollinated varieties. Then, in the course of just a few years, scientists applied genetics to corn breeding—and brought about a transformation of agriculture in this century.
>
> The development of hybrid corn resulted from the exploitation of a phenomenon known as heterosis or hybrid vigor. This increased yield, vigor, and rate of growth of plants comes from the mating of unrelated parents. Many early botanists and horticulturists, including Charles Darwin, had previously observed this phenomenon. But it was geneticist George Harrison Shull who developed the heterosis concept as it is applied today. He and E. M. East, a contemporary whose experiments at the Connecticut Agricultural Experiment Station in New Haven closely paralleled Shull's, were the first to isolate pure strains of corn. These were then crossed to produce the reliable vigor of hybrid corn.[106]

Tom D. Crouch uses the work of the Wright brothers as a background for the origin of the interest of Theodore von Kármán in aerodynamic theory:

> The Wright brothers had triumphed at Kitty Hawk more than four years before and had gone on to an airplane that could fly 24 miles under the pilot's control. Or so rumors circulating in Europe had it, for the brothers refused to unveil details of their inventions without patent protection and purchase orders in hand. But not even the Wrights really understood how airplanes could fly. A science of aeronautics had not yet been born. While aviation could move forward experimentally, only theory could provide a blueprint for rational progress and point to new directions.
>
> This Saturday morning, March 21, 1908, would give Europe its own record book entry, and through good fortune would start aviation down the path to space. By five o'clock, a small crowd had gathered at Issy-les-Moulineaux, a military parade ground on the left bank of the Seine. A handful of street urchins took up positions on the ancient city walls bordering the field, while the cream of Parisian café society clustered around an airplane that had just been rolled from a hangar at the south end of the grounds.
>
> Henry Farman, collar turned up and cloth cap turned backward in what was already appropriate swashbuckling aeronautical fashion, climbed

through the confusion of wire braces and took his seat on the lower wing as a marker 500 yards down the field emerged from the rising mist. Coaxing the clattering biplane along a 50-yard takeoff run and into the air, he made two wobbly circles of the marked course. His distance of 2,004.8 meters in three minutes and 31 seconds established a world record for an officially witnessed flight. More important, Farman gave Theodor von Kármán his first look at a flying machine. The 26-year-old Hungarian, never before interested in flight, was awestruck.[106]

Thomas P. Hughes opens with examples of the problems that arose after invention or discovery:

> Popular accounts insist that the invention of the telephone should be credited to Bell, the automobile to Daimler and Benz, and the electric lighting system to Edison. But today's telephone caller uses components Bell never dreamed of, today's driver depends on systems Daimler and Benz never thought of, and today's homeowner switches on a power and light system that Edison never envisioned. These discoveries have long since been embedded in mammoth networks of technology that no single individual invented.
>
> Technological systems evolve through relatively small steps marked by the occasional stubborn obstacle and by countless breakthroughs. Often the breakthroughs are labeled inventions and patented, but more often they are social innovations made by persons soon forgotten. In the early days of a system such as electric light and power, inventors played the prominent role. Then as the system matured and expanded to urban and regional networks, others came to the fore. Electric light and power systems today are not just scaled-up versions of the Pearl Street station that Edison introduced in New York City in 1882. By the turn of the century, for example, it was the utility manager, not the inventor or engineer, who played the major role in extending round-the-clock service to many different kinds of customers—to the night shift chemical plant as well as the rush-hour electric streetcars.
>
> The automobile evolved similarly.[106]

Phillip V. Tobias goes directly to the point—origin—in "The Child from Taung":

> Blasted out of the hot, dusty Buxton Limeworks near Taung in the northern Cape Province, the skull was the first of Africa's fossil testimonies to the dawn of humankind. The discovery of this ancient African child in November 1924 and all that flowed from it dramatically transformed our conception of human origins. Yet at the time, the dice were heavily loaded against its acceptance.[106]

Allan Sandage opens "Inventing the Beginning" with definition and startles with the recentness of the science despite its ancient origin:

> Cosmology is the science of the universe: its structure, its origin, and its

115

final state. For as long as man could reason, he has wondered about this ultimate order. Yet the basic scientific discoveries in cosmology have been made only within the past 50 years.[106]

In "Broken Eggshells" George M. Woodwell places first an example that leads to origin:

> Ocracoke Island is a sand strand off the North Carolina coast, one of the chain of islands sheltering the mainland in that stormy segment of the middle Atlantic. The island is low and wet; fully one-third of its area is salt marsh. Although it has been occupied for centuries, Ocracoke had always been ruled seasonally by the salt-marsh mosquito—swarms assaulting mouth, eyes, and ears of the human population with maddening persistence. During World War II, however, the U.S. Navy succeeded in making the Island habitable for defenders of our coastlines by spraying dichloro-diphenyl-trichlorethane—DDT—on the marshes.
>
> In the 1950s and '60s, long after departure of the Navy, a municipal fogger traveled the narrow roads of Ocracoke Village blasting DDT and diesel oil into the air. Citizens owned personal foggers: A gasoline-powered lawn mower could be adapted by dripping DDT and diesel oil into the hot exhaust system. Once around the yard, saturating the yaupon and cedar trees at dusk when the air was still, would assure several days' respite from the swarms.
>
> The first of a family of powerful man-made chemicals that have transformed the age-old war against insects, DDT was synthesized in 1874 by Othmar Zeidler, a German chemist. But its importance was first recognized in 1939 by Paul Müller, a research chemist at the J. R. Geigy Company in Basel, Switzerland. Müller had been testing various chemicals for insecticidal properties and discovered the extraordinary effectiveness of DDT. In 1941 the Swiss used DDT successfully to combat the Colorado potato beetle on Swiss farms. Later tests in the United States in 1943 confirmed DDT's effectiveness. In 1948 Müller received the Nobel Prize for his discovery.[106]

Donald G. Fink, who begins "The Tube" with early opinions, follows with the origin:

> Inventors had begun to think about television more than a century ago, but they didn't call it that, and the very notion met its inevitable skeptics. "Shall we ever see by electricity?" wondered an editorial in *The Electrician*, a technical journal of the day. The answer, of course, is that there are now more than half a billion television sets in use, one for every 10 people on Earth. But without the television camera invented in 1923 by Vladimir K. Zworykin, who had arrived from Russia only a few years before, television would have remained a curiosity, captivating but impractical.[106]

"The Making of the Pill" by Carl Djerassi starts with some results and then turns to the origin:

The introduction in 1960 of oral contraceptives was comparable to a sudden switch from horseback to jet plane. Few events during the last 20 years have facilitated women's emancipation more than the Pill, or so quickly secularized the Catholic church in Western Europe and the United States. The fact that contraception has become an accepted form of dinner conversation—something the condom never inspired—is directly associated with the Pill, as is much of the sexual revolution that occurred in the '60s and '70s.

The origin of the Pill, however, is not a "Eureka" tale. The person who synthesized the first oral contraceptive did not say, "Here is the answer to conception control for the rest of the century." The story of the Pill is more circuitous, and while it was not born on a windowsill like Fleming's penicillin, it had its own unusual antecedent in the tropical jungles of Mexico.

The Mexican connection stemmed from the work of one gutsy individualist, Russell E. Marker. In the late 1930s, Marker left the prestigious Rockefeller Institute for Pennsylvania State College, where, as a professor of chemistry, he could pursue his own research. Marker was interested in a group of steroids called sapogenins, which make a soaplike foam in water. While working with the sapogenin diosgenin, he discovered a process that would readily degrade the substance into the female sex hormone progesterone.[106]

Pamela McCorduck discusses the origin of today's computer in her first sentence:

Today's digital computer was born in Great Britain and the United States during World War II, in the frenzy of technological innovation that war invariably spawns. In Great Britain, mathematician Alan Turing helped design an electronic computer named Colossus in absolute secrecy to help crack the ciphers of Enigma, the Nazi code machine. Turned on in December of 1943, it was the first computer to employ vacuum tubes as digital on–off switches—some 2,000 of them—instead of the slow, noisy electromagnetic relays its predecessors had used. Some military analysts say that Colossus was the pivotal factor in Hitler's defeat, for the Germans never knew the computer was breaking codes almost as fast as Enigma was cranking them out.

So tight was the secrecy around Colossus that the U.S. designers of Eniac, the first general-purpose electronic computer, began their work confident that their machine would be first. Physicist John William Mauchly and a 22-year-old engineer named J. Presper Eckert at the University of Pennsylvania started building Eniac (for Electronic Numerical Integrator and Computor) in 1943. It passed its tests in the fall of 1945 and went to work the following February, calculating bomb and missile trajectories for the Army—a job that had been done by as many as 200 people using desktop tabulating machines. The heat thrown off by Eniac's 17,468 vacuum tubes sent the temperature in the room to 120 degrees and the tubes to premature deaths; whenever the computer malfunctioned, teams of technicians would search the racks of equipment looking for burned-out glass corpses. But the monstrous machine did work, cost less than $500,000, and spent nearly a

117

decade calculating trajectories and performing such civilian work as weather forecasting.[106]

"Medicated Minds" by Solomon H. Snyder starts with historical background but moves in the first paragraph to origin:

> Until the middle of this century, psychiatric disturbances were virtually untreatable. Schizophrenics were doomed to their visions and disembodied voices, manic-depressives to their violent swings of mood, and neurotics to their anxieties. Then in a span of about 10 years modern psychotherapeutic drugs provided, though not a cure, the first genuine relief for the mentally ill.
>
> In the late 1940s French neurosurgeon Henri Laborit wanted a concoction to calm patients prior to anesthesia. Laborit felt that histamine, naturally released when the patient is undergoing the stress of an approaching operation, harmed patients. He wanted an antihistamine for his preanesthetic cocktail.[106]

David Baltimore, author of "The Brain of a Cell," also reaches origin before the end of his first paragraph:

> Biology, for all its mystery, is woven into the fabric of our daily lives. We cultivate plants and raise animals to eat; we burn wood to keep warm; we wear leather, cotton, or wool. Over the centuries, man has become increasingly adept at harnessing the biologic world, but before the 1950s our knowledge of how living things work was at best superficial. That has changed. One discovery stands out as the primary generator of our new understanding of biologic systems and our power to manipulate them: the 1953 elucidation of the structure of DNA by James D. Watson and Francis Crick.[106]

Charles H. Townes, using a welcome personal style, begins immediately with origin:

> The laser was born early one beautiful spring morning on a park bench in Washington, D.C. As I sat in Franklin Square, musing and admiring the azaleas, an idea came to me for a practical way to obtain a very pure form of electromagnetic waves from molecules. I had been doggedly searching for new ways to produce radio waves at very high frequencies, too high for the vacuum tubes of the day to generate. This short-wavelength radiation, I felt, would permit extremely accurate measurement and analysis, giving new insights into physics and chemistry.[106]

In the *Harvard Business Review* Thomas D. Morris, writing subjectively, presents origin in his opening paragraph:

> Last year I visited the Pentagon frequently to interview key officials on federal management problems. My itinerary each morning took me along

the third floor "A Ring" corridor, where handsome paintings of past secretaries of defense, from James Forrestal on, hang on the walls. Each morning, I thought back to those whom I had known or worked for—eight altogether—and pondered their records. From my perspective, perhaps half deserve a mark of high excellence. Others earn low grades, perhaps because of short tenure or perhaps because of generally ineffectual performance. Out of those morning strolls has come the idea for this article. The theme: what distinguishes successful leadership in the public sector.[107]

The variety of the uses of origin, its easy combinative nature, and its logical suitability for first place in informational prose have all helped to increase its popularity in the past twenty-five years. Much influence toward more frequent use has come from business and industry, where managers and some writing teachers and consultants have pressed upon scientists and technologists the need and desire of their readers for certain information at the start. Company readers unfamiliar with the work want reports, particularly reports that do not start with final conclusions, to answer in the beginning the questions Why did the company (or division or department) undertake this research? or How did we get into this? Unfortunately, beginning many papers with origin may lead to hackneyed expression: "We are doing this because . . ."; "We started this study because. . . ." If only one person wrote such beginnings, they would be serviceable if not elegant; but when dozens or more versions of this trite beginning reach the same readers every week or every day, the oft-repeated sentences become boring and even annoying.

Writers in business and industry who must begin with the answer to how did we get into this? and have difficulty avoiding a cliché answer can learn much from the preceding examples as well as from the opening sentences on origin in well-written newspapers, journals and magazines, and books. The usefulness of opening with source in orienting a company reader immediately and in keeping scientists and engineers reminded of why they began their research project, while important, is not all. This beginning can also be interest-arousing, emphatic in stressing a writer's points, imaginative, and different from the other beginnings company readers must face.

## POINT OF VIEW

An excellent beginning when used discreetly, point of view may be the worst of all beginnings. An initial statement of point of view is useful in the same way that a statement of scope is; both warn the reader early. An early warning should be given when it is necessary or helpful to the reader or likely to arouse interest. A writer must judge cautiously, because if the initial statement of point of view antagonizes, an angry reader may stop after the first sentence or may read only to disagree.

David A. Burge opens the preface to *Patent and Trademark Tactics and Practice* with a paragraph composed of one sentence on point of view and one on scope:

> I have written this book from the viewpoint of a registered patent attorney who enjoys an active patent and trademark practice. My intention is to outline fundamental principles that should be understood by inventors and business persons engaged in the development, protection, and management of intellectual property.[108]

George S. Dominguez begins Chapter 4 of *Government Relations: A Handbook for Developing and Conducting the Company Program* with a one-sentence statement of his point of view: "The influence, reach, and control of government as it affects business and the marketplace are virtually limitless.[109]

"Tactics to Employ When a Lawsuit Looms," published in the *Harvard Business Review*, also begins with an opinion:

> Whether a company wins or loses an important lawsuit may be determined not by what happens after it has been filed but by what happened before it was filed. Often the critical factor in litigation is the actions of the involved parties after a dispute has arisen but before litigation starts.[110]

In the same issue the section "Realities of Japanese Productivity" begins with a question that is answered immediately by a statement of point of view:

> If Japanese workers are not absent less than U.S. workers, do not have a significantly lower propensity to quit, and don't work harder, why are they more productive? My answer to this question is not very romantic. Superior productivity in Japan does not hinge on the "oriental" style of management or on Japanese corporate culture, but rather on the mundane decisions managers make.[111]

"The Green Revolution Revisited" opens with a much longer treatment of point of view:

> As I left the United States in May 1979 for 3 months of field study in India, I wondered whether my article "The Green Revolution and Social Structure in India," then being published, would be consistent with what I was about to observe. From a home-based vantage point, working with data published up to 1976 covering the period to 1972 or 1973, I had found, first, that small as well as large farmers had gained economically from the new high-yield varieties (HYV)—smaller farms were becoming more viable because of greater productivity of land; second, that agricultural labor tended to gain from the greater volume of crop activity, but this might have been offset by mechanization. To shed light on the latter question there were wage rate data, but income data for agricultural laborers were vir-

tually nonexistent. I concluded from my findings that HYV should be further supported by government policies promoting water and fertilizer use but that mechanization should be limited to necessary support for HYV such as pump sets for irrigation, but not tractorization of farm processes. Further, I thought that resumption of tenanted lands by owners should not be impeded but that a fixed land ceiling of 10 acres should be implemented so that mechanization would be discouraged and the owner-cultivator class enlarged while tenantry and agricultural labor would be decreased. Bold views. Would they be supported by what I was to observe?[112]

Point of view is treated briskly in "Skin—The Ultimate Solution for the Burn Wound" in the *New England Journal of Medicine*:

> When the history of 20th-century burn care is written, topical antibacterial agents are unlikely to receive the credit for contributing to progress in burn care that they have been awarded during the past two decades.[113]

Brief statements of point of view are common openings in book reviews:

> According to Dirckx, "medicalese" is no longer jargon; it is a language. A picky philologist might disagree, but never mind. This book is an important contribution to the understanding of the argot that enshrouds the working of the medical profession.[114]

> In his 1984 Report to the Board of Overseers, Harvard President Derek Bok called for a massive rehumanization of medical education. It is fortuitous that publication of this new book follows closely on the heels of that call. Jay Katz's poetic manifesto, urging replacement of the traditional physician–patient relationship of blind trust (based on silence) with a new alliance of mutual respect (based on conversation), will no doubt long be noted as a milestone in the rehumanization effort.[115]

A review of *Brave New Workplace* in *The Atlantic* starts:

> As America's economic complacency has eroded, its regard for big, powerful machines has increased. Nearly every proposal for improving the economy starts with a lecture about the need to raise the industrial-productivity rate. Inventing and installing new machines are not the only ways to do this, but they come high on most lists. What was the American steel industry's sin, if not its delay in turning to continuous casters? When the Japanese are envied for their high savings rate, it is mainly because their frugality leaves them with extra money to invest in advanced production equipment.[116]

Reviewing a new book by William Trevor, Elizabeth Spencer begins by stating her point of view:

> Ten novels, six story collections and numerous dramatic writings have

121

already come to us from William Trevor, and in this his seventh book of short fiction he proves once more that he is best when writing about his native land. Born in County Cork and educated in Dublin, he now lives in Devon. That England has been his home for a long time can be surmised from the quantity of fiction he has written about the English—country English, village English, London English, the English abroad. Mr. Trevor is a sturdy writer of clear and satisfying prose, who knows how to seek out the fictionally viable character and how to catch the pulse beat in a relationship or the secret strain within a heart. Yet his English stories seem strangely "produced," planned instead of crying to be written. It is the news from Ireland that, wander where he will, he is always returning to give voice to—and these are the stories with flow and power.[117]

In the same issue of *The New York Times Book Review* Wilfrid Sheed also starts by stating his point of view:

People expect an awful lot from their newspapers, and the news isn't the half of it: they also expect the Last Word on practically everything. Thus the poor drudges who blunt their taste at bad plays night after night, and the drones who have to master two books a week—books of a kind that most people put aside a whole summer to read—are assumed to be the most important critics in town, while the hurry-up sages of the editorial pages may be the only voice some people will ever hear on the Middle East or the decline of the yen. (All the above, by the way, is supposed to be done entertainingly.)

It's a lot to ask for 30 cents, but certain hardy souls devote their lives to bucking the odds and making newspapers just that good; and of these, none delivered the goods more consistently than Red Smith, the subject of Ira Berkow's skillful biography.[118]

Many writers for journals also find point of view a suitable beginning. "Slow-wave Sleep: A Recovery Period After Exercise" opens with a paragraph that leads to a statement of point of view in the last sentence:

Recent research has lent credence to the hypothesis that sleep and in particular slow-wave sleep (SWS) is a recovery period for daily metabolism. Evidence in support of this theory includes the synchrony of growth hormone release with SWS in humans, the suggestion that optimum conditions for anabolism prevail during sleep, and studies showing SWS duration to be proportional to preceding wakefulness. Although many other studies have yielded supportive evidence for the theory, the prediction that daytime exercise would increase SWS has produced equivocal results. Possible reasons for these conflicting findings include the variable fitness of the subjects tested, the time during the day at which the exercise is performed, and the absolute amount of exercise. The absolute amount of exercise is relevant since it is the increase in energy expenditure during exercise over and above basal metabolism that would be expected to influence the amount of SWS. To evaluate the theory that SWS is a recovery process and to resolve the question of the effect of exercise on sleep, an experiment was carried out in which the sleep patterns of six subjects were studied after a 92-km marathon. We thought that this extreme event would highlight the effect of a large increase in energy expenditure on sleep.[119]

Journalists writing about science and technology not infrequently use point of view for a lively beginning, especially if they can contrast two points of view as Daniel Goleman does:

> In an editorial in June, the influential New England Journal of Medicine concluded that "our belief in disease as a direct reflection of mental state is largely folklore."
> That statement touched off a furor.
> The American Psychological Association responded with a resolution denouncing the editorial as "inaccurate and unfortunate," asserting that it ignored a substantial body of research findings linking psychological factors and health.[120]

James Fallows writing in *The Atlantic* begins with point of view expressed in one sentence, follows with historical evidence, and ends with a question that his article addresses:

> The biggest surprise about America's defense boom was how long it lasted. Between the end of the Second World War and the inauguration of Ronald Reagan, military spending had never gone up for more than three years in a row—despite the Korean War, despite Vietnam. But in the last two years of the Carter Administration and the first four years under Reagan, military spending rose by an average of eight percent a year, after allowing for inflation. In the late Carter years we were spending roughly $200 billion a year (in 1986 dollars) on the military; now we're spending almost $300 billion. (All dollar comparisons in this article are in constant dollars, to remove the effects of inflation.) By 1983 the annual defense budget had grown larger than it had been in 1968, the peak spending year for Vietnam. By last year it had passed the peak for the Korean War. From 1982 through 1985 the United States spent more money on the military than it had during any four years of the Vietnam or the Korean War. It even spent more to *operate* the forces, now much smaller, than it had from 1951 to 1954, or 1967 to 1970.
> During those earlier periods, of course, the money was going into field hospitals, combat pay, jet fuel, R&R, mortars, grenades, prostheses, induction centers, and all the other raw materials of war. During the past four years larger amounts of money have gone into . . . what?[121]

M. Mitchell Waldrop opens "Machinations of Thought" with a paragraph ending in a prediction. The next paragraph states the point of view of "more exuberant enthusiasts." The third paragraph presents the opinions of those who find the talk of paragraph two "a little premature"; and the fourth paragraph comments on these opinions, states another point of view, and asks a question:

After three decades of frustrating work, artificial intelligence is coming of age—moving out of the laboratories and into the marketplace. Expert systems, computer programs that give advice like a human specialist, are pinpointing mineral deposits and diagnosing diseases. Programs are taking shape that can do a pretty fair job of understanding plain English or French. Robotics will soon benefit from computer vision systems able to store a digitized photograph of an object or scene and "recognize" a good bit of what is there.

As the more exuberant enthusiasts see it, we might soon have machines to advise us about our income taxes or the baby's fever; silicon tutors could help a child master the enthralling possibilities of geometry and numbers; trucks might drive themselves through the night and unload themselves at their destination. In short, we could one day have machines to do almost anything that now requires "intelligence" in a human.

Many AI researchers find this kind of talk a little premature. AI has become commercial, they feel, not because of any fundamental break-throughs but because venture capital is available. They've learned just how far from fulfillment some of the promises really are. "You've got to separate the science from the hype," says John Seely Brown of Xerox Corporation's Palo Alto Research Center.

Besides overselling, those who trumpet the impending triumphs of AI often miss the point. At heart, AI isn't about creating smart computers. The computer is simply a tool, a laboratory for testing out ideas. AI itself is about something much more interesting: intelligence. Mind. The nature of thought. What is it that we *do* up there in our skulls? And how, exactly, do we do it?[122]

James Cornell begins "Science vs. the Paranormal" with a point of view in the first paragraph reinforced by a quotation in the second:

Ours may be an age of reason—or at least high-tech reducibility—but belief in the irrational, the occult and the supernatural seems almost as persistent and pervasive today as it was in the Middle Ages. Indeed, thanks to the wonders of mass communications, public exposure to claims made for paranormal phenomena may be more widespread than in any previous era.

"On the current world scene, belief in the paranormal is fed and reinforced by a vast media industry that profits from it, and it has been transformed into a folk religion, perhaps the dominant one today," says Paul Kurtz, professor of philosophy at the State University of New York at Buffalo.[123]

In presenting a point of view a writer may state an opinion and then attack it. Thus writers can state opposing points of view as weakly or strongly, as completely or incompletely as they wish and make the arguments of others lead to a long, medium-length, or short combating of the points of view of others. This method, known as the *straw man*, is analyzed and illustrated beginning on page 358.

It is not unusual for an author to begin an article with several points of view, as Marlene Zuk does in "A Charming Resistance to Parasites":

> By 1889, British naturalist Alfred Russell Wallace had concurred with many of Darwin's ideas about evolution. He balked, however, at the notion that males of many animal species have evolved bright colors and complicated mating displays simply as a result of females' preference for them, and wrote:
>
>> it may also be admitted . . . that the female is pleased or excited by the display. But it by no means follows that slight differences in the shape, pattern or colours of the ornamental plume are what lead a female to give the preference to one male over another. . . . A young man, when courting, brushes or curls his hair, and has his moustache, beard or whiskers in perfect order, and no doubt his sweetheart admires them; but this does not prove that she marries him on account of these ornaments, still less that hair, beard, whiskers and moustache were developed by the continued preferences of the female sex.
>
> Darwin disagreed and maintained that "the power to charm the female has sometimes been more important than the power to conquer other males in battle." Most evolutionary biologists would now agree with Darwin, but they still disagree over exactly what females find charming and how their tastes could cause the development of such bizarre ornaments as a peacock's tail or the elongated wattles of a turkey. Recently, even the possible role of tiny blood parasites has been brought into the controversy.[124]

That last word, *controversy*, is the clue to the selection of writings that might well begin with point of view. When conflicting views are difficult to present or to organize, stating them at the start clears the air and makes it unnecessary for a reader to skim through half an article to see whether a favorite point of view is included.

A brisk statement of a point of view, on the other hand, is useful for many short reports and for memorandums in business, industry, and government provided that the writer does not exasperate the reader who is to decide the issue. An initial assault with an opinion that the reader is known to oppose or that this reader has already rejected is seldom successful.

Journalists use a brief statement of point of view successfully, especially when writing on business:

> Mexico's economic crisis, which has worsened with the collapse of world oil markets, has handed the Government and the ruling Institutional Revolutionary Party, known as the P.R.I., a challenge that could threaten their strong hold over Mexican life.[125]

Journalists also often use a point of view that is or that includes a prediction. The business pages provide many examples of this combination:

> The chairman of the Financial Corporation of America said today that

problem loans would rise, possibly substantially, and that the company was unlikely to show an operating profit until at least the end of this year.

This bleak assessment by William J. Popejoy, who has been struggling to save the troubled savings and loan company since August 1984, contradicts widespread speculation that the Financial Corporation's worst loan problems were behind it and that profitability was on the rise.[126]

In this example Nicholas D. Kristof presents two main opinions clearly and briefly. Many articles and columns on real estate also open with point of view:

> The idea of giving the customer what he or she wants often works fine in retailing. But applying that adage to 80 exclusive town houses in this Westchester community caused construction to slowly grind to a halt.[127]

## PREDICTION

The forecast beginning will appear more frequently in writing for industry as scientists and technologists discover the advantages of contributing their background and judgment to the making of company decisions. The timid and overly self-protective will avoid forecasting, but other writers will use their information and their experience to interpret and predict coming events. And even the timid may quote the predictions of others.

John Tierney, writing a special report for *Science 85*, begins with a prediction in two short sentences and then develops it to a comment in two short sentences on what happened:

> They'd put up with nature's whims long enough. This time the scientists would do it *their* way. For two decades they had patiently been observing the vast stream of charged particles spewed by the sun into space. This hot "solar wind" produced brilliant auroras over the Earth's poles, disrupted radio signals, even seemed to cause rainstorms. But how it did all this was mostly a mystery. As the solar wind bombarded their planet with 50 trillion watts of electrical energy, scientists could only watch powerlessly from below.
>
> But now they were finally going to do something about the weather in space. Their contribution, however brief and isolated, promised to be spectacular: history's first artificial comet (*Science 84*, December). On Christmas morning a satellite would release a cloud of barium gas that would glow yellow, green, blue, and purple. The solar wind would blow it into a shimmering tail appearing 10 times as long as the moon's diameter. Instead of passively observing, scientists would actively experiment. They would study nature at their convenience.
>
> It was a fine plan. Nature, unfortunately, did not go along with it.[128]

Charles D. Allen, Jr., opens "Forecasting Future Forces" with prediction followed by analysis of the prediction:

126

The Navy has always been a technically oriented service, playing a leading role in the development and application of virtually all the new technologies of the past century. In the face of a seemingly endless acceleration in the rate of technological growth, it will be increasingly difficult for the Navy to retain its historic position as leader and innovator in the use and application of science and technology. The reasons for this prospect are twofold.

First, largely as a result of the enormous advances in data processing, new technological opportunities are presenting themselves at an accelerating pace. The Navy is well past the point where all technically feasible and operationally useful research and development (R&D) options can be funded to the point of feasibility demonstration. Yet, in a competitive world, the Navy must choose wisely from a glittering array of offerings. We have a staggering prioritization problem in undertaking R&D starts.

Second, it has become increasingly difficult to develop a clear picture of the operational need for new systems—the traditional means for establishing priorities. The lead time from initiation of advanced development of critical components to widespread fleet introduction of the new system or platform has grown, typically, to 20 years. Therefore, the environment in which they will be used will be more difficult to visualize. Aegis, for example, was conceptualized in the mid-1960s, but the first Aegis cruiser will not join the fleet until Januanry 1983, and there will not be enough Aegis ships at sea to provide protection to all of the carriers until the end of the 1980s. There are exceptions, of course—of which Polaris is a shining example—but the costs of the redundancy necessary for that sort of time compression are enormous. The Navy cannot afford many such programs. Most of the development programs commonly perceived as going from concept to deployment in shorter periods of time began with previously developed components. Basic point defense, for instance, exploited a missile and illuminator which had been operational in fleet aircraft for years. NATO Sea Sparrow was essentially an engineering effort involving off-the-shelf components. However, to avoid being overtaken by technology, a larger share of our future systems will have to embody new and therefore unfamiliar concepts. Their relationship to future operational needs will be more difficult to identify.[129]

Robert T. Lund in *Technology Review* begins with a briefly presented prediction:

When a durable product such as a machine, appliance, or automobile wears out, most of us simply discard it and buy a new one. But there is an alternative: have the product remanufactured. This approach, used more often behind the scenes in manufacturing than most of us realize, is destined for a greater role as energy and materials become scarcer.[130]

In the same issue Lawrence M. Lidsky starts with two questions that lead to a prediction of "The Reactor of the Future." The following paragraph supports the forecast:

Can a new type of fission reactor solve the nuclear power problems of

127

U.S. electric utilities? Is it realistic to believe that manufacturers will build, federal regulators will allow, and utilities will buy such radically different reactors? A growing coalition of manufacturers, utility planners, and engineers thinks the answer may very well be yes.

A group of professors and students in M.I.T.'s Department of Nuclear Engineering has come to the same conclusion: we believe the modular high-temperature gas-cooled reactor may be a "better mousetrap." We are now trying to determine whether this reactor—which would be much smaller and inherently safer than today's light-water reactors—can and should play an important role in generating the nation's electricity during the next decade and beyond.[131]

"The Real News About Multiple Sclerosis" in *The Harvard Medical School Health Letter* first tempers the hopes raised by recent reports of new treatments:

> Recent reports of two new treatments for multiple sclerosis have raised the hopes of its many victims. The good news is that hope is justified, but the realistic news is that the research therapies cannot yet be applied in patient care.[132]

Medical doctors often face the need to discuss recent discoveries or research. They must learn to comment with great care, as the writer of the preceding example does. Thus "Day Care and Contagion" in the same issue opens with a paragraph leading to a common kind of medical forecast—a warning:

> Day care centers have been the scene of many disease outbreaks, some of them serious. As nearly 2 million American children are in these centers, there is obviously a problem, but it has to be placed in perspective. Some children are there because their working parents have no realistic alternative. Also, many parents have positive reasons, such as the educational programs offered, for preferring day care centers. Whether they are regarded as a virtue or a necessity, day care centers are now a permanent feature of American life. Unfortunately, because child care is receiving a low priority among our national concerns, parents may have to rely on overcrowded and understaffed facilities. Epidemics are but one of the problems that will recur until we as a nation make a serious commitment to quality child care.[133]

In *National Geographic* a medical prediction is stated forthrightly:

> Surgical trauma, the jarring aftermath of the surgeon's knife, may one day be consigned to the annals of primitive medicine—thanks to a procedure called "least invasive surgery" by its growing number of practitioners. Using an endoscope, surgeons can view the interior of the body and operate with the least amount of damage.[134]

An Associated Press story combines prediction and contrast effectively; the subordination of a number of details to maintain stress on the main points is admirable:

> More people will die of AIDS five years from now than were killed in traffic accidents last year, the Public Health Service predicted Thursday in calling for a national commission to guide America's response to the deadly illness.
>
> The agency projected that 54,000 victims will die as a result of AIDS during 1991, most of them people who are infected now but do not know it. By comparison, auto accidents killed 45,700 people in 1985, according to the National Safety Council.[135]

Bob Woletz begins with a quotation a short article composed mainly of the predictions of another and then presents a list of one-paragraph comments on the forecasts. The organization is masterly:

> "Prediction," says John F. Magee, chairman of Arthur D. Little, "is not what we do best." But on the occasion of its centennial, the nation's oldest think tank—set among the oaks on a 40-acre campus in Cambridge, Mass.—has gone out on a limb. Herewith, the future as Little sees it:
>
> • The marriage of plastics and artificially grown human cells will enable scientists to create arteries and organs that will not be rejected by the body. . . .
>
> • The biggest new business opportunity by the year 2000 will be in orbiting industrial platforms, used to develop—even manufacture—such new materials as superstrong cast iron and extremely pure polymer crystals. . . .
>
> • In the near term, Americans' taste for fresh, natural food, born of the current obsession with personal health, will continue unabated. . . .

The stream of forecasts ends in the splash of an aptly amusing quotation:

> All of his firm's predictions, the chairman insists, are tentative, extrapolations of current trends. "The only thing in the future I'm very sure of," he adds, "is that the company's next centennial will pass without me."[136]

## PROBLEM

How to begin with the material of most importance worries some writers, particularly those who wish to start writing before their research is completed or before all their ideas are formulated. Often they solve this by beginning with a problem.

Christopher Lehmann-Haupt opens a book review with a simple statement of a problem:

What moved Jacob Needleman to write "The Way of the Physician" was the perception of a problem that should be familiar to most readers by now. American medical practice is in some sort of trouble these days.[137]

Bruce Fellman also states a problem simply and briefly:

A seed falls, and finding the weather hospitable, it swells, splits its protective coat, and sends a shoot and a root out into the air. Suddenly and without fanfare, the shoot curves upward and heads for the sky, while the root turns downward and angles its way into the soil.

This uncanny ability to sense gravity and act accordingly, called gravitropism, has long fascinated botanists. But so far no one has been able to explain how plants get their directions straight.[138]

In *Technical Communication: Journal of the Society for Technical Communication*, "From Transcript to Minutes: An Editorial Challenge" begins with a major problem and its solution:

A transcript poses a number of challenges for the editor who must prepare minutes from it. Difficulties often emerge from the length of the text, which usually cannot be reproduced in its entirety. A transcript is commonly a record not only of the presentations made but of everything said at a meeting. Along with informative material, it may include repetitious and irrelevant statements, such as natterings about the seating arrangements and coffee. During some sessions, the irrelevant commentary may take more space than does the significant information. In these cases, the editor must reduce the text substantially to clarify the essential material. The editor might have to cut further into the text if the client wants the minutes to include only the most important sections of the presentations given. After satisfying the requests of the client and meeting the anticipated needs of readers, the editor could find that 100 pages of transcript have been boiled down to as few as 20 pages.[139]

An abstract in *Science* begins with a problem stated in one sentence: "Homing pigeons that had never seen the sun before noon could not use the sun compass in the morning; nevertheless they were homeward oriented."[140]

In the same issue another article begins with a longer statement of a problem:

The neuronal mechanisms of information storage remain one of the principal challenges in contemporary neurobiology. Over the past decade, however, the development of effective model systems has significantly advanced our understanding of the cellular basis of nonassociative learned behaviors. This progress has resulted largely from the exploitation of "simple" invertebrate models, but few effective systems are available for cellular analysis of associative learning or of learning in vertebrates.[141]

130

The authors of "Sinus Node Function and Dysfunction" open with a combination—problem leading to scope:

> Sinus bradycardia is a normal cardiac rhythm (Fig. 1). However, in some patients, the heart rate is inappropriately slow for the physiologic circumstances. This is a common problem, estimated to account for approximately half of all permanent pacemaker implantations in the United States. Symptoms that may be due to sinus node dysfunction are also common manifestations of numerous other clinical problems. These symptoms are often sporadic and transient, making their documentation by electrocardiographic recording quite difficult. The electrocardiographic manifestations of sinus node dysfunction blend imperceptibly into variations of normal. Therefore, the approach to the patient with suspected sinus node dysfunction requires a thorough understanding of normal and abnormal sinus node function, careful clinical assessment of symptoms, as well as perseverance in the often frustrating attempt to document the real-time correlation between symptomatic and electrocardiographic events.[142]

"Finessing the Risks of Nuclear Power" combines brief background with a statement of the problem, a frequently used combination:

> Twelve years ago the Atomic Energy Commission faced an unenviable dilemma. It had just suffered through 22 months of public hearings on the safety systems of nuclear power plants, embarrassed by the paucity of data supporting its claim that nuclear plants could survive loss-of-coolant accidents without a major release of radioactivity. How could the agency license new plants or allow operating plants to keep running?[143]

Paul S. Bender opens Chapter 4 of *Resource Management: An Alternative View of the Management Process* with a statement of the problem and one of its effects:

> As a consequence of the exponential changes in demographic, sociopolitical, economic, and technological factors in the world environment, the most pressing economic problem of our time is the need to improve productivity. This problem is especially acute in the United States, but it affects all nations in varying degrees. For this reason, it is essential to look at the characteristics and requirements of management in the years to come not only from a general resource management standpoint directed at trading off informational resources against all other resources. It is also useful to describe how such an approach can improve productivity.[144]

Robert C. Cowen begins "The $CO_2$ Threat: Solutions Welcome" appropriately with problems but soon switches to point of view:

> Carbon dioxide—which accumulates in the air as we burn fossil fuels— provides colorful imagery for climatic alarmists. Theoretically, its tendency to warm the atmosphere could cause various catastrophes. The melting of

131

Antarctic glaciers might raise sea level and flood coastal cities. Shifts in the rain belts could parch North American wheat lands. The news media love to publicize such scary predictions. But between the doomsayers and the critics who denounce such sensationalism, the real carbon-dioxide ($CO_2$) story is often lost. This is the simple but awkward fact that any climatic threat $CO_2$ may present could be ameliorated by timely action. Arguing about the problem seems easier than searching for solutions.[145]

The opening paragraph of *Solar Engineering of Thermal Processes* states and solves a problem:

It is generally not practical to base predictions or calculations of solar radiation on attenuation of the extraterrestrial radiation by the atmosphere, as adequate meteorological information is seldom available. Instead, to predict solar performance, we use past measurements of solar radiation at the location in question or from a nearby similar location.[146]

"New Airbags: Low Tech, Low Price" by David Kennedy presents a problem and predicts a solution, a not unusual combination:

The airbag, designed to protect passengers in car crashes, has so far resulted mainly in heated disputes. Consumer groups and insurance companies favor it, but the airbag costs at least several hundred dollars because of its complexity. Some consumers balk at paying that much, and the automobile industry doesn't want to increase sticker prices. The National Highway Traffic Safety Administration (NHTSA) has been preoccupied for over a decade trying to decide whether to require automakers to install the safety device on all new cars. This dilemma may be coming to an end: a simple, cheap airbag may soon be available that will satisfy all parties.[147]

Chapter 6, "The Thyroid Gland," of *Comparative Endocrinology* introduces many problems:

It is possible to define the thyroid gland in a fairly simple way as the vertebrate tissue that can accumulate iodide in great excess and can combine it chemically in the organic compound thyroxine. Defined in this way, the thyroid gland appears refreshingly simple. It may be contrasted with the pituitary and adrenal glands or the gastrointestinal tract, which produce numerous hormones, some of them complex proteins that, unlike thyroxine, usually vary from one species to the next. Unfortunately, much of the apparent simplicity of the thyroid gland disappears upon closer study, as we shall see, and some of the knottiest problems in comparative endocrinology involve this gland.[148]

"Science and the Citizen" opens with a national problem:

Facing the budget-cutting imperatives of the Gramm–Rudman Act and the steep tilt of the Reagan Administration's 1987 budget toward military research, those government agencies, universities and private laboratories that rely on Federal support for nonmilitary research and development are casting about for ways to cope with straitened circumstances. As they do so they hope attitudes in Congress toward the Administration's budget and certain trends in the economy may make their task easier.[149]

Many journal articles begin by stating a problem:

Although much attention has been given to the construction of $D$-optimal exact (integer replication) experiments, criteria concerned with the variance of the response estimator, such as $G$ optimality, have received scant consideration in the literature. This is inappropriate because, in practice, many experiments are aimed at estimation of the response over the region of interest rather than parameter estimation. Moreover, $D$ and $G$ optimality are not necessarily equivalent for exact designs.[150]

We consider the following problem: Given a set of $n$ data points $\{(x_1, y_1), \ldots, (x_n, y_n)\}$, how can we summarize the association of the response $y$ on the predictor $x$ by a smooth, monotone function $s(x)$? Put another way, how can we pass a smooth, monotone curve through a scatterplot of $y$ versus $x$ to capture the trend of $y$ as a function of $x$? This problem is related to both isotonic regression (e.g., see Barlow et al. 1972) and scatterplot smoothing (e.g., see Cleveland 1979).[151]

Traditional approaches to image restoration or resolution enhancement have typically involved two main steps. First, an image is formed with a given imaging system. Second, this image is then processed by using an image-restoration algorithm. Little attempt has been made to take into account the relationship of these two steps. Better results should be achieved by a procedure in which the image-gathering system is designed specifically to enhance the performance of the image-restoration algorithm to be used. We use the term "image-gathering system" here instead of "image-forming system" because no recognizable image may actually be formed until after the image-restoration algorithm has been executed.[152]

Negative luminance contrast conveys a wealth of information in such varied forms as typescript, backlit objects, shadows, and the myriad of objects that reflect less light than their backgrounds. Nevertheless, the perception of negative luminance contrast has not been studied as intensively as that of positive luminance contrast. Thresholds for the detection of decremental stimuli only tap the beginning point of the suprathreshold realm of negative contrast and the associated perceptual continuum that runs from the light gray to deep black. In this paper we explore this suprathreshold realm by obtaining some measurements of the relation between luminance contrast and *perceived contrast*. We define perceived contrast as the *perceived difference* between a target and its background.[153]

A number of optical systems exist for which it would be desirable to have a diffuser whose directional characteristic of scattering could be shaped.

Imaging with diffraction noise reduction, Fourier transform (FT) hologram recording, and incoherent systems in which the diffusers are used as view screens or reflecting filters of infrared radiation are examples.[154]

Discontinuities in dielectric waveguides are assuming increasing importance in the design and development of optical and millimeter-wave components. A longitudinal discontinuity problem arises in the splicing of two dielectric waveguides and in the design of the associated connector.[155]

As Reve and Stern (1979) indicate, "A main purpose of recent channel research has been to demonstrate how a study of power relationships in marketing channels may provide useful insights into channel functioning and interorganizational interactions" (p. 407). While numerous articles have appeared on this general topic over the past decade, only a small number of major field studies dealing with actual channel relationships have been conducted. Furthermore, some of the results from these studies appear to be inconsistent with what one might expect from existing theory. As a result, the literature on power relationships in channels has become based heavily on theories and research findings developed in the related behavioral disciplines of social psychology, sociology, and political science. Although this related research provides an important base from which to begin, additional field studies are essential to advancing our understanding of interfirm interactions in distribution channels.[156]

Paul Meier illustrates how neatly a writer may handle an inappropriate assignment of topic:

To cover my assigned topic comprehensively would take a book. What I undertake to do instead is to examine some of the differences between the methodology employed in the early 1950's, when clinical trials were first taking hold in this country, and the effect of statistical research on the methodology currently in vogue.[157]

In the same *Memorial Symposium in Honor of Jerome Cornfield*, P. Armitage discusses a problem about which Cornfield was concerned:

During the last few years of his life, Jerry Cornfield became increasingly concerned with the problem of assessing the risk to the general population of exposure to low doses of substances that may possibly be carcinogenic. This is inherently a statistical problem, although by no means exclusively so. It is also an extremely difficult problem, as we shall see. It has exercised the minds of many of the most able biostatisticians in this country, several of whom have been Jerry's associates at NIH and elsewhere, and many of whom are present today. Jerry was an original thinker, unlikely to follow established lines of thought or to expect others to agree readily with his own views. So we shall find his approach to this topic at deviance to that of many of his colleagues. I want to try to place his work in the context of that of his contemporaries, and to identify the particular contributions he made.[158]

134

To begin "Accurate Assessment of the Optical Quality of Infrared Systems" D. R. J. Campbell chose a problem and an advantage of solving it:

> The accelerating use of far-infrared (8 to 12 μm) thermal imaging systems for civil and military applications has required the development of a host of new techniques for the fabrication and testing of such lenses. The sophistication of the systems coupled with the higher materials costs, compared with optical systems, effectively means that any system cannot fail its acceptability tests without incurring a very high cost increase in the same production run. This implies that the effective scrap rate must be negligible to maintain the cost effectiveness of these systems. New materials, mounting techniques, and more complex optical engineering compound the problem, for, coupled with the visual opaqueness of most materials used, classical techniques for establishing the usual constructional parameters are no longer available.
>
> Apart, therefore, from establishing adequate testing methods in this waveband, there is a real advantage in providing tests that give a diagnostic capability. If such tests were available, then any lens failing its acceptance test would have the reason for failure determined, and remedial action could be undertaken and special attention given to that area in the lenses following it. Such a system would also be ideal for evaluating any of the new optical engineering techniques that are being used to make such systems easier and therefore more economical to build.[159]

## QUESTION

What easier way to begin than with a question? Writers find this beginning so satisfactory that they feel lucky when their material allows them to catch the attention of readers with a query.

David H. Hubel begins an article on the brain with questions that he answers immediately, thus clearing away common misconceptions:

> Can the brain understand the brain? Can it understand the mind? Is it a giant computer, or some other kind of giant machine, or something more?[160]

Louis E. Underwood uses inquiries to outline his report on a conference that did not supply all the answers:

> When unlimited quantities of biosynthetic human growth hormone become available, how will physicians decide which children should receive therapy? Is it ethical to administer growth hormone to short children who are believed not to be deficient in the hormone according to current criteria? Will such treatment produce taller (or better) adults? What are the possible adverse side effects of pharmacologic doses of growth hormone, and what should physicians, growth hormone producers, and regulatory agencies do to prevent misuse? These are but a few of the questions that were addressed by a panel of nearly 50 experts convened November 22 and 23, 1983, by

Mortimer B. Lipsett, director of the National Institute of Child Health and Human Development, for a conference on the uses and possible abuses of biosynthetic growth hormone.[161]

Questions are springboards for writers' discussions:

Where are the men and women who, ten years from now, will fill the senior management positions in your organization? Perhaps more importantly, what is the process by which those people will arrive in those positions ten years from now?[162]

How do managers, both individually and as a group, respond to crises?[163]

Are banks and savings and loan associations really that bad? Is there a pattern of across-the-board employment discrimination against minorities and females? Do financial institutions consciously structure their personnel policies to inhibit the hiring opportunities of minorities? Once hired, are minority and female employees channelled into low-end, low-paying, dead-end jobs?[164]

What conditions are necessary in order to attract utilities to invest in improved versions of the light-water reactors that produce most of the nuclear power generated in the U.S. today?[165]

What will happen to the robot industry once its primary market, the auto makers, has enough "steel collar" workers, as is expected to happen in a little over five years?[166]

Some questions require introductory background. Eberhard Gwinner's question is all the more effective for the preceding material:

The sight of a flock of birds migrating south in the fall or north in the spring hardly ever fails to evoke a sense of wonder. The flight may be the orderly aerobatics of a V of Canada geese or the ragtag progress of a group of starlings. Whatever its details, the overwhelming impression conveyed to the observer is that of a powerful inner impulse. The birds do not hesitate in their flight; they travel smoothly and unerringly toward a goal far out of the viewer's sight. Where does the impulse come from that guides the birds toward warmer climates in winter and brings them back to their northern breeding grounds in the spring?[167]

Also, in "The Development of the Brain" by W. Maxwell Cowan introductory background leads to questions:

The gross changes that take place during the embryonic and fetal development of the brain have been known for almost a century, but comparatively little is known about the underlying cellular events that give rise to the particular parts of the brain and their interconnections. What

136

is clear is that the nervous system originates as a flat sheet of cells on the dorsal surface of the developing embryo (the neural plate), that this tissue subsequently folds into an elongated hollow structure (the neural tube) and that from the head end of the tube three prominent swellings emerge, prefiguring the three main parts of the brain (the forebrain, the midbrain and the hindbrain).

It is not on these changes in the external form of the developing brain, however, that the attention of developmental neurobiologists has focused in recent years. More interesting questions intrude. How, for instance, are the various components that constitute the major parts of the nervous system generated? How do they come to occupy their definitive locations within the brain? How do the neurons and their supporting glial cells become differentiated? How do neurons in different parts of the brain establish connections with one another?[168]

Gene Bylinsky, writing in *Fortune*, uses introductory questions briefly:

Factories are humming again, manufacturing employment is up 6% in the past year, and the term "rust bowl"—shorthand for the noncompetitiveness of much of American industry—fading from use. But ominous questions remain. How many more products will lose the "Made in U.S.A." label? Can advanced production technology overcome the high-wage handicap of American industry—of steelworkers, for example, who are paid eleven times as much as South Korea's? Even if colossal infusions of capital can keep old-line industries competitive without protection, will the investment yield an adequate return?[169]

## QUOTATION

"Next to the originator of a good sentence is the first quoter of it. By necessity, by proclivity, and by delight, we all quote"—Emerson. Beginning with a quotation is one of the most popular introductions. A quotation that seems to have been designed just for a writer's need is irresistible to writer and reader. An author can use such a passage with admirable effect, as Catherine Caulfield does, melding gracefully a particularly apt quotation with her own thinking:

"Its lands are high; there are in it many sierras and very lofty mountains. . . . All are most beautiful, of a thousand shapes; all are accessible and are filled with trees of a thousand kinds and tall, so that they seem to touch the sky. I am told that they never lose their foliage, and this I can believe, for I saw them as green and lovely as they are in Spain in May, and some of them were flowering, some bearing fruit, and some at another stage, according to their nature." This is how Columbus described to Ferdinand and Isabella the forests of Hispaniola—the island that is now divided into Haiti and the Dominican Republic. The description contains the seeds of two different but not incompatible views of the rain forest—the romantic and the scientific, each of which still has its adherents.[170]

Don P. Wyckoff uses a quotation both as title and introduction for his article "Let There Be Built Great Ships":

> The order from Prime Minister Winston Churchill was in his usual grand style: "Let there be built great ships which can cast up on a beach, in any weather, large numbers of the heaviest tanks."[171]

Journalists sometimes break a quotation to suit the brevity necessary in leads and reserve more extensive quotation for later paragraphs, as Ronald Sullivan does:

> "Nowhere near enough" doctors are being disciplined by state medical boards, and organized medicine "can no longer abide a brotherhood of silence" on the incompetency of some physicians, Dr. Otis R. Bowen, Secretary of Health and Human Services, said yesterday.
>
> Dr. Bowen's remarks to the graduating class of the New York University School of Medicine coincided with a report by his inspector general that the state and Federal health authorities should have more power to weed out incompetent physicians and to discipline others for misconduct or negligence.
>
> Dr. Bowen said, "Experts have estimated that anywhere from 5 to 15 percent of practicing physicians would be candidates for some kind of disciplinary action, many because of drug or alcohol problems."
>
> But he said the inspector general's report "indicates that nowhere near that percentage of formal actions actually occur in a year" against the nation's 500,000 doctors.[172]

A writer may use a quotation for its flavor, for the style of the expression. Laurence Shames introduces "Wharton Reaches for the Stars" with a breezy quotation, his own style in the rest of the article being less colloquial:

> "When I came here," says Russell E. Palmer, dean of the University of Pennsylvania's Wharton School since 1983, "I tried to look at Wharton from the viewpoint of the marketplace. And I realized that if we asked the question, 'What is Wharton known for?' the marketplace would sing out, 'Finance.' That was the perception."
>
> Settling into his commodious armchair, locking his hands behind his head, Palmer continues. He is 51 years old, with clubby good looks and hair the color of silvery sand. "Finance is fine," he says, "but we're also damn good at turning out entrepreneurs—the Donald Trumps, the Michael Milkens, the Saul Steinbergs—people who went on to make a very personal mark in business. That side of Wharton was never really talked about before. Well, we're talking about it now."[173]

In *Natural History* Richard Wassersug uses a brief quotation effectively to open his article:

As Martha Crump pointed out in these pages several years ago (January 1977), there are "many ways to beget a frog." The vast majority of frog species deposit eggs in water. These hatch into tadpoles and subsequently metamorphose into frogs. Crump showed that this is only one of at least ten distinct reproductive modes for anurans—short-bodied, tailless amphibians such as toads, frogs, and tree frogs.[174]

Indirect quotations, too, are frequently good beginnings, particularly where nothing is lost in color or style. The following leads composed of indirect quotations come from two columns of one page of *The New York Times:*[175]

The Exxon Corporation said yesterday that it planned to withdraw more than $1 billion in surplus pension funds and use the money for general corporate purposes.

The action would have no significant impact on 1986 earnings, the company said.

The Union Pacific Corporation said yesterday that it had reinstated an agreement with Katy Industries under which it would acquire the Missouri–Kansas–Texas Railroad for $110 million.

Viacom International, the diversified media company, said it would report a loss for the second quarter as a result of its previously announced repurchase of stock held by the New York investor Carl C. Icahn. The repurchase will result in a charge of $28 million. In last year's second quarter, the company earned $9.5 million.

Poor communications between the A. H. Robins Company and its Chapter 11 bankruptcy lawyers led company officials to make debt payments without court approval, Robins's attorney, Dennis Drebsky, said today.

The Scott Paper Company, a leading producer of toilet tissues and paper towels and napkins, said it had agreed to sell its Cut-Rite brand of wax paper to the Reynolds Metals Company. The company would not disclose the purchase price, but said that the gain from the sale would "more than offset" an $8 million one-time charge associated with an early retirement program put into effect by Scott during the second quarter.

Spantax, Spain's charter airline, has reached agreement in principle to sell a stake to the Texas Air Corporation, a spokesman, Rafael Chavarri, said. He said the stake to be acquired by Texas Air was still under negotiation. "If Texas Air comes up with the right terms we are not against a majority takeover," he said.

The International Minerals and Chemical Corporation, one of the world's largest fertilizer producers, said it expected to report a fourth-quarter net loss of between $105 million and $150 million, largely because of current difficulties in its industrial and energy sectors.

139

The Western Union Corporation said that its annual stockholders meeting, set for next Thursday, had been postponed. The company said the delay had been caused solely by the amount of time required to complete a complex Securities and Exchange Commission registration process related to its previously announced financial restructuring. There has been speculation on Wall Street, however, that changes may be made in the restructuring plan, including the possibility that an outsider has been found to put equity in the company. Western Union declined comment.

The sixth example follows an indirect quotation with a direct one, a useful combination for writers who wish to give the gist in their own words and still have the credibility of a direct quotation. The eighth example, quoted in its entirety, uses the indirect quotation for the company's view and summary for the speculation on Wall Street. Journalists find quotation very useful as these examples from one page show. The danger is that these writers may overwork this beginning. Other informational writers have yet to exploit this workhorse and might find that it solves their problem if a beginning containing an opinion of their own requires support; if dull information needs a little glitter, like a distinguished source; if a summary opening seems lifeless; if an eye-catcher is needed in the beginning; if their own style is not equal to the task of creating the atmosphere, freshness, authority, or force of an apt quotation.

These eight direct and indirect quotations on financial information and opinions present each subject briefly. Such quotations are useful for memorandums, short reports, and letters of information and may be used to shorten or bend to a writer's own needs wordy, poorly written, too detailed press releases that the writer must summarize. After beginning with a good direct or indirect quotation, a writer usually does not have to write much more.

Some writers paraphrase well-known material instead of using direct or indirect quotation, as Glen Martin does in the opening sentence of "Magic of the Sea of Cortez":

> To paraphrase Lawrence of Arabia, I like the desert because it is clean. More specifically, I like the juncture of the desert with the sea. There the cleanliness seems somehow transcendent.[176]

In considering paraphrase, writers should be careful to avoid it if the paraphrase has less literary or thought value than the original statement. Why make writing worse by paraphrasing?

Quotations as beginnings appear in a variety of lengths and of combinations with other types of beginnings. The quotation opening Chapter 10 of *Deep-Sea Biology* is brief:

> The late E. W. Fager taught that "pattern [of organisms in space] is one of the easiest obtainable clues to interaction in a non-uniform world." The following is an attempt to apply and extend the methods he taught for

deciphering these clues to data from the deep sea, where spatial patterns have been among the few clues accessible to oceanographers.[177]

A stock-market phrase, *the triple witching hour*, suggests the quotation that introduces James Russell's story in *The Miami Herald*:

"Double double, toil and trouble; fire burn and cauldron bubble."

The chant of the witches, as Shakespeare depicted it in MacBeth, wasn't heard on Wall Street until a few minutes before 4 p.m. Friday. When it came, this three-way expiration of stock index futures, index options, and options on individual stocks drove blue-chip shares up sharply and left the rest of the market broadly mixed.

The market pros call this quarterly event the "triple witching hour," and they expect wild action as computer-directed trading programs are triggered.[178]

An Associated Press release neatly combines direct and indirect quotation:

Now that the sort of nuclear accident she predicted for years has come to pass, Dr. Helen Caldicott says she has too little energy left to do much about it.

"I've been doing this for 16 years and I'm very tired," said the anti-nuclear activist who left her teaching job at Harvard Medical School six years ago to work full-time warning the world of the dangers of nuclear power.[179]

Owen F. Devereux opens Chapter 1, "Foundations of Thermodynamics: The First and Second Laws and Their Use," by quoting thoughts very important to his subject:

Gibbs prefaced his classical work with the words of Clausius:

Die Energie der Welt ist constant.
Die Entropie der Welt strebt einem Maximum zu.

These statements may be regarded as two fundamental postulates, or laws, which summarize actual experience and, in fact, constitute the *First and Second Laws of Thermodynamics*. They may be paraphrased in less pithy but more pragmatic terms:

1. The expenditure of a given amount of work, no matter what its origin, always produces the same quantity of heat.
2. The complete conversion of heat into work is impossible without causing some effect to occur elsewhere.

With some qualifications, much of the remainder of thermodynamics is the result of applying simple and established mathematical procedures to these statements.[180]

141

Frank Graham, Jr., begins with four lines of poetry that lead neatly into his book review:

> Lord Byron, in his long poem *Don Juan*, went out of his way to condemn the cruelty implicit even in such a contemplative diversion as angling:
>
> "Whatever Izaak Walton sings or says:
> The quaint, old, cruel coxcomb, in his gullet
> Should have a hook,
> And a small trout to pull it."
>
> Not an orthodox expression of humane sentiments, to be sure (though likely to give umbrage to devoted Waltonians). Yet it serves to illustrate the range of feelings that humans harbor toward their fellow travelers on this planet. Respect, contempt, love, revulsion, awe, irrational hatred, utter indifference—these and more are found in any sampling of opinion about animals, and all of them may be expressed at one time or another by the same person.
>
> This welter of human feelings, this inextricable mix of compassion, cruelty, and soppiness, is the subject of a new book, *Man and the Natural World: A History of the Modern Sensibility*, by Keith Thomas.[181]

The quotation at the beginning of "Saving the Bounty of a Harsh and Meager Land" introduces and enriches the article:

> "This is Pinacate country, the end of the line. It is one of North America's empty places. Everywhere dunes, hog rocks, and craters shimmer in the heat. There are no cities or towns or lonely ranchos. The rain falls less than five inches a year, and sometimes it forgets to fall at all.
> "And all around are beans and squash and corn. A garden grows miles from any gardener or any faucet. The young researcher, Gary Nabhan, checks the weather station and discovers less than two inches of rain have fallen in the month since the garden was planted. The low has been 72 degrees F, the high 116 degrees in the shade.
> "In the next month there will be no rain. In the next month the temperature will range from 100 degrees to 115 degrees. And then an Indian family will come and harvest the fields. And then their friends will come and celebrate the harvest. And something very old in this desert will have been done yet again. A crop will have been made off less than five inches of rain."

This is how Charles Bowden of the *Tucson Citizen* describes the traditional Indian agriculture in the deserts of the Southwest. Over much of that area sporadic summer thundershowers and brief winter rains total only three to twelve inches of moisture annually. Except in times of rain, streams never flow. Heat makes the ground a shimmering furnace. Aridity makes the growing seasons short. Yet in this harsh and meager land, Apache, Havasupai, Hopi, Mohave, Navajo, Papago, Pima, and Yuma Indians developed an agricultural civilization that was most remarkable. They had dozens of useful plants that were adapted to the sun, soils, and sporadic rains. On the wide, shining flat of the desert they grew crops with less direct rainfall than is used anywhere else.[182]

Some good beginnings come from a writer's twisting of a well-known phrase or thought. Fred M. Hechinger opens an article in *The New York Times* with a twist on an advertising slogan and adds two paragraphs illustrating the twist:

> The well-known advertising slogan "You've come a long way, Baby," playfully celebrating women's progress, might better be reworded "You had a long way to go" when applied to the top rungs of the academic ladder. It explains why today only few full professorships are held by women.
>
> Past "theories" about the education of women today sound like silly jokes. In the 1860's reputable physicians and educators feared that higher education would leave women with "monstrous brains and puny bodies," unfit for childbearing.
>
> Outrageous? Well, perhaps more flagrant but not all that different from misconceptions that rationalized more polite discrimination way into the 1960's. Late in the 1950's I interviewed Prof. Cecilia Froehlicher at the City College of New York because she was the first woman to chair a university electrical engineering department. She pointed at a framed diploma that welcomed "Brother Cecilia" to the honorary engineering society.[183]

A United Press International release plays on a familiar phrase: "American medical schools are afraid they won't be able to get enough skeletons for their closets."[184]

Quotations, direct or indirect, combine effectively with the straw-man technique:

> Fewer jobs, it was predicted sadly. More cockroaches, it was hissed desperately. Higher prices, it was warned knowingly. Undaunted, the people of New York went ahead and approved a bottle bill. A little more than a year later, the soothsayers of doom have been proven wrong on all counts.[185]

Background and quotation are also compatible:

> "Always something new out of Africa," wrote Pliny the Younger 1,700 years ago, and Africa is still exerting its enigmatic allure—*Out of Africa*, the Hollywood movie based on Danish writer Isak Dinesen's memoirs, swept this year's Academy Awards and inspired this summer's on-the-savannah approach to fashion.[186]

The unexpected in a quotation can be a welcome different beginning for a subject swamped in articles with traditional openings:

> A century ago Frédéric Auguste Bartholdi set the Statue of Liberty, torch in hand, at the mouth of New York Harbor as a beacon of freedom to all who entered the United States. He chose his site with care: "If any place in the world needs light," he said, "it's New York City."[187]

And a civil engineer effectively combines quotation and questions to provide a personal touch in his paper:

"You can't convince me that it won't pollute our lake." A feeling of help-lessness overcame me as I realized their minds were thoroughly made up.

People were beginning to filter into the Zone Change Hearing in Coeur d'Alene, Id., where I was proudly discussing the obvious advantages of the new sanitary landfill site to early arrivers. Technically, the landfill issue was not part of the zone hearing, but the country commissioners had invited us on an informal basis.

As part of the consulting team charged with siting the new Kootenai County Sanitary Landfill, I had wrestled with some basic questions. How do you tell these people that there is no risk to their lake? That the landfill will be a good neighbor? That the environment will be thoroughly protected?[188]

## SCOPE

The examples below illustrate beginning with the scope of a paper as well as with scope combined with other methods of beginning. A few cautions follow the examples. A statement of the limitation of the subject or of the amount of detail to be presented is sometimes a necessary be-ginning if readers are not to be misled into wasting time. When the open-ing sentences establish that three computers manufactured by X Cor-poration have been surveyed (a) for price, (b) for dates of availability, and (c) for adaptability to the operation of Z Corporation, a reader will not waste time looking through the paper for information about the expe-rience of other companies with these computers or for information about the computers manufactured by Y Company.

Scope is favored by writers as an opening for chapters in books. Such openings are likely to be plain, simple, and direct, particularly in text-books. Chapter 5 of *Patent and Trademark Tactics and Practice* is prefaced by two sentences:

> Conducting a patentability search and preparing a patent application are two of the most important stages in efforts to pursue patent protection. This chapter points out pitfalls to avoid in both stages.[189]

Chapter 3 of *Solar Engineering of Thermal Processes* opens with scope:

> This chapter is intended to review those aspects of heat transfer that are important in the design and/or analysis of solar collectors and systems. It begins with a review of radiation heat transfer, which is often given cursory treatment in standard heat transfer courses. The next sections cover con-vection heat transfer between parallel plates. The final sections review some convection correlations for internal flow and wind-induced flow.[190]

Chapter 7 of *Physical Processes in the Interstellar Medium* begins as follows:

144

Dust grains scatter and absorb light in interstellar space. In addition, these small solid particles emit radiation at wavelengths much greater than those of the absorbed light. Knowledge of the optical properties of grains can be used to interpret the observed effects in terms of the nature and spatial distribution of these particles.[191]

Chapter 13 of the same book begins

The influence of gravitational fields on the motion of interstellar gas is treated here for three idealized situations. First, the motion in the neighborhood of a mass concentration, either a star or some more massive object, will be considered. Second, motions in the galactic plane under the influence of large-scale density waves, such as are presumably associated with spiral arms, will be discussed. Third, a brief analysis will be given of certain processes occurring when gravitational instability sets in, triggering the complex series of events leading to star formation.[192]

In *Dynamics* Chapter 4 combines definition and scope:

A macroscopic object that cannot be deformed by the forces that act on it is called a *rigid body*. As with "particles" or "nonviscous fluids," a rigid body is a mathematical idealization, which is never realized exactly. Nevertheless, the dynamics of a great many commonplace objects, in many diverse situations, is well modeled by a rigid-body description.

Aside from the usefulness of rigid-body dynamics as a physical model, a study of rigid bodies leads to many valuable insights, since it brings together nice examples of Lagrangians, constants of motion, and moving coordinate systems. Many topics discussed in previous chapters find application in a unified context as we study in detail the motion of rigid bodies.

We begin with some general considerations and then discuss the inertia operator and its role in rigid-body motion. Euler's equations for rigid-body motion are studied, both with and without external torques. The energy and angular momentum constants serve as a basis for a thorough study of the torque-free case.[193]

The introduction of Chapter 8 of *Comparative Endocrinology* explains the background and then states the scope:

The efficient digestion of a meal requires the coordination of an amazing number of physiological events. The digestive process is initiated by the meal itself and is integrated locally by the successive segments of the digestive tube. Coordination of digestion is achieved through neural and hormonal responses to the physical and chemical properties of the food within the lumen of the gut. In this chapter we propose to consider mainly the endocrine mode of integration of the digestive process.[194]

The introduction of an article in the *Journal of Chemical Physics* uses recent background followed by the scope:

145

Recent investigations of hydrogen diffusion in potassium tantalate ($KTaO_3$) and the interactive role played by transition-metal impurities have led to renewed interest in the solid state properties of doped tantalate crystals. In particular, work is currently in progress in which the effect of hydrogen on the high-temperature electrical properties of $KTaO_3$ is being investigated. In the specific case of $Fe^{3+}$ in single crystals of $KTaO_3$, it has been shown that the conditions resulting in the formation of OH in $KTaO_3$ also produce a significant change in the relative intensities of the cubic- and axial-site $Fe^{3+}$ EPR spectra. Accordingly, it is of interest to investigate the relationships between the properties of other transition-metal impurities and the diffusion of hydrogen. The purpose of the work reported here is to provide fundamental information regarding the local environments of $f^n$ and $d^n$ electronic impurities in $KTaO_3$ as a basis for future studies of the effects of hydrogen in doped tantalate crystals. In the present study, the EPR spectra of $Yb^{3+}$ and $U^{5+}$ in $KTaO_3$ were observed for the first time, and more complete EPR results are given for $Cu^{2+}$ in $KTaO_3$—a system that was the subject of an earlier preliminary report. Additionally, EPR spectroscopy was used in a reexamination of $KTaO_3$ doped with $Co^{2+}$, $Mn^{2+}$, and $Ni^{3+}$—ions that along with $Eu^{2+}$, $Gd^{3+}$, $Fe^{3+}$, and $Ti^{3+}$ have been observed in earlier work.[195]

In *Statistics for Experimenters* Part 1, "Comparing Two Treatments," briefly details the necessary background and then states the scope of Part 1:

> Experiments are often performed to compare two treatments, for example, two different fertilizers, machines, methods, processes, or materials. The object is to determine whether there is any real difference between them, to estimate the difference, and to measure the precision of the estimate.
>
> In this part of the book we discuss the design and analysis of these simple comparative experiments. We describe pitfalls that await the unwary, for example, how violation of the assumption of statistical independence of successive errors can seriously invalidate standard procedures. By understanding the nature of this difficulty and others one is led to appreciate the crucial importance of randomization, replication, and blocking.
>
> In this context significance tests and confidence intervals are introduced, first for comparison of means, and later for comparison of proportions and frequencies. The ideas introduced are of much more general application and are used in the design and analysis of more complicated experiments in later parts of the book.[196]

Stating the scope—the extent of the subject covered or the amount of detail included or both—is useful if readers are expecting a different scope. But scope is not usually an interesting beginning, particularly as it reminds some readers of the openings of dull classroom lectures. Sometimes it is about as compelling as opening with a summary of the preceding chapter, another dull device that also smacks of academia, where some speakers start every lecture with a review of the preceding one.

146

It is essential, therefore, that writers vary the style to avoid the monotony characteristic of a scope beginning:

> In this chapter we recall briefly the formulation of dynamics given by Newton. Our approach here is mainly one of review. We first focus on dynamics of a single particle, including forces of constraint. We then generalize to many-particle systems. We consider one-dimensional systems in detail and introduce the notions of phase space and phase curves. Phase flow is defined along with the concept of Liapunov stability. We conclude this chapter by considering open systems, that is, systems that may exchange mass or momentum with the surroundings.[197]

Sometimes the tediousness of a poorly expressed beginning can be eliminated by placing the scope in the title and using a different beginning. But please, no dull presentation of scope in the title and also in the first sentences—or worse, in the title, in the abstract, and in the first sentences of the article.

## STRAW MAN

The straw-man beginning, in which a writer opens with an argument and then knocks it down, daunts some writers. They maintain that it gives too much space and emphasis to ideas opposed to theirs and that if writers do not reason effectively readers may accept the introductory statement rather than the counterarguments. They believe also that some readers are annoyed by attacks on opinions dear to them, particularly if their opinions are refuted only by other opinions. Finally, they think that for some papers, some writers, and some readers the contentious atmosphere created by this method is inappropriate. Yet many writers find it an effective opening for the following reasons:

- It disposes of the opposing point of view at the beginning, and therefore as the paper can be less contentious throughout, the reader can give full attention to the writer's view.
- Disagreement even on dull subjects can be lively.
- A writer can clarify the point or points of disagreement early and thus help readers who have incorrect notions about the argument.
- Many writers function best in combat, and argument can create an impression of a forceful, logical, courageous writer.
- The chance to set up an opponent is irresistible.
- Readers enjoy seeing an argument won neatly.

Marilyn Gardner's report of a national conference of young women briefly reiterates the stereotyped "Cinderella fantasy" and then quickly

147

demolishes it by a description of the realities of a woman's life in the eighties:

> Two decades after the women's movement declared him obsolete, Prince Charming lives.
>
> For millions of American girls, the Cinderella fantasy remains a powerful influence as they think about their future lives. Despite the fact that many of their own mothers work outside the home and support families alone, these young women cling to the notion that they need not prepare for a career because they will be provided for—royally.
>
> At a recent conference bringing together 300 delegates from across the country, the Girls Clubs of America took note of the revival of this dangerously dated myth, then tried to sound final curfew on it with disenchanting facts guaranteed to turn the coach into a very unromantic pumpkin.
>
> For every latter-day Cinderella saved from a life of toil by a handsome prince—or at least a husband with an MBA—a hundred others will need to earn part or all of their family's income. Projections show that girls today will work for an average of 30 years. By the end of the century, 80 percent of women will be employed outside the home while raising children, with two-thirds clustered in low-paying, traditionally female jobs.[198]

"The Stress of Being Born" begins with common misconceptions and then corrects them:

> At first thought, being born would seem to be a terrible and dangerous ordeal. The human fetus is squeezed through the birth canal for several hours, during which the head sustains considerable pressure and the infant is intermittently deprived of oxygen (by the compression of the placenta and the umbilical cord during uterine contractions). Then the neonate is delivered from a warm, dark, sheltered environment into a cold, bright hospital room, where some large creature holds it upside down and in many cases slaps it on the buttocks. In addition, during the strains of birth—particularly hypoxia (oxygen deprivation) and pressure on the head—the fetus produces unusually high levels of the "stress" hormones adrenaline and noradrenaline—higher than in such severely taxed adults as a woman giving birth or a person having a heart attack. Adrenaline and noradrenaline, which are major representatives of a chemical class called catecholamines, typically prepare the body to fight or flee from a perceived threat to survival (a stress), and high concentrations of these hormones often indicate that the organism is in danger.
>
> In spite of surface appearances, the stresses of a normal delivery are usually not harmful.[199]

A short article in *Science 84* sets up old beliefs and then offers a new one to replace them:

> To calm a horse, veterinarians still rely on a device dating back at least

to the Middle Ages. It's called the twitch. When this loop of metal or rope is tightened around a horse's upper lip, the animal typically becomes quiet, sometimes even droopy, and meekly submits to a medical treatment or a new set of shoes. Veterinarians have assumed that the pain in the lip merely diverts the horse from reacting to other irritations. Now Dutch scientists have another explanation: The twitch is "a Western example of acupuncture."[200]

Writers who disagree with a previously published article may have to summarize the material with which they disagree at or near the beginning of a paper if they are to argue against it effectively. "The Economics of the Diaspora Revisited" begins with such a summary:

> In a recent issue of this journal, Reuven Brenner and Nicholas Kiefer raised the issue of the effect of the refugee status of an ethnic group on the levels and types of investments made by members of the group. Their hypothesis is that refugees will overinvest in human capital and underinvest in physical capital. They repeat the often heard point that Jews in the Diaspora invest more in human capital than others because of a lingering fear of expropriation and expulsion, a fear well grounded in the historical experience of the past 2,000 years. Brenner and Kiefer then consider the general nature of their hypothesis by examining data for Palestinian Arabs. The data show higher levels of schooling and occupational status for U.S. Jews compared with non-Jews and for Palestinian refugees compared with most other Arab groups in the Arab Middle East. We do not dispute these patterns. What is at issue, however, is the interpretation or explanation of these patterns.
>
> We have three objections to the Brenner-Kiefer article. First, it is ambiguous whether a refugee experience will result in greater investments in human capital than otherwise or greater investments than residents of the host country: ceteris paribus. Second, an examination of data on the earnings of Jews in the United States does not provide support for the hypothesis that Jews overinvest in human capital. The earnings analysis finds that Jews have a higher rate of return on human capital than non-Jews. Third, the higher level of schooling among Palestinians in the Arab Middle East is consistent with the fact that the first refugees from the 1947–48 war were disproportionately from the urban middle and upper classes in Palestine: that prior to 1948 Palestinian Arabs had higher levels of education than Arabs in neighboring countries except Lebanon; that changing attitudes toward the role of women led to a vast increase in female education; and that the United Nations Relief and Work Agency (UNRWA) provided Palestinian refugees with educational facilities superior to those available to others in the area. Hence, we would expect the Palestinian refugees to have a higher level of education and occupational status than the general population of the Arab countries. Moreover, the data cast doubt on the allegations that the fields of study chosen by Palestinian Arabs are significantly different from the general pattern in the area. The Brenner-Kiefer analysis does not demonstrate that the refugee status per se determined the human capital investment decisions of Palestinian Arab refugees.[201]

The opening of Chapter 6 of *Comparative Endocrinology* offers an inaccurate opinion, which the authors oppose:

> It is possible to define the thyroid gland in a fairly simple way as the vertebrate tissue that can accumulate iodide in great excess and can combine it chemically in the organic compound thyroxine. Defined in this way, the thyroid gland appears refreshingly simple. It may be contrasted with the pituitary and adrenal glands or the gastrointestinal tract, which produce numerous hormones, some of them complex proteins that, unlike thyroxine, usually vary from one species to the next. Unfortunately, much of the apparent simplicity of the thyroid gland disappears upon closer study, as we shall see, and some of the knottiest problems in comparative endocrinology involve this gland.[202]

Kenneth L. Woodward combats the set-up argument strongly in *Psychology Today*:

> Is Mao's new China a mental-health utopia? Some Western psychiatrists, psychologists and mental-health professionals would have us think so. The People's Republic of China (PRC), they claim, has not only eliminated much of the emotional conflict seen in Western societies but has found the keys—social solidarity and purpose—to positive mental health. Psychologist Martha Livingston and psychiatrist Paul Lowinger in their recent book, *The Minds of the Chinese People*, say, "Social organization solves most of the people's everyday problems without their having to resort to the mental-health establishment."
>
> Such claims are hopelessly romantic and naive, the modern equivalent of Marco Polo's tales of fabulous Cathay. They are also outdated, based on an already discarded line of propaganda about the virtues of Chinese socialism. Today, Chinese mental-health professionals are quite candid in acknowledging that the PRC is not an emotional Promised Land. Yes, they admit, some of those busy Chinese factory workers do suffer from depression, anxiety and personality disorders. Yes, some of those delightfully expressive Chinese children are hyperactive. Indeed, there are millions of diagnosed schizophrenics in China—though it will be a long time before the Chinese have the epidemiological data necessary for reliable comparisons with the West.[203]

In the same issue Brett Harvey begins a book review with opinions that disagree with those of the author of the book and then summarizes the argument of the book, a fresh use of the straw-man opening:

> Recent critics of the United States have accused us of being a nation of "narcissists" obsessed with our own psyches. Psychology and its practitioners, they say, have led us down the garden path to a national psychosis in which the immediate gratification of every desire has replaced the transcendent values of community, commitment and service to others.
>
> In *The Poverty Of Affluence* (The Free Press/Macmillan), Paul Wachtel

argues that, on the contrary, we are not psychological enough as a nation; that our malaise is more the result of our obsession with economic growth than of our search for self-fulfillment.[204]

In another interesting variation an expert corrects his own misconception. Gina Kolata opens "Proper Display of Data" with the past error:

> A few years ago, William Cleveland of AT&T Bell Laboratories began looking through scientific journals to see how researchers use graphs. There had been a revolution in graphic methods in their own field of statistics as investigators began developing new ways of analyzing and presenting data and Cleveland thought that similar developments might have occurred in other areas of science. "I figured, 'Look, there's this incredible talent in science. Surely, there's all kinds of amazing techniques invented that those of us in statistics don't know about. Let's find out about them and bring them into our tool kit,'" he reasoned.
> But, Cleveland remarks, "It didn't happen. I found just the opposite was true. I saw all kinds of errors and abuse [in the graphical display of data]."[205]

Maya Pines states opposing views briefly in "Resilient Children":

> Their parents are poor, illiterate, alcoholics, drug addicts, or schizophrenics. The children may be neglected or abused or may have severe physical handicaps. One would expect such youngsters to be seriously disturbed and to fail both at school and in adult life. But to the amazement of researchers, a number of these children do remarkably well despite all the strikes against them.[206]

In *Psychology Today* a paragraph of opinions is knocked down immediately:

> The dramatic rise in popularity of the martial arts has resulted in expressions of concern by some psychologists, law-enforcement officials and the general public, based on the assumption that training in the martial arts inevitably leads to increased aggressiveness on the part of the *karatekas* (martial arts practitioners). This assumption is strongly supported by the numerous motion pictures and television movies that sensationalize violence in the form of martial arts expertise.
> Scientific evidence, however, supports the hypothesis that martial arts training results in a decrease, rather than an increase, in aggressiveness. A study by T. A. Nosanchuk of Carleton University, for example, found an inverse relationship between aggressiveness and length of martial arts training.[207]

Conor Cruise O'Brien begins "Thinking About Terrorism" with his point of view, then quotes a long passage of one kind of thinking and uses it as a straw man, thus adding one more variation to the technique:

Terrorism is disturbing not just emotionally and morally but intellectually, as well. On terrorism, more than on other subjects, commentary seems liable to be swayed by wishful thinking, to base itself on unwarranted or flawed assumptions, and to draw from these assumptions irrational inferences, muzzily expressed.

Let me offer one example, typical of many more. The following is the conclusion to a recent *Washington Post* editorial, "Nervous Mideast Moment":

> The United States, however, cannot afford to let its struggle against terrorism be overwhelmed by its differences with Libya. That gives the Qaddafis of the world too much importance and draws attention from the requirement to go to the political sources of terrorism. A principal source, unquestionably, is the unresolved Palestinian question. The State Department's man for the Middle East, Richard Murphy, has been on the road again, cautiously exploring whether it is possible in coming months to bring Israel and Jordan closer to a negotiation. This quest would be essential even if terrorism were not the concern it is. It marks the leading way that American policy must go.

The clear implication is that negotiation between Israel and Jordan can dry up "a principal source of terrorism." Now, nobody who has studied that political context at all, and is not blinded by wishful thinking, could possibly believe that. For the Arab terrorists—and most other Arabs—"the unresolved Palestinian question" and the existence of the State of Israel are one and the same thing. The terrorists could not possibly be appeased, or made to desist, by Jordan's King Hussein's getting back a slice of the West Bank, which is the very most that could come out of a negotiation between Jordan and Israel. The terrorists and their backers would denounce such a deal as treachery and seek to step up their attacks, directing these against Jordan as well as Israel.[208]

## SUMMARY

The popular summary beginning takes several forms: (1) a brief statement of the main ideas; (2) a digest of all sections of the paper; or (3) an abridgment of results, conclusions, or recommendations. This beginning is so useful that many professional journals preface each article with a summary. Some research institutes consider the summary important enough for a writer to spend months reducing a summary of years of research to a single page. At least one institute never sends a customer more than one page, no matter how lengthy the research; and its customers do not ask for more even though upper management bases decisions on these few summary sentences of scientists and technologists. Therefore writers should expend their best efforts on opening summaries. The ability to summarize the main ideas of a report is one of the most important skills writers in industry can acquire.

A brief statement of main ideas is considered by many writers the most helpful beginning of all. We suggest it when a report is organized in the order of climax or reverse climax (Chapter 3), and it is also suitable for reports based on other plans. Although a summary of main ideas at the beginning of a paper is sometimes considered useful only to those who are not experts, it may also help specialists. Specialists say that knowing the main points helps them to read a report more intelligently and more rapidly. And this summary also saves the time of specialists who like to glance over reports to get the gist of each before starting to read the most urgent one. Nonspecialists, many of them in management, may read only the summary or only the summary and one other section. Therefore, many writers supply, in addition to the general summary of main points, informative subheads and for a long report expository tables of contents.

Unfortunately, the useful summary beginning, which is often compared to the first sentence in a news story, is frequently misunderstood. Some writers, thinking of the five W's in newspaper leads (Who? What? Where? When? Why?), try to place too much unimportant material in their first sentences. But a good story answers only the important W's in the lead sentence:

> Millions of dollars is spent each year to ship Pennsylvania coal 3,000 miles to heat American bases in West Germany, and the Pentagon wants it stopped.
> Defense Department officials say the coal is unneeded and the money wasted, and they are stepping up their efforts to end the practice.
> But they face the weighty opposition of members of Congress from this coal-mining state. With the support of allies in the shipping and railroad industries, they have pushed through a series of laws requiring the Defense Department to use coal.[209]

> Defective education, mental instability, easy opportunities to break the law, the excitement of lawbreaking, feelings of alienation and poor parenting—in that order—were the reasons most often cited by 374 British men and women as the most important causes of juvenile delinquency, according to a survey conducted by Adrian Furnham of University College, London, and Monika Henderson of Oxford.[210]

> An outraged mother, who says her son's shooting death was "handled like a traffic case," used her own sleuthing to persuade a judge to throw out the defendant's plea bargain and order him to stand trial for murder.[211]

> Tales of a spider that casts a net over its prey have long been related in South Africa and Australia. The Australian netting spider (*Deinopis subrufa*), shown slightly larger than life-size in this picture sequence, practices this strange predatory method.[212]

> As a probe for examining the structure of the atomic nucleus, the neutron offers several advantages.[213]

A study of such examples should help the writer who begins, "This memo summarizes the discussion that Drs. Jones, Smith, Cadwallader, and I had at the Scientific Trivial Verbosity Plant with Drs. Brown,

153

Greene, and Fitzgibbon on June 11, 1987." He is using valuable space—the position of major emphasis, the main clause of his opening sentence—to say that he is summarizing. Instead of telling what he is doing, he should do it. If the talks at the Scientific Trivial Verbosity Plant disclosed that scientific articles are attenuated by the persistent use of unnecessary passives, he might begin, "Scientific articles are attenuated by the persistent use of unnecessary passives," and tell his reader who and where in a footnote or later in the text of the report. If the plant name is important, he might begin, "The Scientific Trivial Verbosity Plant attenuates scientific articles by the persistent use of unnecessary passives." But he should avoid the incorrect emphasis of a main clause stating that he is writing or summarizing or describing or reporting because this relegates his most important thought to a subordinate construction. Moreover, he insults his own writing and his readers' intelligence when he tells his readers what he is obviously doing. Such a writer should take pity on the executives who have already seen a thousand times or more

This memo summarizes
This report is written in answer to
This memo recommends
This memo reports
This memo covers
This memo concerns
This report gives the results of . . .

Instead of saying, "This paper describes an instrument which is useful to measure the total enthalpy content of high-energy gas streams at temperatures where dissociation effects become important," the writer should subordinate the unimportant and stress the important: "The instrument described in this paper measures the total enthalpy content of high-energy gas streams at temperatures where dissociation effects become important."

A digest of all sections of a paper, called an abstract, is helpful to readers. Examples from *The New York Times* and professional journals show that a digest is useful for a short paper that is not highly specialized. The digest may be written from an outline, and indeed, it often borrows the wording of an outline. Thus it assists readers who wish to find certain sections and librarians who must describe reports and refer them to appropriate readers.

Informative abstracts are discussed and illustrated in Chapter 12. The examples given there of an informative abstract organized (1) in the order of the paper and (2) in an order that emphasizes findings help writers see the difference in two abstracts of the same paper and thus aid them in choosing. The example of an indicative abstract shows how little it accomplishes and serves as a warning to write an informative abstract instead.

An abstract in *Science* preceding a paper reads as follows:

> Monoclonal antibodies to guinea pig T cells and antibodies to guinea pig immunoglobulin G were used in immunofluorescence studies to identify T and B cells in central nervous system tissue from guinea pigs with acute autoimmune encephalomyelitis. T cells appeared before B cells and were distributed within the white matter parenchyma, while B cells remained in perivascular spaces.[214]

If writers of long specialized papers do not have the advantage of writing mainly for technically trained readers, as journal writers do, they will probably find some other beginning easier to adapt for readers. But long reports that are not highly specialized, short reports, and memorandums often begin effectively with a digest of all sections. Beginning with such a summary is also effective in reports treating several unrelated subjects because the digest introduces each subject to the readers. Thus the writer, by establishing at the beginning the separate nature of each subject in a report, avoids the need to create artificial connections between subjects.

An abridgment of results, conclusions, or recommendations in terms intelligible to all readers is a suitable and helpful beginning for many papers, particularly those that recommend action. It may be used even though the complete technical description of results is important to the specialists. They can find that later in the paper.

John A. Osmundsen reporting on interferon in the *Times* summarizes conclusions clearly:

> A husband-and-wife team of Belgian scientists reported experimental results here today that may explain one way that certain chemicals cause cancer.
>
> The chemicals do this, the findings indicated, by suppressing the production of a substance called interferon that is normally put out by cells in response to virus infection. Interferon prevents the spread of the infection by interfering with virus reproduction.[215]

Many memorandums and reports begin quite properly and helpfully with summaries of results, conclusions, or recommendations or even with a full statement of these:

> The Marketing Committee recommends for the reasons detailed in this report that we market by May first a 2.5-mg Blank Drug Suppository.
>
> Preliminary trials in private medical practice and in hospitals demonstrated the efficacy of Blank Capsules as an adjunct in controlling appetite and anxiety.

155

In many papers the initial statements of conclusions present only the most significant features. Detailed results and conclusions are reserved for discussion after the presentation of evidence.

> Decreasing the water content of the feed results in a corresponding decrease in acid consumption for hydrofluoric acid alkylation. Lower corrosion rates and less polymer formation are additional benefits.[216]

> Babies 2 to 4 days old whose mothers received heavy medication during labor were less attentive than those babies whose mothers received light medication.[217]

These articles express the introductory statements of results in terms intelligible to all readers. When readers are concerned about recommendations, conclusions, or results and a writer can summarize them briefly and effectively, they are the best introduction.

Presenting the main points of a report or survey is necessary for many writers today. "Women and Minorities Continue to Grow in Workplace" does this neatly:

> The size and composition of the professional work force is changing, according to a report prepared by the Scientific Manpower Commission. The report, *Professional Women and Minorities*, chronicles the increasing participation of women and minorities in the science, engineering, and professional populations, calling particular attention to the gains made by women.[218]

This introduction is followed by an indented list of trends emphasized by black dots, or bullets.

A short summary of main points introduces "Their Professional Journal" in the *Proceedings of the U.S. Naval Institute*:

> There are similarities between their *Morskoy Sbornik* and the *Proceedings*. Both are aimed at an audience larger—foreign as well as domestic—than only naval officers. Both contain articles by junior as well as senior officers and occasionally, although less frequently in the Soviet journal, by civilian specialists. Dissimilarities? One is less likely to see disagreements aired on the pages of the Soviet journal, although their inclusion is not altogether unknown.[219]

Donald Worster begins Chapter 2 of *Rivers of Empire* with a summary of the importance of water:

> Earth has been variously called the planet of water and the planet of life, the connection between the two attributes being by no means casual. Without water, there simply can be no life. Water flows in the veins and

roots of all living organisms, as precious to them as the air they breathe and the food they eat. It is the lifeblood of their collective body.

Water has been critical to the making of human history. It has shaped institutions, destroyed cities, set limits to expansion, brought feast and famine, carried goods to market, washed away sickness, divided nations, inspired the worship and beseeching of gods, given philosophers a metaphor for existence, and disposed of garbage. To write history without putting any water in it is to leave out a large part of the story. Human experience has not been so dry as that.[220]

In *Scientific American* an opening paragraph summarizes a prediction of future results:

> Advances in biotechnology have brought U.S. agriculture to the brink of its third 20th-century revolution. According to a two-year study by the congressional Office of Technology Assessment (OTA), the tools of molecular biology may change the economics of agriculture as profoundly as they can alter the DNA of crops or livestock.[221]

Sandra Herbert presents in her first two paragraphs some of the results of her study of Charles Darwin:

> One tends to see Charles Darwin today essentially as a biologist, concerned with the origin and evolution of plant and animal species. Yet it was as a geologist even more than as a naturalist that he took part in the famous voyage of the *Beagle* from 1831 to 1836, and it was as a geologist that he saw himself in the years immediately following the voyage. His contributions to geology were significant. His meticulous fieldwork on the voyage produced collections of material that remain valuable. His insights on the origin of coral reefs laid the basis for today's view of the matter. Finally, in large part it was his geological work that led him to his views on evolution, or "transmutation" as the vocabulary of the time had it.
>
> Darwin's interest in geology during the voyage did not stem from long involvement with the subject. Indeed, at the outset of the voyage he was better prepared to collect insects and invertebrate animals; he had been an amateur entomologist since his youth and had been well trained in the study of invertebrates during the two years he spent as a medical student at the University of Edinburgh.[222]

Paul Lewis, reporting on two studies of women's education and careers by the Organization for Economic Cooperation and Development, summarizes the conclusions succinctly:

> Throughout the industrial world, women are catching up with men in education and earning power. But if some current trends continue, they may never achieve equality.[223]

Opening paragraphs may helpfully summarize background. David H.

Hubel and Torsten N. Wiesel open "Brain Mechanisms of Vision" with such an effective paragraph:

> Viewed as a kind of invention by evolution, the cerebral cortex must be one of the great success stories in the history of living things. In vertebrates lower than mammals the cerebral cortex is minuscule, if it can be said to exist at all. Suddenly impressive in the lowest mammals, it begins to dominate the brain in carnivores, and it increases explosively in primates; in man it almost completely envelops the rest of the brain, tending to obscure the other parts. The degree to which an animal depends on an organ is an index of the organ's importance that is even more convincing than size, and dependence on the cortex has increased rapidly as mammals have evolved. A mouse without a cortex appears fairly normal, at least to casual inspection; a man without a cortex is almost a vegetable, speechless, sightless, senseless.[224]

The first paragraph of Chapter 14, "Coexistence in the Technosphere," in *Only One Earth* summarizes a projection:

> The bleak mathematics of population growth, of needed food supplies, of urban pressure, and of the capital sums required to mitigate their effects leave us in no doubt about the threat to humanity in developing lands. By the year 2000, their numbers will have grown by nearly three billion—or more than doubled in thirty years. To keep pace with this growth and to ensure a modest increase of income above the pitiful $100 a year of the great majority, governments must aim at economic growth of at least 6 per cent a year. To do this they must push up annual savings to well above 15 per cent a year of gross national product (the sum of goods and services), and do this in communities where, in many cases, 90 per cent of the people live little above margins of subsistence. Since much of the equipment, the technology, and the skills they need cannot yet be secured locally, parts of the needed investment must be in the shape of foreign exchange.[225]

More common is the use of an introductory summary paragraph for past events and beliefs. Lewis Thomas neatly epitomizes past events in an informal paragraph opening Chapter 13, "The Board of Health," in *The Youngest Science: Notes of a Medicine Watcher*:

> Before I went to medical school my knowledge of the public health profession was limited to the childhood singsong: "Marguerite, go wash your feet, the Board of Health is down the street." I learned very little about public health in medical school, beyond what was called "The Sanitary Survey," a required field study in the summer between the third and fourth years, in which all students were assigned in pairs to a municipal or county health department as observers. I had a classmate friend who lived in Cincinnati, and we arranged to survey that city. It took two weeks to do and another week to write a report. We learned more than we had known about sewage disposal, water bacteriology, venereal disease clinics, premarital Wasser-

mann tests, and public baths, but it seemed a long way from medicine, a long way from science. That was the extent of my training until I was appointed in 1956 as a member of the Board of Health of the City of New York, on which I served for fifteen years.[226]

## WIT AND HUMOR

If, as Meredith said, "The well of true wit is truth itself," then wit is certainly an appropriate beginning for a scientific paper. But stale jokes are not to be confused with dry wit. Although ancient jokes have ancient sanction, it is better if a writer of informational prose does not wonder, like the character in Aristophanes, "Shall I begin with the usual joke/ That the audience always laughs at?" Laughter at oneself or one's subject is so rare in the journals of science and technology that Robert Schmidt, finding some in the "Letters" section of *Science*, was moved to write to the editors:

> If the editors continue to exercise, or, still better, even improve a little, their present judgment in selecting the very best morsels for publication, they might even succeed in overthrowing my pet hypothesis: the principal reason for the abysmal distance between scientists and humanists is found in the inability of the former to laugh at themselves.[227]

In the late 1980s there is more humor in journals and magazines than there was when Schmidt wrote. Still some of the best examples are not new:

> Rodin's sculpture "The Thinker," modified slightly by the addition of a wrinkled brow and a suggestion of anguish in the facial expression, could represent man in a more familiar aspect that might be labeled "The Forgetter."[228]

> The trouble with any measurement is the inevitable mismatch between the messiness of reality and the purity of the abstractions used to describe it. Of this truism, thermometry is the exemplar.[229]

In a recent *Atlantic* Paula Roberts begins "A Ministerial Portfolio" with amusing material about government cabinets:

> One of Parkinson's laws holds that the ideal government cabinet contains just five members. One of them knows the law, one knows finance, one knows foreign policy, and one knows defense. The fifth member, the one who has failed to master any of these subjects, usually becomes the Prime Minister. Over time, however, other officials, representing other interests, will seek to join the cabinet, and it often proves easier to let them in than to keep them out. As a result, cabinets tend to get larger and larger.
> Americans will recognize the truth of this. The first cabinet of the United

159

States, in 1789, had exactly the four secretaries that Parkinson prescribes. The cabinet grew to eight by the end of its first century and stands at thirteen as it approaches the end of its second. A nice case for Parkinson. Still, the United States is cabinet-poor compared with most countries. On average, each of the world's 174 national governments employs twenty-five cabinet-level ministers. These 4,300 officials are identified and paired with their portfolios in a bimonthly publication called *Chiefs of State and Cabinet Members of Foreign Governments*, which is put out by the Central Intelligence Agency.[230]

"Enough of Roads" by Paul Quinnett opens with an anecdote:

> When a friend of mine's grandfather graduated from the University of California with a degree in civil engineering, his old professor remarked to him, "Son, I'm afraid you have wasted your time. They've built all the highways they're ever going to build."
> The year was 1911.[231]

A *Miami Herald* story on hurricanes chooses satire:

> South Florida observed the start of the 1986 hurricane season today in traditional fashion: By the millions, coastal residents failed to buy storm shutters, check their flood insurance, stock up on emergency food and supplies, or make personal evacuation plans.
> In what has become an annual orgy of avoidance, they treated the threat of hurricane with true knock-on-wood spirit, adopting an out-of-sight, out-of-mind attitude as a talisman against the giant storms.[232]

And Laurie Baum covers the opening of a branch of a popular store with a joke that many shoppers relished:

> On opening day at Loehmann's new Kendall store, one manager saw an eager customer begin to change clothes in the middle of the sales floor.
> Another manager heard a man who had escorted his wife to the women's apparel store say that when he died, he wanted to be cremated and have his ashes spread over Loehmann's parking lot. Why? So he could be sure his wife would visit him three times a week.[233]

In "The Joke That Bombed" *The Washington Post* makes the most of presidents not being permitted jokes on a certain subject:

> Life isn't fair, our presidents like to tell us, and especially is it not fair to presidents. There are so many things we can do that presidents cannot. Presidents cannot stand in the rain and wait for a bus, for example. They cannot be told there are no tables available in the restaurant and that it is unlikely that there will even be a seat at the counter for another half-hour. Presidents cannot spend two days on the phone trying to find out why the trash wasn't collected. They cannot miss a plane. A miserable life, really,

160

when you consider its deprivations, and we have not even mentioned the most egregious unfairness of all: presidents cannot make nuclear bombing jokes.[234]

The selection of details in this example lends much to the humor.

"Pet Turtles—Revisited" in *The Harvard Medical School Health Letter* begins with amusing advice:

> Better a pet rock than a pet turtle. Infectious disease experts have known for years that pet turtles are often crawling reservoirs of salmonella bacteria.[235]

Ricki Fulman writing for the New York News Service begins with an anecdote:

> It was one of those chic parties.
> Helene, a newly divorced computer programmer, had completely charmed the appealingly eligible lawyer. Not wanting him to know her age, she tried to erase the terrible fact that she grew up in the '50s by talking about what it was like growing up in the '40s. She meant to say the '60s, but never could get her arithmetic right.
> He backed off in horror, figuring that though she must have had a superb facelift, she was still 10 years too old for him. Actually they were contemporaries.
> Helene is not unusual. The tendency is to lie about one's age.[236]

And an article on software humorously introduces a change in ownership:

> RAMIS II, conceived and dedicated to the proposition that all fourth generation languages are not created equal, is about to have its second new address in three years. It's enough to give a product an insecurity complex.
> Currently residing in Martin Marietta Data Systems' Information Technologies Division (ITD), Princeton, N.J., RAMIS II at press time was all but signed, sealed, and delivered to On-Line Software International Inc., Fort Lee, N.J.[237]

An article in *IEEE Spectrum* achieves humor through the surprise ending of the opening paragraph:

> In most functional testing of electronic systems, engineers hope that as many units as possible will pass the test. But environmental stress screening, or ESS, takes a different tack: its goal is to make as many units as possible fail.
> Unlike the more familiar qualification testing, in which manufacturers check prototypes before they are produced in quantity—by operating them within the range of temperatures for which they were designed, for ex-

161

ample—ESS subjects the units to random vibration and temperature cycling during production.[238]

## SELECTING A BEGINNING

In addition to those illustrated here, there are other ways of beginning, such as dialogue, figures of speech, statement of the plan of the writing (common at the beginning of chapters of textbooks), steps of a process or procedure, purpose. Writers also combine many kinds of beginnings.

The reader of this book has probably noticed that we introduced the discussion of each type of beginning by writing that kind or that kind dominant in a combination. In scientific and technological journals many openings are combinations. Usually, however, one type dominates the combination: an introduction that defines may contain a little background, or a brief opening question may be followed by a long answer.

With such a wide variety of beginnings available, writers need not worry about being unable to think of an opening. Instead, they can think about which opening to choose. Choosing is not a dead end street or a highway to procrastination like, "I can't think of a beginning. If I could think of a beginning, the rest would flow." No, with such a wealth of beginnings available, everyone can find one or many suitable ways to start a paper. For instance, some readers of this chapter may remember beginnings that appeared to them particularly well suited to their subjects. Others may skim the chapter and stop at one or another way of beginning because they feel, "This is the one. This is it." Still others may simply rely on the kind of beginning that they have seen used for similar subjects and leave braver choices for future attempts when the writers will be more experienced.

Therefore many examples have been included for those shopping for beginnings. Instructors using the first edition asked for these examples. Writers in business, government, and industry said that they are no longer tied to starting with history of the art or some other set beginning and need to study what is available. And writers of scholarly papers, enjoying their newfound freedom to choose a beginning, also wanted to examine and practice different methods.

If as seems likely these writers find several appealing beginnings for a paper, how should they choose wisely? They can estimate the quality and suitability of a beginning by considering the following points:

- Is it appropriate for the subject and material?
- Does it stress major ideas?
- Is it suitable for the writers and their style?
- Does it appeal to readers; that is, does it catch and hold their interest?
- Does it lead, or can it be helped to lead, gracefully into the rest of the writing?

Writers should avoid forcing their material into inconvenient or unsuitable introductions. They should not begin with pointless questions, ineffective quotations, or unnecessary details just to get started. The best beginnings emphasize the main thoughts of papers or at least suggest the main thoughts. And the introduction of a paper should be harmonious with the rest of it. Once in a while a journal errs by prefacing with a colloquial or slangy abstract a scholarly paper written in a formal style. The only readers not jolted by the jump in style between the introduction and the paper are those who omit one or the other. The material of an introduction should not lead to disappointment in the subject of a paper; thus it is unwise to begin with material that may capture a reader's interest and then to drop that material. If a paper begins by comparing two plant sites but then studies only the one near Boston, it may disappoint and annoy readers who, while reading the introduction, developed an interest in the site near Chicago. An introduction suitable to the subject matter of a paper prepares readers for what lies before them and enlists their interest in that subject.

In choosing a beginning for their papers good writers weigh their own strengths and weaknesses. If they lack humor, they avoid trying to begin with humor; if they quickly become contentious, they reject a beginning that knocks down an argument; and if they are inclined to be long-winded or overconscientious in discussing background, they may prefer another kind of opening. Although beginners should practice writing a few of the easier kinds of introductions before attempting to write complex combinations, experienced writers may increase their own interest in the work as well as their readers' by experimenting with beginnings that are new to the writers.

But the dominant consideration in the choice is the readers. Careful attention to their interests, background, education, and reading habits is essential. It enables writers to avoid quoting authorities that readers will distrust, using technical language that they do not understand, and reviewing boring familiar background. Consideration of their readers helps writers to answer important questions about beginnings: Should authors use tried-and-true beginnings well known to readers or new, imaginative, daring ones? The answer depends on how conservative the readers are, on how they would respond to striking originality. How many details of background should writers include? Enough to show the thoroughness of their research? Enough to build a beginning that seems impressive to the writers? No, they should include in the introduction the details of background that their readers need at the beginning and that will lead easily into the paper. A beginning is not supposed to be a tower separate from the rest of the structure but an intrinsic part of the whole. (Ways to connect paragraphs and stress coherence appear in Chapter 12.) Whenever writers find an introduction that satisfies the five criteria above, they can rest assured that their choice is a sound one.

If a chosen beginning is received well, as one that suits the five criteria usually is, writers may be tempted to use it in most of their reports. If many such reports are sent to the same readers, they are likely to become bored after a time or even irritated. Variations are necessary in regularly scheduled reporting. Yet some writers and supervisors, having been taught an order for reports in school or university or government or industry, insist that they cannot think in any other sequence and therefore submit reports year after year with the same beginnings (and middles and endings).

Typical is a scientist who said he could not think in any other order and wrote all his reports in one pattern: (1) Work Planned, (2) Work Accomplished, (3) Conclusions. He might better have used mimeographed lists than sentences and paragraphs:

1. Work Planned—last four experiments of Project 998B10
2. Work Accomplished—the same
3. Conclusions—none

But by expanding these three topics into wordy sentences, he managed to fill a single-spaced page with no more than the listed information.

He also used this order for ideas not well suited to the order. A report titled "Literature Survey" opened by stating that he had located and summarized eight journal articles. The bibliography followed and then a summary in technical language of the results and conclusions of each highly specialized article. At the end of the last conclusion of the last article was a sentence stating that the research of A.B. Company was supported by the articles.

When readers were questioned about this report, most of them remembered that it was about eight journal articles and that they did not understand the summaries. Only one reader mentioned that the articles supported his company's research, and he was the chemist's supervisor, a captive reader. Writers and supervisors should never underestimate the influence of opening sentences. They should also remember that each person is in no way a typical reader of the writing of another. Authors who write with only their supervisors in mind may soon find that they have no other readers.

Writing for the wrong readers can confuse; even young writers know that. One very young writer began every school composition by printing "I LOVE MY MOMMY." An incipient critic in the class finally asked, "Why do you always write that? Your mother has nothing to do with our class trip to the zoo." "That's right," chimed in a bored supporter. "She didn't even come." "But she likes me to begin that way, and then I don't have to think of how to spell another beginning." "B-o-r-i-n-g," the critic–classmate managed to inject before the teacher intervened. Papers in business and government and industry and journals might be less b-o-r-i-n-g if authors wrote not to their supervisors but to their noncaptive readers—or perhaps we should say if supervisors encouraged authors to write to their readers and gave them enough information about the readers to enable them to do so.

## NOTES

1. S. Nozette and W. V. Boynton, "Superheavy Elements: An Early Solar System Upper Limit for Elements 107 to 110," *Science* 214 (16 Oct. 1981): 331.
2. Garry A. Rechnitz, "Bioselective Membrane Electrode Probes," *Science* 214 (16 Oct. 1981): 287.
3. J. C. French and P. B. Tomlinson, "Vascular Patterns in Stems of Araceae: Subfamilies Colocasioideae, Aroideae and Pistioideae," *American Journal of Botany* 70 (1983): 756.
4. Stephen J. Pyne, *Introduction to Woodland Fire: Fire Management in the United States* (New York: Wiley, 1984) vii.
5. Fred Hapgood, "Roden's Eye," *The Atlantic* 260 (Aug. 1987): 47.
6. Stuart L. Shapiro and Saul A. Teukolsky. *Black Holes, White Dwarfs, and Neutron Stars: The Physics of Compact Objects* (New York: Wiley, 1983) 512.
7. Cornelius J. Casey and Norman Bartczak, "Cash Flow—It's Not the Bottom Line," *Harvard Business Review* 62 (July/Aug. 1984): 61.
8. Stephen W. Carmichael and Hans Winkler, "The Adrenal Chromaffin Cell," *Scientific American* 253 (Aug. 1985): 40.
9. C. Arden Miller, "Infant Mortality in the United States," *Scientific American* 253 (Aug. 1985): 31.
10. Tim Carrington, "U.S. Agencies Want Nation's Banks to Disclose More Data on Problems," *The Wall Street Journal* 22 Aug. 1984: 27.
11. "High-Resolution Images Yield New Data on Venus," *Geotimes* 29 (Sept. 1984): 10.
12. Walter Pincus, "40 Years of the Bomb," *The Washington Post National Weekly Edition* 5 Aug. 1985: 6.
13. Lawrence D. German and Raymond E. Ideker, "Ventricular Tachycardia," *Symposium on Cardiac Arrhythmias I* (*Medical Clinics of North America* 68 [July 1984]): 919.
14. Jeremy Shapiro, foreword, *Resource Management: An Alternative View of the Management Process,* by Paul S. Bender (New York: Wiley, 1983) ix.
15. Harry B. Gray and Andrew W. Maverick, "Solar Chemistry of Metal Complexes," *Science* 214 (11 Dec. 1981): 1201.
16. Cladd E. Stevens et al., "Hepatitis B Vaccine in Patients Receiving Hemodialysis," *New England Journal of Medicine* 311 (23 Aug. 1984): 496.
17. Maurie Markman et al., "Antiemetic Efficacy of Dexamethasone," *New England Journal of Medicine* 311 (30 Aug. 1984): 549.
18. Walter Adams and James W. Brock, "Integrated Monopoly and Market Power: System Selling, Compatibility Standards, and Market Control," *Quarterly Review of Economics and Business* 22 (Winter 1982): 29.

19. Stuart Diamond "'New Frontier' in Plastics," *The New York Times* 1 Aug. 1985: D2.

20. Walter Pincus, "Can the MX Do the Job?" *The Washington Post National Weekly Edition* 4 June 1984: 16.

21. "Sober Side of Boating Is Safety," *The New York Times* 20 July 1985: 48.

22. Robert Giroux, "Happy Birthday, William Shakespeare, and Keep Those Plays and Sonnets Coming," *The New York Times Book Review* 28 April 1985: 3.

23. Gregory Sandow, "Age of the Conductor," *Saturday Review* (Jan.–Feb. 1984): 46.

24. L. Erik Calonius, "Ambassador's Dilemma: Position Has Prestige but White House Keeps the Power for Itself," *The Wall Street Journal* 7 Sept. 1984: 30.

25. Vicente Navarro, "Hazardous Substances in Western Developed Countries," *New England Journal of Medicine* 311 (23 Aug. 1984): 546.

26. F. Peter Woodford, "Sounder Thinking Through Clearer Writing," *Science* 156 (12 May 1967): 743.

27. Carlo M. Croce and George Klein, "Chromosome Translocations and Human Cancer," *Scientific American* 252 (March 1985): 54.

28. Kosta Tsipis and Eric Raiten, "Antisatellite Weapons: The Present Danger," *Technology Review* (Aug. 1984): 54.

29. Jonathan P. Hicks, "Second-Quarter Profits Off Moderately," *The New York Times* 22 July 1985: D2.

30. Barry Sussman, "Do-It-Yourself Tax Reform: Many Think Cheating Is Okay," *The Washington Post National Weekly Edition* 29 May 1984: 36.

31. James S. Bristow, "Computers: Our Aging Workhorses," *Proc. U.S. Naval Institute* 108/8/954 (Aug. 1984): 56.

32. Robert H. Jeffrey, "The Folly of Stock Market Timing," *Harvard Business Review* 62 (July/Aug. 1984): 102.

33. H. Chew, D.-S. Wang, and M. Kerker, "Surface Enhancement of Coherent Anti-Stokes Raman Scattering by Colloidal Spheres," *Journal of the Optical Society of America* 1 (March 1984): 56.

34. Nevin S. Scrimshaw, "The Politics of Starvation," *Technology Review* 87 (Aug.-Sept. 1984): 18.

35. Daniel Jean Stanley, Terry A. Nelsen, and Robert Stuckenrath, "Recent Sedimentation on the New Jersey Slope and Rise," *Science* 226 (12 Oct. 1984): 125.

36. "Weapons Builders Eye Civilian Reactor Fuel," *Science* 214 (16 Oct. 1981): 307.

37. Gerard K. O'Neill, *Technology Edge: Opportunities for America in World Competition* (New York: Simon, 1983) 84.

38. O'Neill 51.

39. Lewis Thomas, *The Youngest Science: Notes of a Medicine Watcher* (New York: Viking, 1983) 179.

40. Deirdre Carmody, "Peddler Patrol on Fifth Avenue Cracks Down," *The New York Times* 25 Dec. 1986: 33.

41. Carl Rain, "Uncovering Hidden Flaws," *High Technology* 4 (Feb. 1984): 49.

42. "Poor Form, INS," *The Miami Herald* 7 June 1987: 2C.

43. D. H. Kelly, "Disappearance of Stabilized Chromatic Gratings," *Science* 214 (11 Dec. 1981): 1257.

44. Frank Graham, Jr., rev. of *Biophilia* by Edward O. Wilson, *Audubon* 87 (March 1985): 32.

45. "Will Biotechnology Revolution Become Farmers' Bad Seed?" *The Miami Herald* 5 June 1986: 3D.

46. Melvin Levine, "Learning: Abilities and Disabilities," *The Harvard Medical School Health Letter* 9 (Sept. 1984): 3.

47. Michael E. McCormick, *Ocean Wave Energy Conversion* (New York: Wiley, 1981) 28.

48. David A. Burge, *Patent and Trademark Tactics and Practice* (New York: Wiley, 1980) 111.

49. Aubrey Gorbman et al., *Comparative Endocrinology* (New York: Wiley, 1983) 517.

50. John A. Duffie and William A. Beckman, *Solar Engineering of Thermal Processes* (New York: Wiley, 1980) 197.

51. S. Neil Rasband, *Dynamics* (New York: Wiley, 1983) 81.

52. Shapiro and Teukolsky 1.

53. Shapiro and Teukolsky 499.

54. Burge 96.

55. Dana Z. Anderson, "Optical Gyroscopes," *Scientific American* 253 (April 1986): 94.

56. Benjamin Weiser, "Rosebud: Worst of Both Worlds," *The Washington Post National Weekly Edition* 24 Sept. 1984: 6.

57. Kathy Sawyer and Kenneth E. John, "Public Concern Grows over Nuclear Dangers," *The Washington Post National Weekly Edition* 24 Sept. 1984: 38.

58. Jarold A. Kieffer, "Looking Forward to Retiring and Living a Life of Ease? Don't," *The Washington Post National Weekly Edition* 24 Sept. 1984: 23.

59. Peter Richmond, "Open Hopes Are Cast to Wind Today," *The Miami Herald* 12 June 1986: D1.

60. G. Ghiselli et al., "Type III Hyperlipoproteinemia Associated with Apolipoprotein E Deficiency," *Science* 214 (11 Dec. 1981): 1239.

61. Fred W. McLafferty, "Tandem Mass Spectrometry," *Science* 214 (16 Oct. 1981): 280.

62. Robert Rainer, "To Save a Place of Wonder," *Sierra* 70 (July/Aug. 1985): 70.

63. Charles C. Mann, "No Parking," *The Atlantic* 257 (May 1986): 14.

64. Richard F. Shepard, "Metropolitan Math," *The New York Times Magazine* 28 April 1985: 90.

65. William K. Stevens, "Stronger Urban Accents in Northeast Are Called Sign of Evolving Language," *The New York Times* 21 July 1985: 36.

66. Geraldine Fabrikant, "Comcast's Consistent Profits," *The New York Times* 23 July 1985: D1.

67. Fred Hiatt, "The Pentagon's Good News Crowd," *The Washington Post National Weekly Edition* 12 Nov. 1984: 23.

68. Cullen Murphy, "Defining Rembrandt," *The Atlantic* 257 (May 1986): 16.

69. Gale Warner, "Low-level Lowdown," *Sierra* 70 (July/Aug. 1985): 19.

70. Rosalind Resnick, "There's No Place Like Home: Workers Like the Comfort but Employers Are Resisting," Business Monday *The Miami Herald* 13 May 1985: 1.

71. Robert Buxbaum, "Age and Exercise," *The Harvard Medical School Health Letter* 8 (April 1983): 3.

72. James B. Hendrickson, "Synthesizing Chemicals by Computer," *Technology Review* 87 (April 1984): 23.

73. Pochi Yeh, "Photorefractive Coupling in Ring Resonators," *Applied Optics* 23 (1 Sept. 1984): 2974.

74. Shapiro and Teukolsky 267.

75. Marlene Sokol, "Domino Effect: Towns Vie for New Industry," Business Monday *The Miami Herald* 9 June 1986: 7.

76. Jeffrey Weiss, "Irked by Slipping Garbage Bags? Here's a Plastic Can That'll Help," Business Monday *The Miami Herald* 9 June 1986: 33.

77. David Wise, "Campus Recruiting and the C.I.A.," *The New York Times Magazine* 8 June 1986: 20.

78. *The New York Times Magazine* 8 June 1986: 55.

79. Steven Greenhouse," The Rise and Rise of Big Mac," *The New York Times* 8 June 1986: 1F.

80. Deborah Rankin, "When Insurers Won't Pay Medical Bills," *The New York Times* 8 June 1986: 11F.

81. N. R. Kleinfield, "Executive Fun and Games," *The New York Times Magazine* 8 June 1985: 37.

82. Wassily Leontief, "The Choice of Technology," *Scientific American* 252 (June 1985): 37.

83. Lisa H. Towle, "Microsurgery Restoring Form and Function," *New York University Magazine* 1 (Spring 1987): 30.

84. Scott M. Fishman and David V. Sheehan, "Anxiety and Panic: Their Cause and Treatment," *Psychology Today* 19 (April 1985): 26.

85. David Kipnis and Stuart Schmidt, "The Language of Persuasion," *Psychology Today* 19 (April 1985): 40.

86. William L. Allman, "Nice Guys Finish First," *Science 84* 5 (Oct.): 25.

87. Maya Pines "In the Shadow of Huntington's," *Science 84* 5 (May): 32.

88. Wayne S. Wooden, "The Flames of Youth," *Psychology Today* 19 (Jan. 1985): 23.

89. "The Last Headline," *Audubon* 85 (Jan. 1984): 32.

90. Scott Haas, "The Nuclear State: Hope in the Face of Danger," *Psychology Today* 18 (April 1984): 6.

91. Michael Harwood, "Country Musician," *Audubon* 88 (Sept. 1986): 62.

92. Esmond Bradley Martin, "They're Killing Off the Rhino," *National Geographic* (March 1984): 404.

93. "Ozone 'Hole' Stumps Scientists; It's the Focus of Antarctic Study," *The Miami Herald* 23 June 1986: 1.

94. Thomas A. Bass, "Fooling Flamingos," *Audubon* 87 (July 1985): 90.

95. Peter Applebome, "A Boom in Texas: Guard Donkeys," *The New York Times* 6 June 1986: 14Y.

96. "Dark Days for the Panther," editorial, *Florida Naturalist* 57 (Winter 1984): 3.

97. Timothy Ferris, "Physics' Newest Frontier," *The New York Times Magazine* 26 Sept. 1982: 38.

98. Robert Jay Lifton, "The Psychic Toll of the Nuclear Age," *The New York Times Magazine* 26 Sept. 1982: 52.

99. "Snake Oil and Other Remedies," *The Miami Herald* 22 June 1986: 1F.

100. "Out of the Mud of Galilee," *The Miami Herald* 22 June 1986: 4J.

101. "Books," *Sierra* (Sept./Oct. 1985): 74.

102. Sandy Rovner, "Spray Lets You Say Goodbye to Poison Ivy Pain," *The Miami Herald* 23 June 1986: 3C.

103. James McGregor "Custer's Men Win Tribute," *The Miami Herald* 23 June 1986: 1.

104. Shapiro and Teukolsky vii.

105. Clive O. J. Grove-Palmer, foreword, McCormick xi.

106. "Twenty Discoveries That Changed Our Lives," *Science 84* 5 (Nov.): 65, 69, 73, 76, 79, 83, 99, 111, 115, 121, 127, 131–132, 141, 149, 153.

107. Thomas D. Morris, "Taking Charge in Washington," *Harvard Business Review* 62 (July–Aug. 1984): 24.

108. Burge vii.

109. George S. Dominguez, *Government Relations: A Handbook for Developing and Conducting the Company Program* (New York: Wiley, 1982) 103.

110. Mark M. Rosenthal and Robert A. Miller, "Tactics to Employ When a Lawsuit Looms," *Harvard Business Review* 62 (July–Aug. 1984): 42.

111. Andrew Weiss, "Simple Truths of Japanese Manufacturing," *Harvard Business Review* 62 (July–Aug. 1984): 119.

112. George Blyn, "The Green Revolution Revisited," *Economic Development and Cultural Change* 31 (July 1983): 705.

113. Jack C. Fisher, "Skin—The Ultimate Solution for the Burn Wound," *New England Journal of Medicine* 311 (16 Aug. 1984): 466.

114. Glenn C. Christo, rev. of *The Language of Medicine: Its Evolution, Structure, and Dynamics* by John H. Dirckx, *New England Journal of Medicine* 311 (23 Aug. 1984): 544.

115. Marshall B. Kapp, rev. of *The Silent World of Doctor and Patient* by Jay Katz, *New England Journal of Medicine* 311 (23 Aug. 1984): 544.

116. James Fallows, rev. of *Brave New Workplace* by Robert Howard, *The Atlantic* 257 (May 1986): 544.

117. Elizabeth Spencer, rev. of *The News from Ireland* by William Trevor, *The New York Times Book Review* 8 June 1986: 14.

118. Wilfrid Sheed, rev. of *Red* by Ira Berkow, *The New York Times Book Review* 8 June 1986: 1.

119. C. M. Shapiro et al., "Slow-wave Sleep: A Recovery Period After Exercise," *Science* 214 (11 Dec. 1981): 1253.

120. Daniel Goleman, "Debate Intensifies on Attitude and Health," *The New York Times* 29 Oct. 1985: C1.

121. James Fallows, "The Spend-Up," *The Atlantic* 258 (July 1986): 27.

122. M. Mitchell Waldrop, "Machinations of Thought," *Science 85* 6 (March): 38.

123. James Cornell, "Science vs. the Paranormal," *Psychology Today* 18 (March 1984): 29.

124. Marlene Zuk, "A Charming Resistance to Parasites," *Natural History* 93 (April 1984): 28.

125. William Stockton, "Crisis Strains Ruling Party," *The New York Times* 13 June 1986: Y29.

126. Nicholas D. Kristof, "Financial Corp. Chief Sees Bleak '86," *The New York Times* 13 June 1986: Y29.

127. Philp S. Gutis, "Long Delay at Luxury Complex," *The New York Times* 13 June 1986: Y44.

128. John Tierney, "On the Faint Trail of the Christmas Comet," *Science 85* 6 (March): 16.

129. Charles D. Allen, Jr., "Forecasting Future Sources," *Proc. U.S. Naval Institute* 108/11/957 (Nov. 1982): 74.

130. Robert T. Lund, "Remanufacturing," *Technology Review* 87 (Feb./March 1984): 19.

131. Lawrence M. Lidsky, "The Reactor of the Future," *Technology Review* 87 (Feb./March 1984): 52.

132. "The Real News About Multiple Sclerosis," *The Harvard Medical School Health Letter* 8 (May 1983): 1.

133. Donald Goldmann, "Day Care and Contagion," *The Harvard Medical School Health Letter* 8 (May 1983): 3.

134. Alan A. Boraiko, "Beams That Heal" in "Lasers—'A Splendid Light,'" *National Geographic* 165 (March 1984): 346.

135. "AIDS Deaths Will Top Traffic Toll, Report Says," *The Miami Herald* 13 June 1986: 1.

136. Bob Woletz, "2020 Foresight," *The New York Times Magazine* 8 June 1986: 13.

137. Christopher Lehmann-Haupt, rev. of *The Way of the Physician* by Jacob Needleman, *The New York Times* 22 July 1985: C16.

138. Bruce Fellman, "How Do Plants Know Which End Is Up?" *Science 84* 5 (July/Aug.): 26.

139. Abby Arthur Johnson, "From Transcript to Minutes: An Editorial Challenge," *Technical Communication: Journal of the Society for Technical Communication* 31 (Second Quarter 1984): 20.

140. R. Wiltschko, D. Nohr, and W. Wiltschko, "Pigeons with a Deficient Sun Compass Use the Magnetic Compass," *Science* 214 (16 Oct. 1981): 343.

141. M. R. Gold and G. H. Cohen, "Modifications of the Discharge of the Vagal Cardiac Neurons During Learned Heart Rate Change," *Science* 214 (16 Oct. 1981): 345.

142. Steven Swiryn, Timothy McDonough, and David C. Hueter, "Sinus Node

Function and Dysfunction," *Symposium on Cardiac Arrhythmias I* (*Medical Clinics of North America* 68 [July 1984]): 935.

143. James J. MacKenzie, "Finessing the Risks of Nuclear Power," *Technology Review* 87 (Feb./March 1984): 34.

144. Paul S. Bender, *Resource Management: An Alternative View of the Management Process* (New York: Wiley, 1983) 174.

145. Robert C. Cowen, "The $CO_2$ Threat: Solutions Welcome," *Technology Review* 87 (April 1984): 8.

146. Duffie and Beckman 1.

147. David Kennedy, "New Airbags: Low Tech, Low Price," *Technology Review* 87 (April 1984): 78.

148. Gorbman et al. 185.

149. "Science and the Citizen," *Scientific American* 253 (April 1986): 62.

150. William J. Welch, "Computer-aided Design of Experiments for Response Estimation," *Technometrics* 26 (Aug. 1984): 217.

151. Jerome Friedman and Robert Tibshirani, "The Monotone Smoothing of Scatterplots," *Technometrics* 26 (Aug. 1984): 243.

152. W. Thomas Cathey et al., "Image Gathering and Processing for Enhanced Resolution," *Journal of the Optical Society of America* 1 (March 1984): 241.

153. Dwight A. Burkhardt et al., "Symmetry and Constancy in the Perception of Negative and Positive Luminance Contrast," *Journal of the Optical Society of America* 1 (March 1984): 309.

154. Marek Kowalczyk, "Spectral and Imaging Properties of Uniform Diffusers," *Journal of the Optical Society of America* 1 (Feb. 1984): 192.

155. P. G. Cottis and N. K. Uzunoglu, "Analysis of Longitudinal Discontinuities in Dielectric Slab Waveguides," *Journal of the Optical Society of America* 1 (Feb. 1984): 206.

156. Gary L. Frazier and John O. Summers, "Interfirm Influence Strategies and Their Application Within Distribution Channels," *Journal of Marketing* 48 (Summer 1984): 43.

157. Paul Meier, "Current Research in Statistical Methodology for Clinical Trials," *Biometrics*, Supplement (March 1980): 141.

158. P. Armitage, "The Assessment of Low-Dose Carcinogenicity," *Biometrics*, Supplement (March 1980): 119.

159. D. R. J. Campbell, "Accurate Assessment of the Optical Quality of Infrared Systems," *Optical Engineering* 23 (March/April 1984): 141.

160. David H. Hubel, "The Brain," *The Brain* (New York: Scientific American, 1979) 3.

161. Louis E. Underwood, "Report of the Conference on Uses and Possible Abuses of Biosynthetic Human Growth Hormone," *New England Journal of Medicine* 311 (30 Aug. 1984): 606.

162. Robert S. Burnett and James A. Waters, "The Action Profile: A Practical Aid to Career Development and Succession Planning," *Business Horizons* 27 (May–June 1984): 15.

163. Stuart St.P. Slatter, "The Impact of Crisis on Managerial Behavior," *Business Horizons* 27 (May–June 1984): 65.

164. Martin K. Denis, "Financial Organizations and Affirmative Action," *The Bankers Magazine* 167 (July/Aug. 1984): 62.

165. Michael W. Golay, "An Agenda for Improving Present-Day Reactors," *Technology Review* 87 (Feb./March 1984): 49.

166. Tony Baer, "The Repositioning of the Robot," *Mechanical Engineering* 108 (July 1986): 65.

167. Eberhard Gwinner, "Internal Rhythms in Bird Migration," *Scientific American* 254 (April 1986): 84.

168. W. Maxwell Cowan, "The Development of the Brain," *The Brain* 57.

169. Gene Bylinsky, "Can Smokestack America Rise Again?" *Fortune* 109 (6 Feb. 1984): 74.

170. Catherine Caulfield, "The Rain Forests," *The New Yorker* 60 (14 Jan. 1985): 41.

171. Don P. Wyckoff, "Let There Be Built Great Ships," *Proc. U.S. Naval Institute* 108/11/957. (Nov 1982): 51.

172. Ronald Sullivan, "Tighter Reviews of Physicians Urged," *The New York Times* 8 June 1986: Y9.

173. Laurence Shames, "Wharton Reaches for the Stars," *The New York Times Magazine* 8 June 1986: 84.

174. Richard Wassersug, "Why Tadpoles Love Fast Food," *Natural History* 93 (April 1984): 60.

175. *The New York Times* 6 June 1986: 28.

176. Glen Martin, "Magic of the Sea of Cortez," *Audubon* 87 (Jan. 1986): 110.

177. Peter A. Jumars and James E. Eckman, "Spatial Structure Within Deep-Sea Benthic Communities," *Deep-Sea Biology*, ed. Gilbert T. Rowe (New York: Wiley, 1983) 399.

178. James Russell, editorial, *The Miami Herald* 21 June 1986: 4B.

179. Anti-Nuclear Activist's Prediction Came True, but She Takes a Break," *The Miami Herald* 25 June 1986: 3B.

180. Owen F. Devereux, *Topics in Metallurgical Thermodynamics* (New York: Wiley, 1983) 1.

181. Frank Graham, Jr., "Us and Them," *Audubon* 86 (Jan. 1984): 96.

182. Noel Vietmeyer, "Saving the Bounty of a Harsh and Meager Land," *Audubon* 87 (Jan. 1985): 100.

183. Fred M. Hechinger, "Liberation of Women Scholars in the 50's," *The New York Times* 30 July, 1985: C9.

184. "Skeleton Shortage," *The Miami Herald* 25 June, 1985: 2E.

185. "Yes Deposits, Yes Returns," *Sierra* 70 (Sept./Oct. 1985): 12.

186. "To Kenya With Love," *Dial* 8 (July 1986): 28.

187. "Statuesque," *Dial* 8 (July 1986): 26.

188. F. C. Budinger, "Overcoming Opposition," *Civil Engineering* (Feb. 1986): 36.

189. Burge 45.

190. Duffie and Beckman 111.

191. Lyman Spitzer, Jr., *Physical Processes in the Interstellar Medium* (New York: Wiley, 1978) 149.

192. Spitzer 270.

193. Rasband 81.

194. Gorbman et al. 325.

195. M. M. Abraham et al. *Journal of Chemical Physics* 81 (15 Sept. 1984): 2528.

196. George E. P. Box, William G. Hunter, and J. Stuart Hunter *Statistics for Experimenters* (New York: Wiley, 1978) 19.

197. Rasband 25.

198. Marilyn Gardner, "Advice to Today's Cinderellas," *The Christian Science Monitor* 6 May 1985: 41.

199. Hugo Lagercrantz and Theodore A. Slotkin, "The Stress of Being Born," *Scientific American* 253 (April 1986): 100.

200. "It's Nature's Way of Saying, 'Whoa!'" *Science 84* 5 (Dec.): 6.

201. Eliezer B. Ayal and Barry R. Chiswick, "The Economics of the Diaspora Revisited," *Economic Development and Cultural Change* 21 (Summer 1982): 861.

202. Gorbman et al. 185.

203. Kenneth L. Woodward, "The Soul of Confucius Meets the System of Mao," *Psychology Today* 18 (April 1984): 41.

204. Brett Harvey, rev. of *The Poverty Of Affluence* by Paul Wachtel, *Psychology Today* 18 (April 1984): 74.

205. Gina Kolata, "Proper Display of Data," *Science* 226 (12 Oct. 1984): 156.

206. Maya Pines, "Resilient Children," *American Educator* 8 (Fall 1984): 34.

207. Michael E. Trulson, Chong W. Kim, and Vernon R. Padgett, "That Mild-Mannered Bruce Lee," *Psychology Today* 19 (Jan. 1985): 79.

208. Conor Cruise O'Brien, "Thinking About Terrorism," *The Atlantic* 257 (June 1986): 62.

209. Lindsey Gruson, "Military Cites Waste in Law Forcing Coal Use in Europe," *The New York Times* 8 June 1986: 14Y.

210. John Masterson, "Public Perceptions of Delinquency," *Psychology Today* 18 (April 1984): 8.

211. "Outraged Mom Convinces Judge to Drop Bargain in Son's Killing," *The Miami Herald* 13 June 1986: 26A.

212. "Predator Nets a Sugar Ant," *Natural History* 73 (March 1964): 54.

213. Lawrence Cranberg, "Fast-Neuron Spectroscopy," *Scientific American* 210 (March 1964): 79.

214. U. Traugott et al. "Autoimmune Encephalomyelitis: Simultaneous Identification of T and B Cells in the Target Organ," *Science* 214 (11 Dec. 1981): 1251.

215. John A. Osmundsen, *The New York Times* 16 Sept. 1963: 120.

216. M. D. Box A. E. Bynum, and R. J. Schoofs, "Molecular Sieves Dry Alkyl Feed Better," *Hydrocarbon Processing & Petroleum Refiner* 43 (Jan. 1964): 125.

217. Gerald Stechler, "Newborn Attention as Affected by Medication During Labor," *Science* 144 (17 April 1964): 315.

218. Betty M. Vetter and Eleanor L. Babco, "Women and Minorities Continue to Grow in Workplace," *Science* 226 (12 Oct. 1984): 159.

219. Roger W. Barnett and Edward J. Lacey, "Their Professional Journal," *Proc. U.S. Naval Institute* 108/10/956 (Oct. 1982): 95.

220. Donald Worster, *Rivers of Empire* (New York: Pantheon Books, 1985) 19.

221. "Science and the Citizen," *Scientific American* 254 (May 1986): 66.

222. Sandra Herbert, "Darwin as a Geologist," *Scientific American* 254 (May 1986) 116.

223. Paul Lewis, "In Wages, Sexes May Be Forever Unequal," *The New York Times* 21 Dec. 1986: 20E.

224. David H. Hubel and Torsten N. Wiesel, "Brain Mechanisms of Vision," *The Brain* 84.

225. Barbara Ward and René Dubos, *Only One Earth: The Care and Maintenance of a Small Planet* (New York: Norton, 1972): 209.

226. Thomas, 238.

227. Robert Schmidt, Letter, *Science* 143 (31 Jan. 1964): 430.

228. Benton J. Underwood, "Forgetting," *Scientific American* 210 (March 1964): 91.

229. Ronald B. Neswald, abstracting "Measuring Temperature" by Lawrence G. Rubin, *International Science and Technology* (Jan. 1964): 74.

230. Paula Roberts, "A Ministerial Portfolio," *The Atlantic* 257 (June 1986): 22.

231. Paul Quinnett, "Enough of Roads," *Audubon* 87 (July 1985): 18.

232. Stephen K. Doig, "Knock-on-Wood Spirit Greets Storm Season," *The Miami Herald* 1 June 1986: 18.

233. Laurie Baum, "Loehmann's Vets Tackle S. Florida Expansion," *The Miami Herald* Business Monday 27 Feb. 1984: 7.

234. "The Joke That Bombed," *The Washington Post National Weekly Edition* 27 Aug. 1984: 26.

235. "Pet Turtles—Revisited," *The Harvard Medical School Health Letter* 9 (Sept. 1984): 8.

236. Ricki Fulman, "We All Lie About Our Age," *The Miami Herald* 9 April, 1983: C1.

237. Willie Schatz, "Clash of Cultures," *Datamation* 32 (1 Nov. 1986): 22.

238. Wayne Tustin, "Recipe for Reliability: Shake and Bake," *IEEE Spectrum* 23 (Dec. 1986): 37.

# Part Two
# STANDARDS OF CORRECTNESS

# CHAPTER 6

# The Standard of Grammar for the Professions

*The first thing you should attend to, is to speak whatever language you do speak in its greatest purity, and according to the rules of grammar, for we must never offend against grammar.*

*Lord Chesterfield*

> *Let school-masters puzzle their brain,*
> *With grammar, and nonsense, and learning;*
> *Good liquor, I stoutly maintain,*
> *Gives* genus *a better discerning.*

*Goldsmith*

When he connects grammar with schoolteachers, the roistering playboy of *She Stoops to Conquer* voices an association familiar today. Even teetotal writers in industry and government have been known to echo the sentiments of his first two lines. Today grammar is both a sacred subject and the subject of much profanity. In government and industry arguments about grammar generate more heat than does discussion of politics—even office politics. It is almost impossible to discuss any other phase of writing for an audience of business executives, scientists, or technologists without meeting extraneous questions like "But does it matter if my grammar is wrong?" and "Why don't the colleges teach engineering students some grammar?"

## TWO POPULAR FALLACIES

In these discussions two antithetical fallacies are often heatedly supported: "Writing is just grammar" and "Grammar don't matter." (For other fallacies see Appendix A.) These two fallacies have been especially influential in technical writing, because some writers and editors majored in science or technology and others in English or journalism. One ar-

177

gument is that writers do not need instruction in grammar. Unfortunately, many of the supporters of this view show by their statements that they do. Other discussants insist that what a technical writer needs most is a complete study of English grammar, but most of them insist in speech and writing that shows not a deficiency in grammar but a need for basic instruction in style.

This fire and smoke arise from misunderstandings. Grammar is not a set of rules arbitrarily imposed upon writers by dictionary-makers, schoolteachers, or anyone else. The grammar of writings on professional subjects should be the grammar that educated writers use and that educated readers expect. This grammar changes—sometimes rapidly, sometimes slowly. What was regarded as incorrect twenty years ago may not be considered incorrect today. Moreover, the grammar of speech is not necessarily that of writing. In answer to "Who's there?" Americans answer, "Me" or "It's me." "It's I," which was once preferred, now sounds affected, pompous, pedantic. But in writing, educated Americans use "It's I," as in the sentence "It's I alone who recommended this change; the committee said nothing about it."

Incorrect writing annoys readers of functional prose. It is ironical that even those who themselves write incorrectly are quick to complain about the errors of others. Readers want to be free to move along a page without the distraction of errors. They want to grasp information and ideas as rapidly as possible. All readers resent having to rephrase or recast or even reread sentences in order to understand them; they balk at bringing their knowledge to bear in order to discover what a writer meant to say but did not say. Readers want to read, not edit or interpret; they think that they are entitled to clear, correct exposition. They would like it to be brief and interesting also. But it must be clear and correct.

If writing is incorrect, readers hesitate to rely on the research of the authors. They wonder, "Is it likely that the writers' tables, charts, and graphs are accurate and correct if their explanations are not?"

The writers in turn would like to offer correct prose but have difficulty achieving it and meeting the other obligations of their busy lives. In desperation some turn to the secretary who can edit well, and they take the secretary with them from position to position and company to company as long as possible. When the secretary refuses to move or a company refuses to hire the secretary along with the writer, confused collapse is likely to follow. The writer cannot report adequately, and the company does not know why. Sometimes a separation or divorce from an English teacher–spouse who corrected reports along with term themes leaves a scientist or engineer helpless. Indeed, it is not unknown for professionals to become so dependent on relatives who are English teachers or editors that the transfer or death of such persons leaves the professionals afraid to submit their uncorrected work. Good writers and editors have been known to support themselves royally through graduate school or through their first years as writers by charging outrageous fees for being always on call for desperate professionals. (One editor even had to carry a beeper.)

## ADVICE ON REVISING GRAMMAR

Fortunately most writers in the professions are independent enough to try to help themselves. Some of them start to do so by undertaking a complete and detailed study of English grammar. We consider such a study unnecessarily burdensome for professionals, who have better ways to spend their time. Worrying about dozens of rules that one has never broken may be harmful. It is like teaching adults to correct their walking by having them study in detail all the muscle and nerve action involved in walking. Writers who never had trouble with *who, which,* and *that* memorize all the rules for the use of those pronouns and then find *who, which,* and *that* stumbling blocks whenever they write. A busy professional does not have time to learn all the rules and to practice applying them until the correct use becomes automatic.

Writers should concentrate on the rules they need—those they break. The rules in this chapter, which have been selected carefully to spare writers the tedium of a complete study, concern the errors and weaknesses that appear most often in the writing of accountants, of lawyers, of scientists, of technologists, and of other administrators in government and industry. These are the errors that editors and supervisors grow weary of correcting. They are the errors that writers in our classes ask about most often. They are the errors that professionals are likely to make. Professionals who do not make them now are in danger of making them later, for they will read them many times in the writing of their colleagues. And errors are contagious.

Writers are fortunate if they can ask a competent editor or supervisor or teacher to mark the errors in five or six pages of their work. Then writers can study the correction of the errors, look for the errors when they revise their next papers, and correct the mistakes. After a time the errors will no longer appear in their first drafts, and the authors will be ready to ask their mentor to correct for them again. When the corrector finds no errors, the writer is ready for the next steps in revision. But an annual check on errors is advisable because writers are likely to acquire new bad habits or revert to old ones.

Many professionals using this book do not have available the help of competent editors or supervisors, and some writers prefer to work without such assistance. They should read the sections on grammar that follow and should mark for study any rules that they break or might break. Those that writers might break will be revealed by their experience with the examples: the correct examples sound wrong, and the incorrect examples occur in their reading or writing or both. After studying the explanations of the rules that they break or might break, writers should search for examples of broken rules when they revise their next papers, not while they are writing first drafts. Very likely they will find some errors that they thought they never made. They should practice correcting these errors until correct usage becomes automatic, as shown by the disappearance of mistakes from their first drafts. After that an annual skimming through the following examples of common errors should enable them to spot the mistakes in their writing.

## THE AGREEMENT OF SUBJECT AND PREDICATE VERB

Errors in agreement between subject and predicate are common in the prose of information. True, one seldom finds "These methods is recommended" or "The mixture are boiled"; but writers—and editors too—often err in less simple constructions. For example, a medical journal published this opening sentence: "Experience with two antidepressant agents which are known as monoamine oxidase inhibitors are presented in the following." *The New York Times* printed "From this work has come improved antibiotics." And writers in industry often publish sentences like "The recommendation for salary increases were examined."

Because blunders in agreement occur most commonly in certain contructions, a study of those constructions helps writers to avoid the errors.

### Separated Subject and Predicate

Many writers become confused about number when a subject is separated from its predicate verb, particularly if the intervening words introduce another number, as in the following incorrect examples:

The effective *use* of these machines *require* study and practice.
A *tabulation* of the animal responses *have* been presented elsewhere.
*Estimation* of the pressure drop and fluidization characteristics *are* desirable.

In these examples the subjects are clearly singular—*use, tabulation, estimation*—but the plural words *machines, responses*, and *characteristics* misled the writers, who would not otherwise write or say *use require, tabulation have*, and *estimation are*. Any writers who think that because of the intervening word they too would have made these mistakes should remember that one step of their revision should be a check of the agreement of subjects and verbs that are separated.

### Subject Following Verb

Errors are also frequent when subjects follow verbs. The journalist who wrote, "From this work has come improved antibiotics . . ." would not write *antibiotics has come*, but the reversed order confused the writer. "Out of even the best college comes many writers ignorant of the basic

180

principles of English," wrote one of our students and unconsciously offered proof of the generalization by stating it. This student would have looked condescendingly on anyone who wrote *many writers comes*. If the order of subject and predicate is reversed, a writer should place the subject before the predicate to find the correct number. The primary-school device of asking, "Who or what come or comes?" leads to the correct answer, "Writers come." This device is helpful to anyone who would otherwise read the sentence and like the sound of it.

Some writers become confused when a subject follows a verb in a sentence beginning with *there is, there are, here is, here are,* or *it is*. The last of these is the simplest. *It is* may be followed by a singular or plural noun:

> It is this book that I need.
> It is these books that I need.

But when a sentence begins with *there is, there are, here is,* or *here are,* a writer should take special care. Usually the question "Who or what is or are?" will help a writer who has difficulty distinguishing the subject.

> There (*is* or *are*) at least three treatments for this condition.
>     (Q. What is or are? A. Three treatments are.)
> There (*is* or *are*) an *x* missing in the equations.
>     (Q. What is or are missing? A. An *x* is missing.)

Many indirect statements have poor emphasis; they should be direct statements. The second example, for instance, is a stronger sentence without *there is*:

> An *x* is missing in the equations.

The first example is better if the writer omits *there are* and says something about the treatments instead of just pointing to their existence:

> At least three treatments for this condition were demonstrated at the conference.
> At least three treatments for this condition are inadvisable because of the patient's allergies.

## Words Attached to a Subject

Writers should remember that words attached to a subject by such connectives as *accompanied by, along with, as well as, including, in addition to, no less than,* and *together with* do not affect the number of the subject in formal English.

181

His secretary, as well as many other members of the division, is attending the meeting on computers.

The singular *is* has as its subject *secretary*; *members*, introduced by *as well as,* does not make the subject plural.

## The Collective Noun

A collective noun (the name applied to a group), such as *band, class, family, jury,* takes a singular verb when the group is regarded as a unit and a plural verb when the members of the group function as individuals.

> The committee *is* in favor of a holiday on June first.
> The committee *are* leaving for vacations.
> The class *is* presenting a scholarship to the college.
> The class *are* going to graduate schools or taking jobs.

Those who do not like the sound of the plural verb after the collective noun should remember that they may dislike the sound because the correct form is strange to them. If they think their readers will be distracted, they may rephrase by substituting *members of the committee* and *members of the class.*

Writers on finance, science, and technology and others who frequently use amounts should check their manuscripts to make sure that they use a singular verb whenever the expression of quantity is thought of as a unit and a plural verb otherwise.

> Three grains diluted in sterile water *was* injected.
> Ten and a half ounces *is* added to the solution.
> Two teaspoonfuls *is* a large dose.
> Add water until there *is* three pints of solution.
> Twenty stamps *are* scattered on the table.
> Twenty stamps *is* a small number to collect in a year.

## Foreign Singulars and Plurals

Some singulars and plurals have been taken into English from other languages: *desideratum, desiderata; erratum, errata; seraph, seraphim.* Some plural nouns, for example, *agenda* and *candelabra,* have been accepted as singular nouns; others, like *criteria* and *media,* are often mistaken for singular nouns.

| Singular | Plural | Singular | Plural |
|---|---|---|---|
| agenda | agendas | media | mediae |
| alumna | alumnae | media | medias |
| alumnus | alumni | medium | media |
| appendix | appendices | memorandum | memoranda |
| basis | bases | minimum | minima |
| cherub | cherubim | parenthesis | parentheses |
| continuum | continua | phenomenon | phenomena |
| crisis | crises | radius | radii |
| criterion | criteria | seraph | seraphim |
| curriculum | curricula | series | series |
| datum | data | species | species |
| desideratum | desiderata | stamen | stamina |
| erratum | errata | stimulus | stimuli |
| formula | formulae | stratum | strata |
| fungus | fungi | syllabus | syllabi |
| index | indices | | |
| maximum | maxima | | |

Because the English language has a strong tendency to impress its own methods upon borrowed words, many foreign words acquire English plurals: *appendixes, cherubs, curriculums, formulas, funguses, indexes, maximums, mediums, minimums, memorandums, phenomenons, radiuses, seraphs, stamens, stratums, syllabuses.* Writers should choose these plurals consistently; for example, they should use either *indexes* or *indices* throughout a work, not both for the same meaning. For the Latin singular *curriculum vitae* consistency dictates the Latin plural *curricula vitae.* (The adjective is *curricular.*)

Often a foreign plural is used in a technical sense. One can see this most clearly in the words used in the specialization of others. *Cherubs,* for example, means beautiful children, but the Hebrew plural *cherubim* in theology means one of the nine orders of angels. Either may be used for a representation of beautiful children with wings. The distinction for *seraphs* and *seraphim* is similar. Less common is the practice of adding an English plural to the foreign plural: *cherubims, seraphims.*

*Formulae* is used in mathematics and chemistry, although even in these sciences *formulas* appears. *Funguses* is used more often than *fungi,* although *fungi* is common in medical writing. In mathematics *indices* is preferred, but elsewhere *indexes* is common. *Maxima* occurs frequently in mathematics and astronomy, but *maximums* is the choice elsewhere. Although *memorandums* is commonly used, some formal writers and some scientists and technologists prefer *memoranda. Phenomena* appears in philosophy with one meaning and in the sciences with two, but exceptional persons, things, or occurrences are *phenomenons. Radii* is more frequent in the sciences and mathematics than is *radiuses,* but the use of *radiuses* seems to be increasing. *Strata* is preferred in biology and geology; *stratums* sometimes appears in general usage. *Syllabuses* is preferred and might just as well be because the word came into English from the Latin through a misreading.

A writer who understands the use of the plurals of these words is not likely to make such common errors as the following:

> This formulae is . . .
> The maxima is . . .
> This fungi are found . . .
> The syllabi was used . . .

*Datum* is seldom used in English except in such phrases as *datum line, datum plane,* and *datum point.* But *data,* which is frequently used, perplexes many writers. With more persistence than patience, teachers and supervisors have trained scientists and technologists to use *data* as a plural. No point of grammar or style has received so much emphasis, with the possible exception of the banishment of *ain't.* Some who inveigh most strongly against *data* as a singular have no true plural concept of *data.* They change *This data is . . .* to *This data are. . . .* They write, "Smythe's data are accurate. It proves. . . ." The use of *this* and *it* reveals that they think of *data* as singular despite their preachments that it is plural. Making *data* both singular and plural as in *This data are* and *Smythe's data are accurate; it proves* is worse than the consistent use of one number—even the wrong number.

But the scientists who write, "This data is helpful to us," are not using the wrong number. They are thinking in a collective sense of *data;* to them it means evidence, not separate points of evidence. When *data* is thought of as functioning as a unit, then the singular is sensible and correct: "This data supports my theory"; "The data appears in the appendix." If *data* is thought of as units operating individually, then the plural is sensible and correct: "These data are useful in a number of fields"; "Few data have been accepted as widely as these."

Writers likely to use such constructions as "Much data are . . ." and "Little data are . . ." do not have a concept of *data* as either singular or plural, for they use singular modifiers and plural verbs. They may test their constructions by substituting *information* for *data* when they wish to indicate that the points are to be considered as a unit. Then they will write, "Much (information) data is . . .," and "Little (information) data is. . . ." When they mean that the points are operating individually, they may experiment by substituting *facts*: "Many (facts) data are . . ." and "Few (facts) data are. . . ." Writers who would never use "Much facts are . . ." become confused by *data* unless they test with another noun—a singular noun when they want a singular concept (*information* or *evidence*) and a plural noun when they want a plural concept (*facts*). When these writers err, they are confused not so much about English grammar as about their concept of *data*.

## Company Names

Although the British use company names and the official titles of other bodies in the plural, Americans prefer them to be singular:

> Corroon, Carroll, Curran & Cooney, Incorporated, *is* mailing the policies this month.
> The Government Printing Office *issues* this document; *it* charges fifty cents a copy.
> The Research Division *submits* quarterly reports. *It* also prepares special reports on request.

Difficulty arises when the company or division does not function as a unit. Then the writer should not awkwardly switch to the plural. "The Research Division is investigating the proposed method. Some of *them* think. . . ." *The Research Division,* established as singular by the verb *is,* may not serve as the antecedent of the plural *them.* The writer may correct by substituting a plural subject word: "The Research Division *is* investigating the proposed method. *Some of the members* think. . . ." The error of switching incorrectly from one number to another is illustrated by the company that advertises in newspaper type nearly an inch high: "A big office furniture company like Itkin doesn't have to be an authority on decorating too. But they are." Is Itkin or are Itkin?

Whether one decides to use the plural or the singular with the titles of companies and other bodies is not important. But being consistent in maintaining the plural or the singular is helpful to readers. If the chosen number is difficult to use for a particular concept, a writer can easily substitute another word. Although the point seems obvious, many writers of company reports do not realize that because one starts to write about "The Medical Illustration Department, which *is* located on the eleventh floor," one does not have to retain the singular throughout the report. Any writer can think of dozens of escapes to the plural, such as *the medical illustrators, the members of the department, the opinions of the department, the methods, the policies, the executives, the artists,* etc.

## Mathematical Expressions

In mathematical calculations either a singular or a plural verb is correct:

> Three times three *is* (or *are*) nine.
> Three and three *is* (or *are*) six.
> Three and three *makes* (or *make*) six.

An equation, regardless of the number of terms on either side, is singular:

> In assuming that $x + y(3a/b) = 46$ *is* valid for every case. . . .
> He was certain that $a^2 + b^2 = c^2$ *was* correct.

## Singular Indefinite Pronouns as Subjects

*Another, anybody, anyone, anything, each, either, everybody, everyone, everything, neither, nobody, no one, nothing, one, somebody, someone,* and *something* require singular verbs.

> Neither of the treatments *is* safe.
> *Is* anyone listening?
> Each of the machines *has* advantages. (But when *each* is an adjective, it does not affect the number of the verb: They each *have* qualifications for the nomination.)

## *None* as Subject

When *none* is the subject, the verb is usually plural. *None* may be used with a singular verb if a writer wishes to stress a singular concept, but in such instances *no one* or *not one* is generally chosen.

> *None* of the animals *are* responding.
> *Not one* of the animals *is* responding.

## Subjects Connected by Correlative Conjunctions

Singular subjects connected by *either . . . or* or *neither . . . nor* take singular verbs.

> *Either the Purchasing Division or the Manufacturing Division is* responsible.
> *Neither crystallization* caused by pressure *nor slippage* at the capillary walls *causes* this discontinuity in flow.

When the subjects connected by *either . . . or* or *neither . . . nor* differ in number, the verb agrees with the subject nearer it.

> Neither polymeric *substances* nor their physical *behavior is* understand thoroughly.
> Either you or *I am* going. (A writer who does not like this sentence may rephrase: Either *you are* going or *I am.*)
> Neither the doctor nor the *nurses are* to blame.
> Neither the nurses nor the *doctor is* to blame.

# THE PRONOUN

Although most writers on professional subjects know that a pronoun agrees with its antecedent in number, gender, and person, many of them have difficulty with the following constructions.

## The Collective Noun as Antecedent

Once writers have decided whether the collective noun is singular or plural, they must use that number not merely for the verb but also for any pronouns that refer to the collective noun and for any words that modify it.

> INCORRECT: The committee *is* planning to submit *their* report before January.
>
> CORRECT: The committee *is* planning to submit *its* report before January.
>
> INCORRECT: *These* data for copper deposition *has* been given previously. *They have* been used to prove. . . .
>
> CORRECT: *This* data for copper deposition *has* been given previously. *It has* been used to prove. . . .
>
> CORRECT: *These data* for copper deposition *have* been given previously. *They have* been used to prove. . . .

(The singular or plural forms are used with *data* according to whether the writer is indicating that the data is functioning as a unit or that the data are functioning individually. Discussion of this point appears under The Collective Noun (page 182).

## The Sexist Pronouns: *He, She; His, Her, Hers; Him, Her; Himself, Herself*

Until about a quarter of a century ago the masculine pronoun in the third person singular (*he, his, him*) was used for nouns that meant in the general sense both masculine and feminine genders, for example, such words as *artist, athlete, doctor, engineer, musician, pilot, voter,* and *writer.* Only a few nouns, like *nurse* and *elementary school teacher,* were usually followed by the feminine pronoun (*she, hers, her*). This assignment of gender to particular jobs or vocations was most noticeable to readers and writers familiar mainly with English and not accustomed to the separation of gender and sex in a phrase like *la plume de ma tante,* which assigns the feminine gender to both *pen* and *aunt,* or to nouns and articles like *das Weib* or *das Mädchen,* which illustrate the use of the neuter gender (*it, its, it*) for the words *woman* and *girl.* In the United States, therefore, the use in English of the masculine gender for nouns representing both masculine and feminine gender was rejected strongly by men and women fighting for the equality of women.

Newspaper columnists who wrote on the subject and the editors of columns of letters to the editor received overwhelming responses, many of them containing suggestions, such as the use of new pronouns *thon* and *hesh* or the use of *he/she, she/he, s/he, he or she, she or he, he and she, she and he*. Jokes on the subject were almost as numerous as suggestions.

Yet the serious side dominated. Men and women pressing for the equality of women made readers keenly aware of the injustice of referring to a doctor by the third person singular masculine pronoun and to a nurse by the third person singular feminine. The same unfairness occurred in the use of the masculine for many other desirable occupations—architect, artist, engineer, manager, professor, school principal, scientist, supervisor, for instance—and the use of the feminine pronoun for subordinate positions—elementary school teacher, nurse, secretary. Change was demanded for fairness in employment and for the establishment of correct concepts in the young.

Most writers and editors made the easy change—avoiding gender by using the plural followed by *they*, which does not indicate gender. The following examples are typical of the change:

SINGULAR FORM: The doctor is to receive his Form B after the nurse has entered her comments in the first two boxes.
PLURAL FORM: The doctors are to receive their Form B's after the nurses have entered their comments in the first two boxes.
SINGULAR FORM: An engineer can easily use the active voice effectively in his engineering reports.
PLURAL FORM: Engineers can easily use the active voice effectively in their engineering reports.
SINGULAR FORM: A chemist benefits from learning to use more than the one report plan required in his undergraduate chemistry courses.
PLURAL FORM: Chemists benefit from learning to use more than the one report plan required in their undergraduate chemistry courses.
SINGULAR FORM: The school principal is to post the notice in his office. The teacher is to post the notice on the bulletin board of her classroom.
PLURAL FORM: School principals are to post the notice in their offices. Teachers are to post the notice on the bulletin boards of their classrooms.

The change to the plural pronoun to avoid sexist implications was initiated to help establish the equal rights of women, but it also benefited male nurses, male secretaries, male teachers, and males in other occupations traditionally followed by pronouns in the feminine gender. Disliking the stereotype that such nouns as *nurse, secretary,* and *teacher* were referred to by the feminine pronoun, the men welcomed the change that eliminated this sexual discrimination.

### He or She, She or He, He and She, She and He

One use of the pronouns for the third person singular did not require change because of sexist implications. The combination *he and she* and *he or she* has traditionally been used to refer to an antecedent composed of one masculine and one feminine member: "Every man or woman should give his or her full attention to the warnings about walking alone at night on the streets just south of the plant." This form is useful for emphasizing that both sexes are meant. A plural noun has a different effect: "All employees should give their full attention. . . ." The plural for both nouns satisfies many writers: "Men and women should give their full attention. . . ." The combination of pronouns often used now is *she or he, she and he*. In any case, the combination should not be used so frequently that it becomes awkward, as it easily does.

### Singular Indefinite Pronouns as Antecedents

The sexist implications of the singular indefinite pronouns have not been as easy to remove as those of the singular personal pronouns in the third person. Indefinite pronouns name or refer to a group or class but not to any specific member or members. The following singular indefinite pronouns are commonly used:

| | | |
|---|---|---|
| another | everyone | no one |
| anybody | everybody | one |
| anyone | everything | other |
| anything | neither | somebody |
| each | nobody | someone |
| either | | something |

Some of these indefinite pronouns, for example, *everyone* and *everybody,* are used frequently with a third person masculine pronoun referring to the indefinite pronoun:

SENTENCE 1: Everyone must submit his monthly reports by the first Wednesday of each month.
SENTENCE 2: Everybody complained about his new union benefits.

Particularly in sentences where *everyone* and *everybody* are employers or employees, many readers consider the use of *his* to be sexual discrimination. To avoid the objectionable usage writers have revised in one of the following ways:

I. *Rephrasing to Avoid the Personal Pronoun*

SENTENCE 1: Everyone must submit monthly reports by the first Wednesday of each month.
SENTENCE 2: Everybody complained about the new union benefits.

This is satisfactory in Sentence 1 but changes the meaning of Sentence 2. If the writer means that everybody complained about all the new union benefits, the rephrasing solves the problem. But if the writer means that everybody complained only about the new union benefits affecting that individual, then the rephrasing has changed the meaning and the writer may have to substitute one of the methods below.

### II. Using the Plural Form of the Third Person Pronoun

*Everybody had expressed their* thoughts before the meeting ended.
Everyone who buys the full package will be given three days of free instruction at *their* convenience.

Although those whose ears are sensitive to any breaking of the old rules may deplore the use of *they* to refer to the singular indefinite pronoun, many speakers use it, especially television newscasters. The English language lacks a third person singular pronoun for persons that does not state gender, and popular usage seems to have assigned the role to *their,* the plural with no sexist discrimination. This raises the question of whether to use the singular or plural for words associated with the pronoun. Fowler wondered in this connection whether the users had "made up their minds between *Everyone* was *blowing their noses* and *Everyone* were *blowing their noses.*"[1] Speakers in the United States seem to have chosen *Everyone was blowing their nose, their* now being widely used in speech as the pronoun for the third person singular in this construction as well as for the third person plural otherwise. We do not recommend a crusade for *they* and *their* as singular, but we do recognize the common sense of speakers who accept this use to help themselves out of awkward and difficult sentence constructions that are not easily avoided any other way. Moreover, *everybody* and *everyone* may indicate a plural number although used with a singular verb:

Everybody has passed the test. (30 students)
Everyone has some faults. (the world population)

Like many changes in language, this one carries a penalty: speakers, writers, and editors must be careful to avoid using *they* to refer to a singular indefinite pronoun if an intervening plural noun or pronoun may be mistaken as the antecedent or if a plural verb makes the use sound awkward:

INCORRECT: Everyone brought their misconceptions to the meeting, and they made the meeting dull and useless.
CORRECT: By bringing their misconceptions to the meeting, all (the employees, managers, students, etc.) made the meeting dull and useless.
INCORRECT: Each employee is to sign the posted notices before noon even though they may not be counted.
CORRECT: Each employee is to sign . . . even though the signatures may not be counted.

190

### III.  Using the Second Person: You, Your, Yours in Informal Writing

FORMAL: "To Cafeteria Employees: Each employee is to report for his medical examination according to the following schedule. . . ."

INFORMAL: "To Cafeteria Employees: Please report for your medical examination according to the following schedule. . . ."

If the style of writing is sufficiently informal (casual, conversational), *you*, whether understood, as in "(you) report," or expressed, as in "your medical examination," solves the problem neatly. Without advocating informal style in formal communications, we consider some insistence on formal style unfortunate: communications to fellow employees, especially to those in the field or in other locations of the company; invitations to informal events; instructions. Simple material thus made complex is a sign of lack of adjustment to readers and thus of poor style.

### IV.  Using the Passive Voice

Although we do not like to recommend recasting even an intractable sentence into the passive voice, doing so may be the answer to some problems caused by sexist pronouns.

One writer editing her rough draft of an instruction guide found "Every manager and field representative must submit his monthly report on the first day of each month except January." The managers and field representatives of her company were no longer all males. Most of the readers were conservative and would not welcome the informal *you* or even the use of *his or her.* In fact, they probably liked the sentence as she had written it, but upper management might not accept the sexist use of the masculine pronoun. A possible change was a shift to the passive voice: "The monthly reports of managers and field representatives must be submitted on the first day of each month except January" or "Monthly reports must be submitted by managers and field representatives on the first day of each month except January."

Writers need not fear that if they use the passive voice for one or two sentences, they must use it for all sentences. Only similar or parallel sentences need be recast in the passive voice; others may remain in the active voice. It is, of course, usually unwise to shift voices within a sentence unnecessarily.

In summary, then, the easiest way for writers to avoid sexist pronouns is to use the plural. In the few sentences where this use seems impossible, writers may choose one of the following methods: recasting the sentence to avoid singular personal pronouns, choosing for informal style *you* instead of *he* or *she,* rewriting in the passive voice.

There will always be writers who use personal pronouns in the traditional way; whether this choice is due to literary or sexist reasons or to a dislike for any change, each reader may decide. When the passage in which the pronoun occurs relates to employment, sexual discrimination

191

may result regardless of the intentions of the writer. Editors, managers, lawyers, supervisors, and officers of companies usually discourage this use of pronouns in such passages and should be sure that the offending writers know the choices available to them if they are asked to revise sexually discriminatory passages. (The subject of nouns and sexism is discussed in Appendix B under **sexist nouns**.)

### Ambiguous or Vague Antecedents

Every pronoun should refer unmistakably to a noun or pronoun or noun phrase. *Unmistakably* means that the antecedent of the pronoun should be obvious to the reader, who should not have to select from two or more antecedents:

> INCORRECT: Miss Jones told Miss Smith that she might receive a raise.
> CORRECT: Miss Jones told Miss Smith, "You may. . . ."
> CORRECT: Miss Jones told Miss Smith, "I may. . . ."
> CORRECT: Miss Jones told Miss Smith that Miss Smith might. . . .
> CORRECT: Miss Jones told Miss Smith that Miss Jones. . . .
> CORRECT: Miss Jones told Miss Smith that she expected a raise. (Miss Jones probably did not know what Miss Smith expected and so was referring to herself.)

Avoid reference to an idea that is not expressed or is not expressed as a noun:

> INCORRECT: The executive secretary has suggested that I request your files and minutes of Board meetings. I hope it will be convenient for you to do this.
> CORRECT: I hope that it will be convenient for you to send them to me.
> INCORRECT: In infections it becomes acidic, the degree of which depends on dilution by exudate.
> CORRECT: In infections it becomes acidic, the degree of acidity depending on. . . .

### One

Writers should avoid overworking *one.* In the most formal writing, where *one* must be followed by *one* and *one's,* the pronoun becomes monotonous. This error is frequent in technical reports, where it occurs in the awkward, wordy statements of writers who are striving to be impersonal.

> AWKWARD: One can see that it is obvious that one can use the tiny David Computer with more speed and accuracy than one can use the massive Goliath Computer.
> IMPROVED: Obviously, the tiny David Computer offers more speed and

accuracy than the massive Goliath Computer. (The improved sentence avoids personal pronouns and *one*.)

AWKWARD: If one supplies $x$ in the first step of one's equation, one will find that one has to spend less time solving and one will find oneself less confused by the complexity of the solution.

IMPROVED: Supplying $x$ in the first step speeds and simplifies the solving of the equation.

A common error is confusing *one* with *I, you, we,* or *they.* An unnecessary shift in pronouns is incorrect whether it occurs in a single sentence or in a paragraph.

INCORRECT: As *one* tours the factory, *one* notices the cleanliness, achieved, you suspect, by little effort.

CORRECT BUT AWKWARD: As *one* tours the factory, *one* notices the cleanliness, achieved, *one* suspects, by little effort.

IMPROVED (informal): As they toured the factory, they noticed the cleanliness, achieved, they suspected, by little effort.

INCORRECT: When *one* applies for employment, *they* should expect long delays.

CORRECT (formal): When *one* applies for employment, *one* should expect long delays.

CORRECT (formal): When applying for employment, *one* should expect long delays.

The use of the nominative *he, she, it*; the possessive *his, hers, its*; the objective *him, her, it*; and the reflexive *himself, herself, itself* to refer to a preceding *one* is common in American informal style.

*One* has a blot on its title page.
*One* oils *itself* for six months.
*One* must keep *one's* opinion to *oneself* while working for Scrooge. (Formal.)
*One* must keep *his* opinions to *himself* while working for Scrooge. (Informal, sexist.)
*They* must keep *their* opinions to *themselves* while working for Scrooge. (Informal and free of sexist discrimination.)
*No one* working for Scrooge should express an opinion.

Many writers have difficulty when a construction containing *one* is followed by a relative clause.

He is one of the competent students who have been hired for summer work.

The antecedent of *who* is *students*; the meaning is that he is one (but not the only one) of the competent students hired for summer work. When the meaning is *the only one,* the antecedent of the relative pronoun is *one.*

193

He is the one of the students who is incompetent.

Usually the meaning is *one,* not *the only one,* and the relative pronoun is plural in agreement with its antecedent, as in the following sentences:

> She is working on one of those research projects that are sponsored by the government.
> He is one of those overambitious executives who are always seeking promotions.
> He made one of those personnel decisions that are hard to change.
> This patient is one of those who suffer nausea and vomiting all through their pregnancies no matter what drugs they are given.
> He made one of those errors in English that are noticed at once.

## Cases

Writers sometimes allow a parenthetical expression to attract a pronoun into the objective case.

> INCORRECT: The employees whom he thought were to be assigned to the project were all working on other research.
> CORRECT: The employees who he thought were to be assigned to the project were all working on other research.
> INCORRECT: He rejected the applicants whom he judged might know more than he did.
> CORRECT: He rejected the applicants who he judged might know more than he did.

In the first sentence *who* is the subject of *were to be assigned.* In the second sentence *who* is the subject of *might know.*

A pronoun takes its case from its own function in its own clause, as the following sentences illustrate:

> 1. INCORRECT: They objected because he recommended *whomever* was his friend for the promotion.
>    CORRECT: They objected because he recommended *whoever* was his friend for the promotion.
> 2. INCORRECT: The report stating *whom* should be held responsible for the poor shipping record was discussed in detail.
>    CORRECT: The report stating *who* should be held responsible for the poor shipping record was discussed in detail.
> 3. INCORRECT: She has a pleasant word for *whomever* comes to her office.
>    CORRECT: She has a pleasant word for *whoever* comes to her office.

In the first sentence *whoever* is the subject of *was.* In the second sentence *who* is the subject of *should be held responsible.* In the third sentence *whoever* is the subject of *comes.* In all three sentences the relative pronouns are in the nominative case because they are subjects. The fact that in each sentence the entire relative clause is the object (of *recommended* in the first sentence and of *stating* and of *for* in the second and third sentences) does not affect the case of the relative pronoun; it is still in the nominative case because it is the subject of the relative clause.

The case of a pronoun following *as* or *than* can be determined by completing the construction in one's mind.

> Her secretary is taller than she (is).
> My supervisor is more interested in causes than I (am).
> I like his substitute better than (I like) him.
> I like his substitute better than he (does).
> My secretary works harder for him than (for) me.

Some writers have difficulty distinguishing participles and gerunds because they look alike: *coming, going, seeing, summarizing*. Participles are used as adjectives, and gerunds, as nouns. In *I dislike his scheming ways, scheming* is a participle modifying *ways*. In *His scheming gained him no advantages, scheming* is a gerund used as the subject of *gained*.

A noun or a pronoun modifying a gerund is in the possessive case.

> When *their* sleeping improved, patients reported their delight.
> After examining the animals she said that she had no doubt about *their* being satisfactory for her experiment.

There is a difference in meaning between a phrase containing a gerund and one with a participle, as well as a difference in grammar.

> GERUND: I observed their *finding* of the lever.
> PARTICIPLE: I observed them *finding* the lever.
> GERUND: I remember his *smashing* of the apparatus.
> PARTICIPLE: I remember him *smashing* the apparatus.

The use of a gerund and of a noun or pronoun in the possessive case emphasizes the act. The use of a participle and of a noun or pronoun in the objective case emphasizes the actor.

> GERUND: I have been worrying about his scheming ever since I hired him.
> PARTICIPLE: I have been worrying about him scheming ever since I hired him.

## COMPARISONS

If comparisons are correctly employed, they clarify and vivify. In functional prose, they are frequently the foundation of the thought. The subject of a report may be a comparison of two chemicals, of methods of controlling pressure, of several computers, of several designs for a bridge, of the side effects of three drugs. And even when comparison is not the central idea of a paper, comparisons frequently appear. Errors unfortunately are also frequent: ambiguous and incomplete comparisons, omission of necessary restrictive words, omission of a standard, use of a vague standard, confusion of classes and constructions.

Incorrect comparisons are published frequently. An advertisement for a book states: "She has seen further into this difficult and rewarding man than anyone else." Does this mean that she has seen further than anyone else has or that she has seen further into this man than into anyone else? Another advertisement for a book compares earnings with professional men: "Today her earnings are comparable to the top doctors, attorneys, and professional men of the nation."

## Ambiguous and Incomplete Comparisons

When *than* completes a comparison, careful writers check to be sure that they have used the words necessary for clear construction.

> INCOMPLETE: The shop steward condemned the supervisor more than the operator.
>
> COMPLETE: The shop steward condemned the supervisor more than the operator did.
>
> COMPLETE: The shop steward condemned the supervisor more than he condemned the operator.
>
> AMBIGUOUS: In the overlapping areas Wilip's results were given more weight because they were in better agreement with Beattie's than those of Ramsey and Young.
>
> CLEAR: In the overlapping areas Wilip's results were given more weight because they were in better agreement with Beattie's than were those of Ramsey and Young.
>
> CLEAR: In the overlapping areas Wilip's results were given more weight because they were in better agreement with Beattie's than with those of Ramsey and Young.

Each of these sentences labeled *ambiguous* or *incomplete* has two possible meanings; each revision has one unmistakable meaning. Writers are obligated to say exactly what they mean. They should not expect readers to stop, consider two or more possible meanings, and select the correct one. Even if readers have sufficient specialized knowledge to interpret an ambiguous sentence, they resent having to interrupt their reading to do so. And sometimes it is difficult—or even impossible—to interpret an incomplete comparison, for example, one like the following:

> INCORRECT: The errors in the sampling technique are greater than Table 5.
>
> CORRECT: The errors in the sampling technique are greater than the errors in Table 5.

CORRECT: The errors in the sampling technique are greater than Table 5 shows.

## Omission of Standard of Comparison

Vagueness often results from failure to state a standard of comparison. When advertisers do this, they may be avoiding the standard because the only one that they can state without risking a suit from their competitors is unsatisfactory. Writers of functional prose should avoid this vagueness, and readers should view with suspicion products that are advertised, intentionally or unintentionally, by omission of a standard of comparison.

INCOMPLETE: The cocktails at the Knaves' Corner are better. (better than water? better than poison? or just better than no cocktails at all?)

INCOMPLETE: This system is more accurate. (more accurate than what?)

COMPLETE: This system is more accurate than Professor Delusive's.

COMPLETE: This system is more accurate than that of Professor Delusive.

COMPLETE: This system is more accurate than I expected (if the context makes clear how much accuracy was expected).

VAGUE: This drug is recommended because it is so safe.

IMPROVED: The Research Committee recommends this drug for marketing because experiments have proved that it has only one minor side effect.

IMPROVED: Dr. Jones recommends this drug to his patients because he considers it safer than any other on the market.

## Omission of Necessary Restrictive Words

When objects of the same class are compared, some restrictive word such as *other* or *else* is necessary.

INCORRECT: That American woman is richer than anybody in the United States.

CORRECT: That American woman is richer than anybody else in the United States.

INCORRECT: These filters are better than any filters.

CORRECT: These filters are better than any other filters.

## Confusion of Classes and Constructions

The terms of a comparison should be of the same class and in the same construction. (See Parallel Construction, below.)

INCORRECT: The molecular weight compared favorably with Butol.

CORRECT: The molecular weight compared favorably with that of Butol.

INCORRECT: The temperature of Monkey A is the same as Monkey B.

CORRECT: The temperature of Monkey A is the same as Monkey B's.
Monkey A and Monkey B have the same temperature.

197

INCORRECT: He attempted to separate the components rather than testing the combination.

CORRECT: He attempted to separate the components rather than to test the combination.

He attempted separating the components rather than testing the combination.

In a double comparison using both *as* and *than,* the terms of comparison should be completed after *as*.

INCORRECT: This machine is as fast if not faster than ours.

CORRECT: This machine is as fast as ours if not faster.

CORRECT BUT AWKWARD: This machine is as fast as if not faster than ours.

## Superlative Adjectives

When a superlative adjective appears twice and modifies singular and plural forms of a noun, the plural form should appear after the first use of the adjective, and the singular form may be omitted.

INCORRECT: One of the greatest, if not the greatest difficulty, is the injection.

CORRECT: One of the greatest difficulties, if not the greatest, is the injection.

There are several other ways of phrasing this sentence correctly, according to the emphasis the writer desires:

One of the greatest, if not the greatest, of difficulties is the injection.
One of the greatest difficulties, if not the greatest of all, is the injection.
One of the greatest difficulties, perhaps the greatest, is the injection.

## Choice of Comparative or Superlative Degree

The comparative degree is used for comparing two persons or things or ideas; the superlative degree is used for three or more.

COMPARATIVE: She is the more reliable of my two assistants.
This typist is faster than the other.
The second page has more errors than the first.

SUPERLATIVE: He is the most reliable supervisor in the company.
This typist is the fastest I have ever seen.
Page twenty has the most errors.

## PARALLEL CONSTRUCTION

Sentence elements that are not parallel in thought should not be placed in parallel structure (Chapter 11).

POOR: The patients received three injections on the first day, four on the second day, and on the third day they were discharged.

IMPROVED: The patients received three injections on the first day and four on the second day. They were discharged on the third day.

INCORRECT: Many accidents are caused by failure to use safety equipment, careless inspection of laboratories, and we need to warn our workers about these.

IMPROVED: Many accidents are caused by failure to use safety equipment and by careless inspection of laboratories. We should warn our workers of these dangers.

IMPROVED: We should warn our workers of the many accidents caused by failure to inspect laboratories carefully and by failure to use safety equipment.

INCORRECT: This drug will be rejected as there is no proof that it is safe, for it has not been tested on human beings for it does not have good marketing possibilities.

THE WRITER'S MEANING: Because this drug does not have good marketing possibilities, it has not been tested on human beings. It will be rejected for lack of proof of safety.


## THE PLACING OF MODIFIERS

English word order is important. The sense of a sentence may depend on the position of the parts. If English were a more highly inflected language, the position of words might be less vital. In Anglo-Saxon, Latin, and German, for example, one can tell case by endings, but in English one tells it by position except for the pronouns that are inflected. In *The girl hit the boy, girl* is the actor, or the subject, and *boy* is the recipient of the action, or the object. In *The boy hit the girl, boy* is the subject and *girl* is the object. Switching the position of the words changes the meaning.

Good writers of English prose take pains, therefore, to place their words correctly so that they convey the meaning precisely. Even when the context makes the meaning clear, a good writer does not misplace a modifier. A misplaced modifier may delay one reader, distract another, and confuse a third. It is courteous as well as accurate to place modifiers where they belong. Writers invite misunderstanding when they place a word incorrectly and expect readers to determine the meaning.

In fact, American readers pay more attention to position than to case. If in an experiment they are asked to correct the sentence *Him feared she*, more than eighty percent retain the position, not the cases (*He feared her.*) They change *him* from the objective to the nominative case and *she* from the nominative to the objective. It would certainly be as easy, or even easier, to change the positions and leave the cases alone: *She feared him.* But revisers are more influenced by position than case. Those who change the position rather than the case are the students who have studied foreign languages that are more highly inflected than English. Amer-

199

icans who know mainly English are strongly influenced by position, as shown by their changing the case of a word to suit it to the position in which it stands, for example *him* to *he* if *him* is in the usual position for a subject and *she* to *her* if *she* is in the usual position for an object.

## Misplaced Modifiers

Grammars call participial phrases, gerundial phrases, infinitival phrases, and elliptical clauses that do not refer unmistakably to the words they logically modify *dangling modifiers*. Confused, exasperated readers call them something worse, especially when the modifiers seem to be not merely dangling in a sentence but floating in outer space or when modifiers placed next to the wrong words become firmly attached to them. Sometimes the writer of a misplaced modifier unconsciously creates humor; sometimes the writer invites Freudian interpretation:

> Injected daily with this drug, the technicians were able to train the animals in three days instead of three weeks (misplaced participle).
> Being slow and stupid, I cannot train Joe Gulch to operate this machine (misplaced participle).
> After daydreaming and napping, Dr. Medical gave Joe Brownstone a thorough examination and many tests (misplaced gerund phrase).
> While chasing around the cage, Dr. Jones observed the rats (misplaced elliptical clause).
> Deliver the machine to the Purchasing Department as soon as properly greased (misplaced elliptical clause).
> To test effectively three grains of this drug must be injected into the monkeys daily (misplaced infinitive).

A writer correcting these errors may choose from several methods:

1. Placing the misplaced modifier next to the word it should modify:

> To test the drug effectively, the experimenters must inject three grains into the monkeys daily.

2. Changing the subject of the sentence so that the introductory modifier will be followed immediately by the actor to whom it relates and that word will be the subject of the sentence:

> After daydreaming and napping, Joe Brownstone was given a thorough examination and many tests by Dr. Medical.

3. Eliminating unnecessary or misleading words or changing the construction in some sentences:

Being slow and stupid, Joe Gulch cannot be trained to operate this machine.
Joe Gulch is too slow and stupid to learn to operate this machine.
Dr. Medical gave Joe Brownstone many examinations and tests because Joe was daydreaming and napping.
An effective test of the drug requires that three grains be injected into the monkeys daily.

4. Completing the elliptical construction:

Dr. Jones observed the rats while they were chasing around the cage.
As soon as the machine is properly greased, deliver it to the Purchasing Department.

Misplaced modifiers occur most frequently when a modifier begins a sentence. An introductory participle, gerund, or infinitive modifies the subject of the principal verb and should be followed by that subject, which should be the word it logically describes:

INCORRECT: Finding many discrete wavelengths differing completely from the wavelength predicted by classical electrodynamics, the information was used to formulate quantum mechanics.
CORRECT: Finding many discrete wavelengths differing completely from the wavelength predicted by classical electrodynamics, Heisenberg, Schrödinger, and others used the information to formulate quantum mechanics.

When writers do not wish to follow the modifier with the subject, they should rephrase:

INCORRECT: Instead of using a standard textbook, a pamphlet prepared in this laboratory may be substituted.
CORRECT: A pamphlet prepared in this laboratory may be substituted for a standard textbook.
A pamphlet prepared in this laboratory may be used instead of a standard textbook.
INCORRECT: Based on error the Accounting Department declared the estimate of taxes fraudulent.
CORRECT: The Accounting Department declared the incorrect estimate of taxes fraudulent.
The Accounting Department erred in declaring the estimate of taxes fraudulent.

Modifiers are also often misplaced before an *it* that is the subject of a main clause.

> INCORRECT: As chief of this department, it has been my pleasure to supervise research.
> CORRECT: As chief of this department, I have supervised research.

Many participial phrases at the ends of sentences modify the wrong nouns:

> INCORRECT: The formulation was applied directly on the lesions starting on the third day of the infection.
> CORRECT: On the third day of the infection and thereafter the formulation was applied directly on the lesions.
> INCORRECT: These data indicate that MWD is affected only slightly using these control agents.
> CORRECT: These data indicate that MWD is affected only slightly by these control agents.
> These data indicate that MWD is affected only slightly if these control agents are used.

An exception to the rule for introductory infinitives and participles is that a phrase stating a general action rather than the action of a specific person or thing may stand before a word it does not describe:

> Taking everything into consideration, the experiment was not too lengthy.
> To sum up, the proposal is now ready for presentation.

## Squinting Modifiers

Careless placing of a modifier between two parts of a sentence confuses readers, who cannot tell whether the modifier belongs with what comes before or with what comes after it; therefore writers should avoid placing between two parts of a sentence a modifier that may refer to either part. An adverbial modifier at the conjunction of a main clause and a dependent clause may be a squinting modifier:

> SQUINTING MODIFIER: After the committee decided that the work must be completed by Monday, in spite of other commitments, it adjourned immediately.

A reader cannot tell whether the work must be completed by Monday in spite of other commitments or whether in spite of other commitments the committee adjourned. A comma will sometimes help, but English word order has a much stronger effect than punctuation; therefore a careful writer places the phrase where it belongs and does not depend on a comma.

CORRECT: The committee decided that in spite of other commitments the work must be completed by Monday. Then it adjourned.

or

After the committee had decided that the work must be completed by Monday, it adjourned immediately in spite of other commitments.

POOR: Although salaries were increased slowly the workers grew dissatisfied.

IMPROVED: Although salaries were slowly increased, the workers grew dissatisfied.

IMPROVED: Although salaries were increased, the workers slowly grew dissatisfied.

POOR: The rats becoming lethargic gradually required an increased dosage.

IMPROVED: The rats gradually becoming lethargic required an increased dosage.

IMPROVED: The rats becoming lethargic required a gradually increased dosage.

## Single Words and Phrases

Writers must place with special care words that can function as two parts of speech. The position of *almost, even, ever, hardly, just, mainly, merely, nearly, not,* and *only* should be checked carefully because they can modify many parts of a sentence. The meaning of the following sentence changes drastically as *just* is moved from one position to another:

*Just* he thought that he would be promised a promotion.
He *just* thought that he would be promised a promotion.
He thought *just* that he would be promised a promotion.
He thought that *just* he would be promised a promotion.
He thought that he would be *just* promised a promotion.
He thought that he would be promised *just* a promotion.

When it is a promotion that is in question, writers can usually see the different meanings clearly. They should train themselves to understand them in other sentences, such as the following:

1. *Only she said that her supervisor introduced errors into her written work.* (She was the only one who said it.)
2. *She only said.* . . . (Saying was all she did: she did not write it, report it, prove it, etc. In this placement of *only,* a second meaning is possible: she said it but did not believe it. In speech an emphasis of the voice on *said* indicates the second meaning; in writing the first meaning is usually taken.)
3. *She said only that.* . . . (The statement after *that* is all that she said.)
4. *She said that only her supervisor.* . . . (MEANING ONE: She said that just her supervisor, no one else, introduced errors. MEANING TWO: *Only her supervisor* means no one else's supervisor. Here too the voice may indicate Meaning Two. In writing *only* is usually taken as modifying the

phrase *her supervisor,* not just *her*; but *only said* and *only her supervisor* may take Meaning Two if the context suggests it.)

5. *She said that her only supervisor.* . . . (She had just one supervisor.)
6. *She said that her supervisor only introduced errors.* . . . (The supervisor did nothing with errors but introduce them. The supervisor did not seek errors or correct errors or remove errors, just introduce them.)
7. *She said that her supervisor introduced only errors into her written work.* (The supervisor did not introduce anything else, just errors.)
8. *She said that her supervisor introduced errors only into her written work.* (The supervisor did not introduce errors into any work but *her written work.*)
9. *She said that her supervisor introduced errors into only her written work.* (MEANING ONE: into no work but hers. *Only* modifies *her.* MEANING TWO: *into her written work,* nothing else. *Only* modifies the phrase *her written work.*)
10. Placing *only* at the end of the sentence may create confusion in an already ambiguous construction. The reader does not know how far back to move *only*: to the position before *written*? before *her*? before *into*?

Sentences that treat technical material show even more drastic changes in meaning when *only* is moved, as the first set of examples shows:

Only I calculated the value of $x$ in the equation.
I only calculated the value of $x$ in the equation.
I calculated the only value of $x$ in the equation.
I calculated the value of only $x$ in the equation.
I calculated the value of $x$ only in the equation.
I calculated the value of $x$ in the only equation.

I nearly accepted his new theory.
I accepted his nearly new theory.

He would not consider even one experiment.
He would not even consider one experiment.

He fired almost all the members of his division.
He almost fired all the members of his division.

Writers who can read these sentences accurately can usually correct their own once they are alert to the different meanings possible. They must remember that meaning may change with the shifting of a word or phrase and should check carefully the placement of phrases because it is easy to overlook even a ridiculous error like *The manuscript was received in the mail from Dr. Simon* instead of *Dr. Simon's manuscript came in the mail* or *The manuscript from Dr. Simon was in the mail.*

The following sentences illustrate the correction of common errors:

INCORRECT: In contrast to the rats, the writers reported that testosterone fails to increase this rate in mice.

CORRECT: The writers reported that testosterone fails to increase this rate in mice although it increases the rate in rats.

INCORRECT: This patient is progressing as well as can be expected under the treatment of Dr. John Doe.

CORRECT: This patient, who is under the treatment of Dr. John Doe, is progressing as well as can be expected.

INCORRECT: Long arguments often occurred about treatments during the conferences of specialists.

CORRECT: During the conferences of specialists long arguments about treatments often occurred.


## IS WHEN, IS WHERE, IS BECAUSE

Explanations and definitions should not be introduced by *is when* or *is where*; these ungrammatical phrases sound childish and illogical.

INCORRECT: A broadband antenna is where certain specific characteristics of radiation pattern, polarization, or impedance are retained.

CORRECT: A broadband antenna is one that retains specified characteristics of radiation pattern, polarization, or impedance.

INCORRECT: Experimental teratology is when drugs and other chemical agents are applied to developing embryos.

CORRECT: Experimental teratology is the application of drugs and other chemical agents to developing embryos.

Statements in a construction like "The reason . . . is because," are neater and briefer if they are expressed more directly.

INCORRECT: The reason nuclear physicists tend to ignore the chemical bonding of the atoms they investigate is because the energies in nuclear reactions are larger than the energies in chemical bonding.

CORRECT: The reason nuclear physicists tend to ignore the chemical bonding of the atoms they investigate is that the energies in nuclear reactions are larger than those in chemical bonding.

CORRECT: Because the energies in nuclear reactions are larger than those in chemical bonding, nuclear physicists tend to ignore the chemical bonding of the atoms they investigate.


## PRINCIPAL PARTS OF VERBS

Many verbs in English are irregular. Their principal parts must be learned, as there are no rules for them. Some frequently misused irregular verbs, with their principal parts, follow. (The present participle always ends in *ing*.)

| Infinitive | Past Tense | Past Participle |
|---|---|---|
| bring | brought | brought |
| | She *brought* the books to us. | They have *brought* the car back. |
| buy | bought | bought |
| | He *bought* the boat. | We have *bought* the property. |
| fall | fell | fallen |
| | He *fell* whenever he skied. | They have *fallen* on hard times. |
| fly | flew | flown |
| | He *flew* solo yesterday. | She has *flown* around the world. |
| go | went | gone |
| | They *went* home. | He had *gone* to the office. |
| lay (to place or put; takes an object. See also Appendix B.) | laid | laid |
| | He *laid* the book on the shelf. | They have *laid* their coats on the bed. |
| lead | led | led |
| | He *led* the way. | They have always *led* the procession. |
| lend | lent | lent |
| | They *lent* their apartment to us. | He has *lent* his brother money for years. |
| lie (to rest or recline; does not take an object. See also Appendix B.) | lay | lain |
| | She *lay* on the couch. | The tools have *lain* there all night. |
| set (takes an object) | set | set |
| | John *set* the package down. | The men have *set* the plants around the pool. |
| sit (does not take an object) | sat | sat |
| | He *sat* on the grass. | The women had *sat* there for three hours. |
| take | took | taken |
| | She *took* the train to the West Coast. | They have *taken* too long. |
| write | wrote | written |
| | He *wrote* rapidly. | They have not *written* since they left. |

## *SHALL* AND *WILL, SHOULD* AND *WOULD*

American usage of *shall* and *will* illustrates change in usage. In speech in the United States, distinctions between *shall* and *will, should* and

206

*would* have disappeared (possibly because of the use of the contractions *I'll, you'll, he'll, we'll, they'll, I'd, you'd, he'd, we'd, they'd*). Whether to observe these distinctions in writing is a controversial issue that provokes handbook writers to present both views and then to emphasize their own view or even to yield in resignation and advise readers to skip the discussion if they are not interested in the controversy.

Skipping may suit college freshmen, but writers for government and industry must know what they are saying when they use these words. Writers of patents and contracts, of technical directions and safety regulations, of legal papers, and of directions for patients find the study of these verbs more important than some writers of handbooks do; therefore we shall describe the forms as they appear in formal style and advise writers for the professions how to choose the most suitable usage.

Until recently most handbooks instructed writers to use *shall* and *should* with *I* and *we*, and *will* and *would* with all other persons for the simple future tense. For the emphatic future tense—expressing the speaker's determination, threat, command, promise, or willingness—they instructed writers to reverse this practice.

### SIMPLE FUTURE

| | |
|---|---|
| I shall or should | We shall or should |
| You will or would | You will or would |
| He, she, it will or would | They will or would |

### EMPHATIC FUTURE

| | |
|---|---|
| I will or would | We will or would |
| You shall or should | You shall or should |
| He, she, it shall or should | They shall or should |

Then "You will see the patient tomorrow" means that the speaker expects you to see the patient. "You shall see the patient tomorrow" means that

1. The speaker is determined that you see the patient, as in

    "You shall see the patient tomorrow even if I have to take you to him myself."

2. The speaker promises that you are to see the patient, as in

    "You shall see the patient tomorrow; I have arranged for you to be admitted."

3. The speaker commands you to see the patient, as in

    "You shall see the patient this week because I am going on vacation."

4. The speaker is willing that you see the patient, as in

207

"You shall see the patient tomorrow if you wish."

5. The speaker is determined that you are to see the patient, as in

"You shall see the patient tomorrow if you do not send his refund today."

Some formal writers use in a question the form of *shall* or *will, should* or *would* expected in the answer.

Shall I send this order now? (Answer: *You shall*—a command)
Will you endorse my check? (Answer: *I will*—willingness)

Other writers consider this stilted, particularly in the second person:

Shall you leave early today? (Answer: *I shall*—expectation)

But they do use *shall* (with the first and third person) in requests for permission and commands.

Shall I leave the book on your desk?
Shall he leave the book on your desk?

These distinctions between *shall* and *will* are still used in formal writing. Some experts recommend them for all writing, a few instructors label those who attempt to retain them "purists," and other experts suggest more mildly that because *shall* and *should* in the first person sound pedantic to Americans, writers need not hesitate to use *will* and *would*. The choice affects the tone of writing. The traditional usage has a formal flavor. The newer forms—*will* for simple futurity in all persons and *shall* for emphatic future—seem modern and informal.

A period of change often places a burden upon the reader, and the shifting in the use of *shall* and *will* is no exception. Readers may have to discover how a particular writer is using them if they are to understand the writer properly. As this may be a time-wasting process, one cannot blame companies or divisions that order writers to use one method for the sake of clarity. If company usage is not established, a writer should choose the forms that seem most appropriate to the taste of the readers. In business and government, which are conservative in language, the customary choice is formal usage, which is also recommended for contracts. Advertising departments, however, are more likely to choose informal usage, which may appear also in business and personal letters and in memorandums. In speech the informal forms are usual; however, addresses, presentations, and other dignified speeches are formal and suitably employ the traditional usage. Above all, writers should be consistent, for a reader can make little sense of a paper that uses both.

In communicating with those in other countries, writers who prefer the informal forms of *shall* and *will* should consider that many English readers use and expect the formal style, as do those who have been taught by the English. It is, therefore, easy for writers to cause misunderstandings by using one set of meanings for *shall* and *will* while their readers are using another set. One of our students wrote to his counterpart in England, "I will be in London Wednesday of the first week in September, and so you may be able to schedule the meeting of the managers from five countries for one day of that week and make the necessary arrangements." He meant that he expected to be in London that week; to his English correspondent the sentence meant that the writer was making a firm commitment when he wrote *I will*. Accordingly the Englishman took care of all the notifications to the five European managers, travel arrangements, and hotel reservations for a meeting on Wednesday of the first week in September. But on that Wednesday the American was at a meeting in California, his notification was in New York, and five managers and his correspondent were confused in London. The six of them reviewed the correspondence, and to each of them "I will be in London . . ." was the emphatic future, a firm commitment. To one executive in California, "I will be in London . . ." meant only "I expect to be there." Fortunately an ocean and a continent separated the writer and his counterpart when they realized what had happened. Each blamed the other. Both, of course, were to blame, the writer for not considering the English interpretation of his words and the reader for not considering the American meaning of "I will."

An equally avoidable but much more colorful misunderstanding occurred in a court, as a student in a Saudi Arabian course reported. An American landlord wrote a lease, which a Saudi tenant signed; it stated, "The tenant will pay a monthly rental of $3000." The Saudi judge considered the document and explained carefully to the American landlord, "This tenant is an honorable man. He has expressed his willingness to pay the rent, and he will pay it when he can. He has not signed any commitment to pay. The document does not say that he shall pay." The judge had learned English from British teachers, who taught the traditional rules.

Some special uses of *shall* and *will* and of *should* and *would* are less controversial.

*Shall* is used in laws, motions, resolutions, contracts, etc.

> LAW: The employer shall provide adequate safety equipment.
> MOTION: The fee for the Demosthenes Lecture shall be five hundred dollars.
> RESOLUTION: Resolved: That the Speech Division shall award a Euphues Prize for Elegant Verbosity annually as a tribute to Professor Abel Webster. (The *shall* may be omitted.)
> CONTRACTS: The car shall be delivered on Monday, January 31, at 9 a.m.
> The car shall be tested at a speed of two hundred miles an hour for one hour.

The machine shall operate under the conditions described in the following paragraph.

The paper shall have 20% rag content.

*Should* or *would* is used to express a writer's doubt or uncertainty.

He should be a good merchandise manager. (This expresses less confidence than the simple future does.)

Would you recommend me for the position? (*Would* is more tentative than *will* in such a sentence.)

The sentence "This car should do two hundred miles an hour" expresses less certainty than "This car shall do two hundred miles an hour."

*Should* used as an auxiliary verb to express obligation is weaker than *ought*. "You should recommend him for promotion" has less force than "You ought to recommend him for promotion."

In some sentences *should* is ambiguous; it may mean probability or obligation. A writer should rephrase such sentences.

AMBIGUOUS. The typists should complete your manuscript before they leave.
ONE MEANING: The typists are to complete your manuscript before they leave.
ANOTHER MEANING: The typists will probably complete your manuscript before they leave.

*Should* is used for all persons in subordinate clauses expressing condition.

Notify your supervisor if you should care to go.
If I should miss the train, do not wait for me.
If they should miss the train, do not wait for them.

Many writers use the present tense in such clauses except when they wish to intensify the conditional element.

If the dyed fabric should look spotty, repeat the procedure.
If the dyed fabric looks spotty, repeat the procedure.

The first sentence does not suggest as strongly as the second that the dye will look spotty.

*Would* may be used in all persons for a wish, but most writers are likely to use some less formal expression.

Would that the founders had lived to see this achievement!
If only the founders had lived to see this achievement!

*Would* is used in all persons for a habitual action.

210

Before this process was introduced, patients would complain of the chalky taste.

We would always have some explanation for our lateness.

You would complain of incorrect billing every month.

## THE SUBJUNCTIVE MOOD

The subjunctive is used to express a wish, a condition, an improbability. It also appears in some clauses introduced by *that*. Usually these clauses introduce recommendations, resolutions, suggestions, and demands in a formal context.

In modern English the simple subjunctive form of a verb can be detected only in the third person singular of the present tense and in the forms of the verb *to be*.

### PRESENT TENSE—THIRD PERSON SINGULAR

| Indicative | Subjunctive |
|---|---|
| the plant succeeds | the plant succeed |
| he retires | he retire |
| she withdraws | she withdraw |

### TO BE—PRESENT SUBJUNCTIVE

(Seldom used today except in resolutions, motions, and other stereotyped expressions and in some *that* clauses)

| I be | we be |
|---|---|
| you be | you be |
| he, she, *or* it be | they be |

### PAST SUBJUNCTIVE

(The form used most often today)

| I were | we were |
|---|---|
| you were | you were |
| he, she, *or* it were | they were |

*Examples*

WISH: It is my wish that he retire.

CONDITION OR HYPOTHESIS: If the temperature were higher, the results would be the same.

IMPROBABILITY: If the estimate were as wrong as that, I would not send the proposal.

*That* clauses expressing recommendations, hopes, prayers, wishes, requests, or commands are expressed in the subjunctive:

The supervisor ordered that the machine be stopped.

The supervisor ordered that the operator stop the machine.

I recommend that Robert Smith be given an annual increase in salary of $2,000.

211

I requested that the window cleaner report every Monday at 9 a.m.
Resolved: That the services of the Trumpery Publicity Agency be terminated by January 1, 1989.
The committee suggests that the investigation of the side effects of this sedative be continued.
It is necessary that every worker wear safety glasses during this entire experiment. (Most writers prefer the direct statement: Every worker must wear safety glasses during the entire experiment.)

Writers should know that many ideas expressed by verbs in the subjunctive may also be expressed by the auxiliaries *may, might, should, would, let, must,* etc.

I hoped that she would sing for us.
If the chemical should be pure, the method would succeed.
If the estimate should be as wrong as that, I dare not send the proposal.
Let the company use the formula immediately.
The services of the agency must be terminated by January 1, 1989.

Sometimes the auxiliaries give meanings slightly different from those of the subjunctive mood. This allows the writer an opportunity to select the precise sense from fine shades of meaning.

## TENSES

### Present Tense

The present tense is used for permanent truths.

Many of the experimental techniques that have been applied to single-phase flow *are* applicable to two-phase flow.
H. N. McManus, Jr., wrote that many of the experimental techniques that have been applied to single-phase flow *are* applicable to two-phase flow.
Newton *is* the originator.

A verb in the present tense that expresses a timeless truth should not be changed to another tense because of a preceding or following verb.

INCORRECT: He taught us that the earth moved.
CORRECT: He taught us that the earth moves.
The fact that the earth moves was accepted in the seventeenth century.
He explained that neurotic patients respond this way.

The use of the present tense for permanent truths presents a scientific decision to experimenters, who may conclude that under the given circumstances a chemical *changes* or *changed* its structure. If they use *changed*, they state only that it happened in their experiment. If they use *changes*, they state that it always happens in the given circumstances.

It is important that writers and readers understand this difference in the meaning of the two tenses.

## Present Progressive Tense

The progressive tense usually indicates action in progress: *I am working in New York. We are collecting for the United Fund.*

> INCORRECT: I am living in New York for five years.
> CORRECT: (speaker is still living in New York) I have been living in New York for five years.
> (speaker no longer lives in New York) I lived in New York for five years.

## Present Perfect and Past Tenses

The past tense refers to past time, to completed action; the present perfect tense connects a past occurrence with the present time, as when (1) the action continues into the present or (2) the action has taken place just before events happening now.

> CORRECT: I have worked harder today than ever before. (The day is not over.)
> I worked harder yesterday than ever before.
> I have worked hard today, and I am glad to be going home at last.

The present perfect tense should not be used for completed action which has no specific or implied relation to the present.

> INCORRECT: I have completed that study ten years ago.
> CORRECT: I completed that study ten years ago.
> (for a student) I have studied physics in college, and I dislike it.
> (for an alumnus) I studied physics in college and dislike it.

## Tenses of Infinitives

The perfect infinitive is used for action completed before the time of the main verb; otherwise the present infinitive is used.

> INCORRECT: We should have liked to have seen the factory before the bidding.
> CORRECT: We should have liked to see the factory before the bidding.
> We should like to have seen the factory before the bidding.

## Tenses of Participles

A present perfect participle indicates action that occurred before the action of the governing verb; a present participle indicates action occurring at the time of the action of the governing verb.

PRESENT PERFECT PARTICIPLE: Having attended meetings all day, we were ready for some sightseeing.

PRESENT PARTICIPLE: Talking as we looked at the sights, we became acquainted.

But if *after* or some other word indicates that the action is past, a present participle is used.

After attending meetings all day, we were ready for some sightseeing.

## Shifts in Tense or Mood

A careful writer avoids unnecessary shifts in tense or mood. This does not mean that all main verbs in a paper must be in the same tense or mood; it means that actions occurring at the same time should be in the same tense—past actions in the past tense, future actions in the future tense, present actions in the present tense, etc.

INCORRECT: When we examined the first products, we find crystals for x-ray analysis. We determined their structure.

CORRECT: When we examined the first products, we found (both past events should be in the past tense—*examined, found*) crystals for x-ray analysis. We determined (past event in past tense) their structure.

INCORRECT: Dr. Wrightwell read his paper so poorly that I hear only one word in three.

CORRECT: Dr. Wrightwell read his paper so poorly that I heard (both past events are in the past tense) only one word in three.

INCORRECT: Dr. Wrightwell read his paper so poorly that during his presentation I find it necessary to consult the copy that I made before the meeting.

CORRECT: Dr. Wrightwell read his paper so poorly that during his presentation I found (both past events are in the past tense) it necessary to consult the copy that I had made before the meeting. (The event in the more distant past is in the past perfect tense: *had made.*)

The results of the third run are about the same. (Present tense is used for historical present.)

INCORRECT: If the results should be positive and the reactions are poor, the patients will be examined.

CORRECT: If the results should be positive and the reactions (*should be* is understood; if one hypothesis is conditional, the other must be also) poor, the patients will be (future tense is used for future time) examined.

INCORRECT: First heat the mixture in the tank. Then you should add the contents of the flask.

CORRECT: First heat the mixture in the tank. Then add the contents of the flask.

CORRECT: First you should hear the mixture in the tank. Then you should add the contents of the flask.

## ELLIPSIS

In formal writing, a necessary verb or auxiliary should not be omitted.

> INCORRECT: The recommendations are sensible and widely used.
> CORRECT: The recommendations are sensible and are widely used. (The first *are* is a main verb; the second *are* is an auxiliary verb.)
> These sensible recommendations are widely used.
> INCORRECT: They never have and never will find the answer to that question.
> CORRECT: They never have found the answer to that question, and they never will find it.
> INCORRECT: The computer never has and never can be used for such work.
> CORRECT: The computer never has been and never can be used for such work.
> The computer never has been used for such work and never can be.

When two or more words require different prepositions, the prepositions should not be omitted.

> INCORRECT: It is difficult to estimate whether the new tax laws will add or subtract from the profits.
> CORRECT: It is difficult to estimate whether the new tax laws will add to or subtract from the profits.
> INCORRECT: The chairman commented on the parliamentarian's knowledge, interest, and respect for the law.
> CORRECT: The chairman commented on the parliamentarian's knowledge of, interest in, and respect for the law.

But when the preposition may serve for two or more words, it need not be repeated: *He works and plays with John; He was prevented and prohibited from working; He had no interest or confidence in the promotion plan.* Sometimes the preposition is repeated for euphony, clarity, or emphasis: *He spoke against, acted against, and fought against change.*

## INCORRECT GRAMMAR IN PROFESSIONAL WRITING

Incorrect grammar sometimes interferes with clarity. Misplaced modifiers, incorrect comparisons, and wrong tenses, for example, may convey the wrong meaning to a reader. A writer who states, "This dye was reliable for all fabrics," when he should state, "This dye is reliable for all fabrics," makes a costly error because he does not understand that the present tense indicates permanent truth. Other mistakes in grammar and construction, like pronoun errors and incorrect parallelism, make writing fuzzy. Writers in the professions must be more exact than other writers. A writer of advertising may be purposely vague to avoid lawsuits or may purposely use street English to flatter the uneducated. But those who are responsible for transmitting information in the professions should write clearly and appropriately.

Although clarity is the minimum expected of a writer in the professions, many scientists and technologists fight the editors and supervisors who change their English and who would train them to write better. I have known an engineer so muddleheaded as to pay for reprinting an entire article to correct an obvious typographical error in arithmetic but to balk at changing in a typed manuscript, "Little data are available." And a chemist who objected to erasures in his letters argued heatedly against changing the position of a misplaced *only*. "Only these three chemicals change the color of the drug," he wrote and insisted that it meant, "These three chemicals change only the color of the drug." He told the supervisor who wanted to correct this, "Leave it the way it is. My readers know enough about chemistry to know what I mean." But those readers are the first to ask why he did not know enough about English to write what he meant.

Other writers err by depending on their ears for correct English usage. Trained as business administrators, scientists, or technologists, they attempt what even a person trained in language and literature hesitates to do by ear. They fail to consider that the ear accepts and likes what it is used to. The ear may reject usages merely because they sound strange to it; therefore an ear accustomed to the weaknesses common in writings on science and technology is an untrustworthy guide for a scientist or engineer. Professionals must learn to beware of the errors that their ears like. If their ears accept an error, that is the mistake they must search for when they revise. This does not mean that writers must use what they do not like. There are innumerable ways of expressing a thought effectively, and writers are free to choose any that suit them. Nor does it mean that the ear is useless. For the more subtle harmonies and rhythms of language, writers must use their ears. But they should not consider their ears authorities on grammar.

## NOTE

1. H. W. Fowler, *A Dictionary of Modern English Usage,* 2 ed. (New York: Oxford UP, 1965) 404.

# CHAPTER 7

# Standards of Diction

*Reading maketh a full man, conference a*
*ready man, and writing an exact man.*

*Bacon*

In the prose of information the right word, like the right number, is of exceeding importance. Scientists, technologists, and business executives cannot afford to be satisfied with something close to what they mean. They must write exactly what they mean. Fuzziness, ambiguity, or misleading incorrectness may be dangerous. Doctors who write, "Often these patients are depressed," when they mean, "Many of these patients are depressed," may lead readers into serious errors.

## COMMON USE OF VAGUE AND OLD-FASHIONED TERMS IN BUSINESS

One cause of many misunderstandings is the vague diction popular today in poor business writing. A director of medical TV programs wrote to the chairman of a program, "If any arrangements for this program have to be changed, contact me." When the three panelists scheduled for the program suddenly found it impossible to appear, the chairman reported this in a letter to the TV director, but the director, who had been two thousand miles away from his office, heard nothing until his plane landed at the airport three hours before the program.

"Why didn't you call or send a telegram?" he wailed. "The office could have reached me in California, and I would have come here a day earlier."

"But you told me to write," the chairman said.

To him *contact* meant *write*; to the director it meant *telephone* or *telegraph*. We have seen many other business executives face difficult and unnecessary emergencies because of similar ambiguities.

Some business clichés are objectionable because they create an old-fashioned atmosphere just as inappropriate as paper cuffs and a celluloid collar would be if worn by someone using the latest computer. A phrase like "your esteemed favor" suggests Bob Cratchit sitting on a high stool in Scrooge's office. One does not expect antiquated business phrases such as "in reply beg leave to state," "under separate cover," and "hoping to hear from you soon, we remain," to come from a word processor.

One writer of a college textbook on technical reporting said that he was omitting discussion of old-fashioned business language because his students had never met it and did not need to be warned against it. But anyone familiar with modern business and industry could have told him that they would meet it soon. A supervisor would encourage them to use it, or young graduates would read it in correspondence and sprinkle it in their own letters to lend a "business tone." Some antiquated phrases are nearly indestructible because supervisors teach them to beginners who become supervisors and teach them to other beginners. Why? Largely because they are easy substitutes for thought. "Yours of the 14th inst. received and contents duly noted" flows from a writer without thought. To express some meaning about a particular inquiry or problem requires more effort. Often the user of business clichés is postponing as long as possible the labor of thinking—sometimes until the final sentence of a letter or report. All the time-worn phrases that precede the last sentence are just procrastination. Eliminating antiquated phraseology and business clichés may cost such a writer a little revision time and some effort at first, but later it will become automatic. The result will be briefer and more effective writing.

## INCORRECT DICTION

Incorrect diction is a different fault, but it also distracts and interrupts and thus creates a poor impression. Frequent distractions of this sort may make busy readers throw down a report before they have completed their reading. Writers who use incorrect diction appear careless, poorly educated, unreliable. Their readers wonder whether they dare trust the scientific information or the professional judgment of writers who make elementary mistakes in the use of English words. To avoid such doubt, distraction, and distrust, careful writers correct their diction. And because usage is highly contagious, they guard against the language errors that they see and hear frequently. Good writers do not confuse similar words like *accelerate* and *exhilarate; accept* and *except; affect* and *effect; adverse* and *averse; alternate* and *alternative; amount* and *number; common* and *mutual; continual* and *continuous; exploit* and *explore;* and *company, concern, firm, corporation.* (See also Appendix B.)

Writers who wish to be correct must learn to be critical of the words they read. Business executives, scientists, and engineers are often berated for careless use of words, the assumption of the critic being that their work should make them accurate in all ways. But they are children of their time, which offers them little encouragement, for they find many poor examples in their professional or other reading.

A signed article in a good newspaper, for example, notes that the home of a bride's family is "unprepossessing." A slum family? No, socialites who have an *unpretentious* home. The producer of a prizewinning documentary film is quoted as saying when the prize was withdrawn, "I'm very disappointed. I think it's unsolicitous of the Academy to publicly announce a nomination and later withdraw it." Unsolicitous? And the following misuse of *compunction* by a journalist is typical of the careless errors found in even the best newspapers: "An F.A.A. official said that the dike and all other aspects of La Guardia's operation met or exceeded Federal regulations and that pilots were under no compunction to land there if they believed the field was unsafe."

A publisher's advertising of a book on human behavior reads:

A Book of Fundamental Insight into the Forces Between Body and
Psyche
which explains in simple non-technical verbiage
the meaning, purpose and promise of
PSYCHIATRY

Well, maybe it is verbiage, but who expected the publisher to admit it? And in an advertisement mailed by an institute that gives seminars on technical writing, *Principals of Clear Writing* is spelled that way twice. The advertisement states that a technical writing association has approved the advertisement "for distribution to its members as an outstanding example of one of the techniques for improving writing ability." The engineers and scientists to whom the advertisement was directed are able to distinguish between *principles* and *principals*. But the writer of the advertisement could not. A reputable publisher advertises as follows a book designed to improve business writing: "The book contains exercises to improve proficiency of grammar, spelling, and vocabulary usage. Interspersed throughout the book are various checklists for inventory and self-evaluation on writing a letter, a report or a memorandum." The misuse of *of* after *proficiency* and *on* after *self-evaluation* and the poor use of *inventory* suggest that the writer of the advertisement needs a few checklists.

Because professionals encounter such mistakes and weaknesses in their reading, even in the very material supposed to help them to write better, it is clear that they must plan to write more accurately and correctly than other writers, even apparently than some of those who would instruct them. To do so they must become alert to words: their uses and abuses.

Because usage changes, every writer faces the delicate question of whether to adopt a new usage. In advising writers in science, technology, and administration, we usually recommend a conservative choice. We have noticed that professionals in those fields tend to resist changes in language, even at times to scorn them. Writers should therefore select conservative usage for such readers. A writer in an advertising divison may well strive always to be the first to use the new, but other writers for business, government, and industry choose well when they prefer established forms. The scientist and the technologist, moreover, have an interest in maintaining established usage whenever it is more precise. They may advantageously cling to usage which their accuracy of expression requires, for thus they help to retain necessary forms that would otherwise disappear. When established usage is more helpful to them than the new is, they should stubbornly be "the last to cast the old aside."

## INTRODUCTION TO PROBLEM WORDS AND PHRASES

The problem words and phrases in Appendix B are those we have found misused most often in business, industry, and government, as well as the words and phrases that our students and clients have questioned us about most frequently. Since the incorrect forms are so common in the reading of most people who use this book, the errors may sound correct to them. That is a danger sign indicating a likelihood that the errors will not be detected in editing. Writers whose ears betray them should mark the word or phrase that is causing the trouble, study the correction, and repeat the correct form aloud until it sounds familiar. Other forms may require just routine study—concentrating on the correct forms that a writer should use to replace the incorrect, for example, or learning to distinguish between two similar words.

Correct diction is the minimum expected of a writer in the professions. Good style requires not merely that a word be correct but that it be the right word in the right place. Choosing the right word from a multitude of possible words is the subject of Chapter 10. Writers who consider their diction correct but not so effective as it might be should consult that chapter for suggestions on how to make words work for them.

# CHAPTER 8

# Punctuation and Other Marks

*Sally: "Show me where you sprinkle in the little curvy marks."*
*Charlie Brown: "Commas."*
*Sally: "Whatever."*

*Charles Schulz*

Standard punctuation eases reading. If, for instance, writers use a period at the end of each declarative sentence, readers know automatically that the sentences are not questions or exclamations, and they expect the sentences to be complete.

Correct punctuation clarifies ideas. When competent readers see, "My daughter, the doctor," they know that the phrase *the doctor* does not limit or identify "my daughter." But "My daughter the doctor" tells them that *the doctor* does limit or identify, that the daughter or daughters who are not doctors are not meant.

Just one or two small marks by their presence or absence convey much meaning, particularly in law courts. "I leave six million dollars to be divided equally among my nephews, who were most attentive during my long illness," writes a wealthy man and leaves his money—as he wishes—to his six nephews. If he forgets the comma after *nephews,* however, the money must be awarded to fewer than six nephews because the absence of that comma makes *who were most attentive during my long illness* restrictive.

The virgule, used in *and/or,* also has meaning in courts of law. A check made out, for example, to Clover Construction/J. M. Sapreton has been declared in some courts cashable by either, the money to belong to that one. If a taxpayer writes a tax-return check to K. X. Brillanton/Tax Collector" and K. X. loses his position as tax collector before the check is cashed, he may cash it as K. X. Brillanton and fight in the courts to keep the money. One canny tax collector instructed taxpayers to make out their checks in his name followed by */tax collector* and had boxes of them in his home waiting to be cashed when he lost his job.

Even when the rules are simpler than those for nonrestriction and the

221

virgule, incorrect punctuation may have amusing or expensive consequences, as the following sentences show:

> The chairman called the officers names.
> The chairman called the officers' names.
> The scientist told the technician to cut the rods into twenty foot lengths.
> The scientist told the technician to cut the rods into twenty-foot lengths.
> The receptionist said Joyce Bellair stole the boxes of cosmetics.
> The receptionist, said Joyce Bellair, stole the boxes of cosmetics.

Many errors in punctuation do not cause such misunderstandings as those just listed but do something nearly as harmful. They force readers to reread and read again sentences that the readers cannot understand until they supply or correct the punctuation. When this happens to students, they almost howl. "We can't understand this textbook," they complain. "What is it supposed to be saying?" Readers in business and industry and government howl too when important meetings are delayed through misunderstandings due to missing or incorrect punctuation or when correct punctuation has not been used as it should be to ease or clarify reading.

Writers need not memorize all the rules of punctuation, but they should be sufficiently familiar with the kinds of rules to look them up if they are needed, and they should avoid freakish punctuation that they base on how they breathe or where they pause when they read. Let them consider the breath pauses of an opera singer and an asthmatic, the variations in thought pauses when a celebrated actor or a semiliterate pupil reads a passage. Standard punctuation avoids these individual variations and thus makes reading easy and clear for readers. The standard rules that follow will help writers to help readers.

## THE PERIOD

### 1. Periods at the Ends of Sentences

A sentence or sentence fragment that is not interrogative or exclamatory is followed by a period. Two spaces follow this period in typewritten material.

> Use examples to clarify rules.
> A declarative sentence is one that makes a statement.
> Questions are called interrogative sentences.
> Sentence fragments should be used sparingly.
> The issue: whether to budget for a loss.
> Yes, twelve dozen.

NOTE: If a sentence ends with a question mark or an exclamation point, no period is used after the mark:

An important question about the significance of research on DNA is will it equate man with God?

Some investigators ignore those who ask, "What is the use of your studies?" (See also Quotation Mark, Section 5.)

An indirect question requires a period.

An important question about research on DNA is whether it will equate man with God.

Some investigators ignore those who ask what use the studies have.

A request is followed by a period or a question mark according to the circumstances. (See Question Mark, Section 1b.)

## 2. Periods with Lists

Items in a list are seldom followed by a period unless they are complete sentences. A period at the end of a list that completes an introductory clause is optional but old-fashioned.

The list included
metal desks
tables
armchairs
files
a drinking fountain
plastic wastepaper baskets

## 3. Periods with Abbreviations

*a.* Common abbreviations take periods.

a.m.    St.    Mr.    M.D.    Esq.    Dr.    Oct.    Ph.D.
(*Ms* referring to a woman often is not punctuated.)

If a sentence ends with an abbreviation, only one period is used.

She asked to be listed as a Ph.D.
The meeting was called for 10 a.m.

NOTE: Periods within abbreviations are not followed by a space.

Abbreviated names, although seldom used today, take periods (*Chas.*, *Wm.*), as do initials standing for given names (*Franklin D. Roosevelt, H. L. Mencken,* but *Harry S Truman*). The periods are followed by a space. In the case of initials, it is wise to follow the preference of the person whose initials they are; many people recommend using periods with all initials.

b. Technical writing usually omits periods after abbreviated units of English measurement but does use periods to avoid confusion: for example, *in.* (for *inch*). Abbreviations of SI units (International System of Measurement) do not take periods, nor do chemical symbols and abbreviations.

| 10 lb | 16 yd | 1,000 ft | 60 m | 25 km | 500 mg |
|-------|-------|----------|------|-------|--------|
| $CO_2$ | He | DNA | PVC | | |

c. Acronyms (words made usually from the initial letters of the name of a committee or department or corporation or other entity) do not take periods. Initial letters that do not form words may be followed by periods, but the current style is to omit these too.

| NASA | ERA | OPEC | AIDS | |
|------|-----|------|------|----|
| USA | CBS | USSR | AMA | TV |

Shortened words, such as *phone, memo, auto, lab, gym,* which are appropriate only in colloquial use, are not followed by a period.

d. For geographical names the practice is varied. The names of most nations are spelled out, but *USSR* is common, and *USA* or *US* is almost always used in newspaper writing. Other abbreviations for country names seen fairly often are *UK* and *NZ.*

Abbreviations for states used to be followed by periods, but in the current postal abbreviations periods are omitted.

MA     NY     FL     CA

e. Latin abbreviations, which should be avoided, require a period if a writer must use them. There are adequate English words for the common Latin expressions, which remain in our language from the time when education meant training in Latin. Nowadays the use of *viz., i.e., e.g., etc.,* and other Latin abbreviations does not denote learning; it suggests unawareness. (See entries in Appendix B.)

## 4.  Periods for Ellipsis

When words, phrases, clauses, or sentences are omitted from quoted material, the omission is indicated by three periods, called *ellipsis dots* or *points.* If the omitted material follows a complete sentence or ends a sentence, the final period is used with the ellipsis dots.

ORIGINAL:
One would wonder, indeed, how it should happen that women are conversible at all, since they are only beholding to natural parts for all their

knowledge. Their youth is spent to teach them to stitch and sew or make baubles. They are taught to read, indeed, and perhaps to write their names or so, and that is the height of a woman's education. And I would but ask any who slight the sex for their understanding, what is a man (a gentleman, I mean) good for that is taught no more?

Daniel DeFoe, The Education of Women

WORDS OMITTED:

One would wonder, indeed, . . . that women are conversible at all, since they are only beholding to natural parts for all their knowledge. . . . They are taught to read, indeed, and perhaps to write their names or so, and that is the height of a woman's education. And I would but ask any who slight the sex for their understanding, what is a man . . . good for that is taught no more?

If quoted material enclosed in quotation marks is run in with the text, ellipsis points are not used before or after the quoted material, although they may be necessary within the quotation.

Defoe summed up women's education in the eighteenth century when he wrote, "They are taught to read . . . and perhaps to write their names or so."

## 5. Periods with Quotation Marks, Parentheses, Brackets

*a.* In American usage the period appears inside the quotation marks (single or double).

The proof is detailed in his article "Some Hazards of High Pressure."
She urged, "Read his article 'Some Hazards of High Pressure.'"

*b.* If the entire sentence is within parentheses, the final period should be inside the closing parenthesis. If the parenthetical material ends a sentence, the period goes after the final parenthesis. The same rule holds for brackets. (See also Parentheses and Dashes, Section 8.)

Our results are tabulated in Appendix 3. (Additional copies are available.)
The laboratory orders most chemicals in bulk but prefers small quantities of some (those that are expensive or that have a brief shelf life).
Previous reports of our work appeared earlier [volume 451, number 6, of this journal]. (*The bracketed material in this and the following sentence is an editor's interpolation.*)
As we described earlier, the preparations for this program were extensive. [Dr. Smith and Dr. Brown have been publishing reports on their work as it progresses. See volume 451, number 6; volume 460, number 2; volume 468, number 3.]

## 6. Periods in Lists and Headings

*a.* In lists containing items distinguished by numbers or letters, a period usually follows the introductory number or letter.

The meeting will consider the following problems:

1. The selection of a representative to attend the conference in Toronto.
2. Three key people's requests for the same vacation period
3. Repeated violations of security rules
4. A ban on smoking in the office

*b.* Similarly, formal outlines also take periods after the various classifications.

I. Interest Rates
  A. Mortgages
    1. Commercial
    2. Residential
      a. Owner occupied
      b. Fully rented

(See also the foregoing Section 2 and Parentheses and Dashes, Section 4.)

*c.* Headings for sections of long reports, of other papers, or of books are frequently identified by numbers or letters or both, usually followed by a period in a typescript, although the period is often omitted in printed material.

1. The Rise and Fall of Interest Rates
*d. Mortgage Interest*

## 7. The Period as a Decimal Point

The decimal point is a period with no space following.

$25.75    $0.35    6.25%    0.85%    .000783

NOTE: For amounts less than 1 a zero usually precedes the decimal point unless there is no likelihood of a misreading. If the words *cents* is spelled out, the decimal point is not used: *35 cents.*

## THE COMMA

Commas are used within a sentence to clarify meaning or to aid in reading.

# 1. The Comma with a Coordinating Conjunction

When two or more main clauses are connected by a simple coordinating conjunction (*and, or, but, for, nor*), a comma is placed before the conjunction.

> Most aquatic mammals are insulated from their environment by blubber, but the sea otter is an exception.
>
> There is some danger of an explosion, for the gauges are registering unusually high pressures.
>
> The physician prescribed the medication immediately, and by noon all the patients at risk had taken it.
>
> This radial thrust of several thousand pounds should not last more than a few seconds, nor should it cause any difficulties.

NOTE 1: A comma does not separate a coordinating conjunction from a following clause unless other components of the clause call for a comma:

> The manager arbitrarily allotted the work, and the project was finished ahead of schedule.
>
> The manager arbitrarily allotted the work, and, despite some disagreements, the project was finished ahead of schedule.

NOTE 2: Compound predicates (two or more verbs with the same subject) connected by a coordinating conjunction should not be separated by a comma:

> Sea waves are generated by undersea earthquakes and often cause more damage than hurricanes.
>
> This unusual sparrow spends the spring on an island but inhabits the mainland the rest of the year.
>
> (In the first sentence there are one subject, *waves,* and two verbs: *are generated* and *cause.* In the second sentence *sparrow,* one subject, *spends* and *inhabits.*)

NOTE 3: Two dependent clauses joined by a coordinating conjunction do not take a comma before the conjunction if both clauses complete the thought of the preceding main clause:

> When we were about halfway through the project, we were certain that our hypothesis was justified and the results would lead to economic development.
>
> The research was canceled because the budget had been cut or no discernible benefits had resulted from the project.
>
> It is easy to sell a product if it is fairly priced and consumers think they need it.
>
> (Many writers would recast these sentences to make them more precise. In

227

the first sentence a second *that* might be inserted before *the results*. In the second sentence *either* might be inserted after *because*. In the third sentence *if* might be added before *consumers*.) In all the sentences the last clause can be called *dependent* only because the reader realizes that it is introduced by the conjunctive adverb following the main clause (*that* in the first sentence, *because* in the second sentence, and *if* in the third).

## 2. Commas after Short Parallel Independent Clauses in a Series

A comma is used after each short main clause in a parallel series whether a conjunction is used before the last member or not.

> He suggested, he argued, he ruled.
> He suggested, he argued, and then he ruled.

## 3. The Comma after an Introductory Dependent Clause

A dependent clause preceding a main clause is followed by a comma.

> When satellite television receivers are used, viewers have a wide choice of channels.
> Although it is designed to profit a trader, a "put" option anticipates a falling market.

NOTE: Some writers omit the comma after short introductory clauses if there is no likelihood of misreading:

> As soon as he arrived the meeting began.
> *But* As soon as he started, the car stalled.

## 4. The Comma after an Introductory Word or Phrase

After an introductory word or phrase a comma is necessary unless the introductory element denotes time, location, manner, or other closely related idea.

> Without these safeguards, the FDA will prohibit the sale of the drug.
> Citing these examples, we hope to prove our case.
> Generally, only a few items fall within a given span of attention.
> The question having been resolved, the research continued.
> Buying or selling, he is always on the floor of the Exchange.
> Here the soil had been eroded for generations.
> Hopefully he examined the result of his experimentation.
> Within these walls all confidences are respected.
> Seldom can an intense stimulus be overlooked.
> In the second place, all tools should be listed in the upper corner.

Yes, messages of each kind should be presented consistently.
Carefully he precipitated the unusual substance.
Then the salaries will be reviewed.

NOTE: If a sentence is likely to be misread without a comma after the introductory element, the comma should be used:

In recent decades the idea of space commuting has come to be accepted. Before, it was in the realm of science fiction.
Sales figures have plummeted; however, we try to cut our employment figures only by attrition.

## 5. Commas around Parenthetical Expressions

Words, phrases, and clauses inserted within a sentence for explanation, clarity, emphasis, or transition are set off by commas. (See also Parentheses and Dashes, Sections 1 and 2.)

Some bats, dependent as they are on echolocation, manage to migrate hundreds of miles.
The branched rhizomes of the plant are, because of their resemblance to the undersea formation, responsible for the name Coralroot.
Several other oxides, such as manganic-manganous oxide and manganese trioxide, are less commercially successful.
The new executive, it was generally believed, planned to reorganize the current strategy.
Many industries in the United States, unfortunately, have established their manufacturing plants abroad.
Blocking the ledger, therefore, may be necessary when it is difficult to uncover an error.

## 6. The Comma in an Elliptical Construction

A comma is used to indicate the omission of a word or words that are understood from the construction of the sentence.

Our losses are up; our profits, down.
A few will remain to the end; most, to the end of this day.

NOTE: In very short expressions usually not containing a verb, the comma may be omitted:

The more the merrier.
The sooner the better.
The older the wiser

## 7. The Comma with Contrasted Elements

Words or phrases that express contrast are set off or separated by commas. (One element in the contrast is often introduced by *not*.)

> The chairman was witty, not comical.
> Not just perseverance, but intelligence, is required for this assignment.
> The true stage efficiency, not the adiabatic, should be considered in this design.
> The harder she worked, the easier the job became.
> The greater the demand for profits, the less enthusiasm he felt for the project.

## 8. Commas with Nonrestrictive Material

Nonrestrictive modifiers (modifiers that do not define or limit) are set off by commas.

> His right arm, which is longer than his left arm, is difficult to fit.
> My twin sister, who values her individuality, hates being taken for me.
> My brother, who is an engineer, likes living in California.
> My brother who is an engineer likes living in California.

> (The commas in the third sentence indicate that I have no other brother; therefore *who is an engineer* does not restrict the meaning of *my brother*. The absence of commas in the fourth sentence indicates that I have more than one brother and that I am restricting *my brother* to the one who is an engineer.)

> I consult my sister who is a medical doctor just as often as I consult my sister who is a lawyer.
> Girls who wear glasses are considered more stylish and attractive than they were in Dorothy Parker's day.
> Only employees who have passes may park in this lot.
> Employees, who have passes, may park in this lot. [This punctuation indicates that all employees have passes.]
> The section of Lake Warner that is located in Wayne County is being tested.
> Living reefs, which are rare, should be protected.

(For the use of *that* and *which* with restrictive and nonrestrictive clauses, see Appendix B.)

## 9. Commas with Words or Phrases in a Series

Words and phrases in a series are separated by commas unless the material is displayed (centered or otherwise set off on a page rather than run in with the rest of the sentence).

The company offered bonuses, medical and dental plans, a credit union, a
  pension plan; but he preferred a high salary to benefits.

(See The Period, Sections 3a, b, c, d for displayed material.)

## 10. The Comma before *and* or *or* in a Series

Words, phrases, or clauses in a series require a comma before a connecting *and* or *or*. Omitting the comma may mislead if the last two items
in a series may be read as a unit. Even when confusion is unlikely, the
comma is used today except in some newspapers and magazines. (See
also preceding Section 2.)

Proprioception indicates what various parts, such as the eyes, head, and
  body, are doing at a given time.
Any changes in pressure, volume, or temperature must be noted.
The building materials were classified as wood, brick, cement or concrete.
The most popular pies are lemon, pecan, apple and raisin.

(In each of the last two examples there seem to be three classifications; if
  the writer means four categories, a comma before *and* and *or* is
  necessary.)

English pounds, American and Canadian dollars, and French and Swiss
  francs are accepted here.
The company has voted to give bonuses to those who have been with the
  company more than two years, to those who have completed at least
  two continuing-education courses, and to those who volunteered to work
  overtime during the transportation strike.

## 11. Commas with Coordinate Adjectives

If each adjective preceding a noun modifies the noun, commas are
placed between the adjectives. If one or more of the adjectives form an
entity with the noun they precede, there is no comma between the entity
and the preceding adjectives.

The hot, heavy mass was in danger of liquefying.
The membrane was described as a thin, stable film.
The old Hemingway house is a tourist attraction.
The many small particles formed a familiar pattern.
The tall white stone building stood well above the surrounding wooden
  structures.
The viscid, malodorous, ugly effluent was difficult to work with.

One common test to determine whether to insert a comma is to insert it if it could be replaced by *and*: the *hot and heavy* mass but not the *old and Hemingway* house.

## 12. Commas with Modifying Expressions Preceding a Noun

If more than one expression modifies a following noun, commas should be placed around the second and subsequent expressions, particularly if the first expression merely describes and the following expressions comment.

> Her impressive, slightly ostentatious, office was calculated to express her new standing in the firm.
> The chart on page 7 shows the steady, albeit unexciting, progress of the company under his management.
> Few, if any, grasses can withstand the intense heat and long dry periods of the island.
> The recent outpouring, not to say deluge, of computer programs for the analysis of multivariate data may not lead to the expected advantages.

(The second modifier is parenthetical and not essential to the sentence.)

## 13. Commas with Words or Phrases in Apposition

When a word is followed by a word or phrase that explains or supplements the first word, the second expression is enclosed in commas. If the second expression is at the end of the sentence, it is preceded by a comma. Sometimes the supplementary expression is introduced by *or*.

> The committee deferred to the treasurer, John Collier.
> Readers often misunderstand the meaning of the virgule, or slash.
> Computer programming was widely held to be a cause of the market panic, or "meltdown," of October 1987.

(See also Section 8 above.)

## 14. The Comma after *Namely, That Is, for Example,* and Similar Expressions

A comma follows words or phrases that introduce examples, explanations, supplementary information, and the like. (See also The Colon, Section 2.)

> Three officers of the company—that is, the CEO, president, and treasurer—threatened to resign.
> Today's demand for a quick return on investment would make difficult any repetition of the success of the pioneers of the late nineteenth and early twentieth centuries: for example, Bell, Edison, and Ford.

Not the toy industry, but two of the most serious pursuits in the world—
namely, economics and warfare—depend on the theory of games.

NOTE: *Like* and *such as* are not followed by a comma:

Many occupations, such as warfare and finance, depend on the theory of
games.
Inventors like Bell, Edison, and Ford might have a difficult time in today's
financial world.

## THE COLON

### 1. The Colon before Elements Developing a Stated Idea

A colon separates an independent clause from the sentences, clauses,
phrases, or words that develop the idea of the main clause.

Two laws of physics which most of us learn are Boyle's Law and Charles's
Law: these laws relate the volume of a gas to pressure and temperature.
One benefit of modern medicine is questionable: people are living longer,
but many old men and women say that they are enjoying life less than
their parents did.
Three factors require the consideration of axial thrust in this equipment:
high pressure, the use of many stages, and the consequent cumulative
effect.
Three methods are suitable for this application: suspension, emulsion, and
solution.

NOTE: If the colon is followed by more than one sentence, the sentences begin with capital letters:

The Fig Shell and the Pearl Whelk look alike, but there are differences:
The Fig Shell is light and fragile. The Pearl Whelk is heavy and fairly
tough.

### 2. The Colon after *Follows, as Follows,* or *the Following*

*a.* A colon precedes an enumeration or explanation introduced by *follows, as follows, the following.*

Our order follows:
  6 dozen pencils (Write Right, Number 3)
  10 reams of paper (style 731)
  2 dozen cassettes (model 34)

*But*

Our order is for
  6 dozen pencils (Write Right, Number 3)

233

10 reams of paper (style 731)
2 dozen cassettes (model 34)

She rewarded her staff as follows: long weekends for some, bonuses for others, and substantial salary increases for a few.

Among requirements for valid opinion polls are the following: sufficient respondents, a representative sampling of the community, careful phrasing of questions, adequate gradation of responses.

*b.* If *for example, namely,* and similar expressions introduce several clauses or a number of phrases separated by semicolons, a colon, instead of a comma, follows the introductory expression.

World War II was responsible for the rapid development of a number of processes that ordinarily would have taken years to complete. For example: synthetic rubber was required because transportation of crude rubber from the tropics was difficult; penicillin, needed by field hospitals, became available within months although it had been discovered years before; dried blood plasma, also essential in the field, was processed in vast quantities that could travel long distances and last indefinitely.

Data for this project were gathered by several research groups—namely: ABG Associates, responsible for phase 1; Grant University, supplying much of the work of phase 2; and the Senior Research Society, also engaged in the development of phase 2, as well as in overseeing the entire project.

*c.* Occasionally a colon is placed before a word, phrase, or clause summarizing or emphasizing the preceding part of a long sentence.

Chairmen, chief executive officers, and general managers are all striving to a greater or lesser degree toward one goal: power.

*But*

He had only one goal, power.

## 3. The Colon with Bibliographical and Scriptural Citations and in Formulas for Time and Proportion

Lester, James D. *Writing Research Papers: A Complete Guide,* 3d ed. Glenville, Ill.: Scott, Foresman, 1980.

1 Corinthians 13:6      Genesis 1:3,5,6

10:30 a.m. to 3:15 p.m.

a ratio of 3:9      $x:9 = 4:12$

## 4. The Colon after the Salutation in Business Letters

The prevailing style is to omit punctuation after the inside address, the salutation, and the complimentary close of the business letter; how-

234

ever, some writers still punctuate the salutation and close. In such a case a formal salutation requires a colon. (A comma follows the complimentary close.)

Dear Mr. Smith:
Ladies and Gentlemen:

## THE SEMICOLON

### 1. The Semicolon with Independent Clauses

*a.* A semicolon should be placed between independent clauses in a sentence (Exception: Comma, Section 2). Separating the clauses with a comma is a childish error known as a *run-on sentence* or a *comma fault.*

> The rise of the service society threatens to enslave an entire generation of the undereducated; these "have nots" will be without hope or sustenance.
> The terms *insecticide* and *fungicide* are often confused; obviously, their purposes are quite different.

*b.* When an independent clause following an independent clause is introduced by a conjunctive adverb, such as *however, therefore, moreover, so,* a semicolon precedes the conjunctive adverb, and a comma usually follows it.

> The average citizen overlooks pollution from waste disposal; however, garbage uncollected in the street renders him vociferous and may open his wallet.
> On its introduction Keynesian theory was rejected; nevertheless, when recently it again lost favor, it had been the bible of many economists for half a century.
> White is commonly considered the absence of color; therefore the layman is surprised that to the physicist white is all hues diffusely reflected.

### 2. Semicolons in a Series

When commas appear within phrases or clauses used in a series, the elements of the series (coordinate elements) are separated by semicolons.

> Microsurgery, which many people accept uncomprehendingly; lasers, which many more regard as science fiction; and organ transplants, which some find repugnant, are all modern developments that are generally perceived as extending and improving human life.
> Ants, nearly all classes; some wasps; and bees, at least many families of them, nest underground.

235

### 3. The Semicolon before a Coordinating Conjunction

When one or both of the independent clauses connected by a coordinating conjunction (*and, but, or, for, nor*) are very long or contain several commas, a semicolon is used before the coordinating conjunction.

> He would not volunteer; nor, unless there were extreme pressure from top management, would his fellow chemists, however much they needed his ideas, consult him about their problems.
>
> It is hard to say whether failure to observe, which suggests poor training, or failure to act on an observation, which suggests an inability to reason well, is the greater fault in a scientist; for there is little doubt that constant vigilance and the willingness to risk disappointment are essential to the advancement of science.

## THE QUESTION MARK

### 1. The Question Mark at the End of a Sentence

*a.* A question mark follows a direct question.

> Is carbon monoxide really odorless?
>
> What practical applications have resulted from the experiments conducted in space?

*b.* An indirect question takes a period at the end. (See The Period, Section 1.) A request couched as a courteous question takes a question mark or a period, depending on the situation.

> Will you let me know when you place the order. (This polite request takes for granted that the recipient of the request will do as requested.)
>
> Will you return the material when you finish with it? (The writer does not take the recipient's response for granted.)

### 2. The Question Mark within a Sentence

*a.* A question contained within a declarative sentence takes a question mark at the end of the question.

> Why did the plane crash? remained unanswered for years.
>
> Would the new computer immediately speed our reports? spin out more data? reduce the number of employees in the department? were the questions that concerned most of the staff.

EXCEPTION: A series of numbered or lettered interrogatives may take a question mark at the end of the last item:

Your supervisor should ask (1) did you unpack the tape yourself, (2) load it, (3) initial it?

*b.* If a declarative sentence ends with a question, a question mark is used, never a question mark and a period.

> The question answered most easily was would the computer speed the paperwork?
> The subject of the symposium is What are the prospects for individual space flight?

NOTE: A capital letter may be used to introduce the question.

## 3. The Question Mark with Parentheses to Express Uncertainty

If information, such as a date, is not definite, it may be followed by a question mark enclosed in parentheses.

> When Leonardo da Vinci went to Milan in 1483(?), he described himself chiefly as a military engineer.

NOTE: Some style guides omit the parentheses after an uncertain date:

> Ayatollah Ruholla Mussaui Khomeini (1900?–    )

Many writers avoid using the question mark enclosed within parentheses to indicate doubt or uncertainty about a statement.

> POOR: Her successful (?) career seemed to have gained her few tangible rewards.
> His thirty-ninth (?) birthday occurred last week.
> IMPROVED: That her career was really successful seemed doubtful as she had received few tangible rewards.
> His last birthday, according to him, was his thirty-ninth.
> His birthday last week was supposed to have been his thirty-ninth.

## PARENTHESES AND DASHES

### 1. Parentheses and Dashes with Parenthetical Material

Parenthetical material within a sentence may be set off by commas, parentheses, or dashes. (See also Comma, Section 5.) Supplementary material not closely connected to its subject, such as explanations or comments, should be enclosed in parentheses. Supplementary material that is inserted for emphasis or elaboration is set off by dashes. In typescript a dash is formed by two hyphens without any spacing between or around them (*cold--really freezing--weather*) or by a hyphen with a space before and after (*cold - really freezing - weather*).

> The active voice, as most stylists advise, should generally be preferred to the passive.
>
> We must, in the words of I. A. Richards, "look upon a correct understanding as a triumph against the odds."
>
> If he thinks (as well he may) that the problem is not worth correcting, don't employ him.
>
> Biochemical engineering is a fairly new subject (or a new branch of an old subject).
>
> People who know their own worth—and this cannot be stressed too often—can usually command their own salaries.
>
> His territory embraced three states—one hundred and twenty-three cities—and had to be covered once a month.

## 2. Dashes with Supplementary Material Containing Commas

When there are commas within the supplementary material, it is usually set off by dashes even though it may be closely related to its subject.

> Aphasia—which has different causes, among them acute alcoholism—will be the topic of three different seminars.
>
> Trying to understand American politics—a difficult feat for Asians and Africans, most of whom find it chaotic—is crucial to nearly all foreign governments.

## 3. Dashes for an Abrupt Shift or Break in Thought

Dashes mark abrupt breaks in sentence thought.

> The committee reports—especially the one read by Sally Rogers—made a favorable impression.
>
> I ask the chair to rule on—oh, never mind, I yield to the preceding speaker.

## 4. Parentheses with Numbers or Letters before Items in a Series

When a sentence or paragraph contains itemized elements distinguished by letters or numbers, the letters or numbers are enclosed in parentheses.

> The following topics will be discussed: (1) inflation, (2) unemployment, and (3) the consumer price index.
>
> The laboratory is concentrating on the development of a new vaccine: (*a*) the technicians must cultivate the agent; (*b*) they must then prepare the agent for serological tests; (*c*) most important, they must guard

against accidental contamination. Item *b* will be explained in detail later.

When itemized material is presented in list form, the letters or numbers are followed by periods or sometimes by a single parenthesis.

The laboratory technicians
   *a.* cultivate the agent
   *b.* prepare the agent
   *c.* guard against contamination

The laboratory technicians
   *a*) cultivate the agent
   *b*) prepare the agent
   *c*) guard against contamination

## 5. Parentheses with Years of Birth and Death

When the years of a person's life span or the year of birth or death follows the name, the year or years are enclosed in parentheses.

Sir Isaac Newton (1642–1727) and Adam Smith (1723–1790), although both lived several centuries ago, are names commonly recognized today.
Konrad Adenauer (b.1876) was a young man at the turn of the century; yet he reached his greatest prominence after World War II.

## 6. Parentheses with Numbers Supplementing Spelled-out Amounts

Although in simple matters it is not necessary to repeat spelled-out amounts in numerals or to follow numerals with spelled-out amounts, it sometimes is advisable to do so, particularly in writing to someone to whom English is not completely familiar or in drafting legal documents. The explanatory material should be in parentheses.

Please send us two hundred and fifty thousand (250,000) fasteners.
The price of these motors is thirty-seven hundred dollars ($3,700) each.
The term of the contract shall be thirty-six (36) months, beginning on April 1, 1992.

## 7. Parentheses with Abbreviations and the Corresponding Spelled-out Terms

When an abbreviation is followed by or supplements the word or words abbreviated, the second term appears in parentheses.

He is a specialist in the disposal of PVC (polyvinyl chloride).
She is a specialist in the disposal of polyvinyl chloride (PVC).

239

## 8. Parentheses and Dashes with Other Punctuation Marks

*a.* A complete sentence within parentheses takes a period or question mark or exclamation point before the closing parenthesis. If the sentence is within another sentence, however, a period is not necessary, and the parenthetical sentence may begin with a capital or small letter.

> The company has declared an extra dividend. (This decision was not arrived at lightly.)
> Humpty Dumpty (we consider him one of the most delightful of the *Alice* characters) said, "When I use a word, it means just what I choose it to mean—neither more nor less."
> Visual readers (Doesn't that describe most good readers?) see not words but symbols of ideas.

*b.* Punctuation that is not part of the parenthetical expression should be placed after the closing parenthesis.

> When you see her secretary (Mary Jones), ask for the films.
> He worked with a number of chemists (Roy Arnold, Rosalind Burns, George Steele, Moses Vladik).
> When confronted with compulsive personalities (those obsessed with rules and behavioral codes), physicians often find some indication of coronary disease.
> With all due respect to technological achievement (and lack thereof), these people remain convinced that only ethical principles will hold back Armageddon.

*c.* Commas do not precede or follow dashes even though, as in the next example, a comma ordinarily would be used to express nonrestriction:

> Disabled persons—especially those in wheelchairs—who ask only for reasonable consideration from designers of public buildings, often find access to the buildings difficult, if not impossible.

## 9. Parentheses for References within the Text (See also Brackets, Section 2.)

*a.* When figures and tables are included in a paper, reference to them is preferably within parentheses.

> The data are too scattered (Figure 6) to allow a valid conclusion to be drawn.
> The patients, who were divided into three groups which were given varying amounts of the drug at varying intervals for ten days, showed markedly similar reactions within each group (Table 5).

NOTE: This form is preferable to *Table 5 shows patients divided into three groups. . . . The reactions within each group were. . . .*

*b*. When equations are numbered, the numbers appear at the right-hand margin and are enclosed in parentheses. Reference to the equations requires brackets if the reference is in parentheses.

Therefore

$$ad = 5y^2 \tag{7}$$

Equation (7) may be used for the next step.

The next step (see Eq. [7]) is easily determined.

## BRACKETS

### 1. Brackets for Editorial Interpolation

Brackets are used to enclose editorial comment or explanation within quoted material.

> "They [the members of the Board] will vote tomorrow on the new salary scale," the foreman announced.
> The rule read, "A guiding principal [*sic*] of this organization is 'A fair day's pay for a fair day's work.'"

NOTE: The Latin *sic* ("thus"), without a period, is permissible if it is essential to point out an error in quoted material.

### 2. Brackets with Parentheses

When parenthetical material ordinarily enclosed within parentheses appears within other material enclosed in parentheses, brackets are substituted for the inside pair of parentheses.

> Without expression the secretary read the report (it had been edited by the president, who believed in "smoothing" [eliminating] any matters that would invite dissension).
> Spontaneous remission of a disease (the unaided recovery from a terminal condition [supported but unexplained by the textbooks]) encourages those striving to find a cure for the disorder.
> The final report on this project (which will be read in its entirety by three people: [1] the manager who initiated it, [2] his supervisor, who has to account for the funds, and [3] some remote researcher who is madly looking for something to bolster his own conclusions) will be carefully typed and bound before being consigned to the files.
> Earlier investigators (Smith [1985], Brown and Jones [1987], Carson [1988]) overlooked the suggestive work of Graham (1975).

### 3. Brackets with Other Marks of Punctuation

The rules that apply to other punctuation with parentheses apply to other punctuation with brackets.

# QUOTATION MARKS

## 1. Quotation Marks with Quoted Material

*a.* A direct quotation (the reproduction of someone's exact words) must be enclosed in quotation marks.

> "We have nothing to fear but fear itself," Franklin D. Roosevelt's reassuring statement, has been repeated by his successors and would-be successors.
>
> Martin Luther King Jr.'s simple words "I have a dream" dramatized the hopes of many.
>
> This formula, according to Smith and Brown, "may be used without question."

*b.* An indirect quotation does not require quotation marks. It must, however, be properly attributed, not passed off as the writer's own words or idea.

> Smith and Brown were the first to say that this formula works.
>
> Martin Luther King said that he had a dream, and millions gained hope.

*c.* Quoted material of more than one paragraph takes quotation marks at the beginning of each paragraph and at the end of the last paragraph. Instead of quotation marks, writers sometimes indent long passages or, if the rest of the manuscript is double-spaced, use single spacing. In printed material such long passages are usually printed in smaller type than is the rest of the work. (See the examples in Chapter 5.)

## 2. Quotation Marks with Dialogue

Dialogue, which is seldom used in technical or business writing unless one is presenting case histories, must be set off from other material by quotation marks. Each speech is usually a new paragraph.

> "For several years," the patient said, "my friends have been avoiding me."
>
> The doctor asked, "What makes you think so? Can you give me a specific instance?"
>
> "There have been many instances," the patient replied. "I can't think of one now."

NOTE: Dialogue is separated from the speaker by a comma, as a rule, but see the immediately preceding example. A semicolon might have been used in place of the first period.

## 3. Quotation Marks with the Titles of Short Works

Titles of articles, short poems, and subdivisions of books are enclosed in quotation marks. Titles of books, magazines, newspapers, long poems (unless included in a collection), and long reports are italicized (underlined in manuscripts).

> "From Newton to Einstein," which appeared in *The New York Times Magazine* some years ago, is well worth reading.
> Chapter 4, "Overcoming Dyslexia," should be helpful to her.
> Ogden Nash's "A Lady Thinks She Is Thirty" is a charming birthday tribute.

## 4. Quotation Marks with Words Used in a Special Way

Quotation marks should be placed around words to which the writer gives a special meaning.

> The "loss" reported here would be welcomed by many financial officers.
> Their "land planning" strongly resembles land plundering.
> Their so-called land planning strongly resembles land plundering. (Quotation marks would be redundant after *so-called.*)

> NOTE: Technical terms, trite phrases, slang, proverbs do not require quotation marks:

> One consideration in the design of solar collectors is beam radiation.
> The machine ground to a halt.
> By the time the default was discovered, those responsible had flown the coop.
> Facing his first major failure, the manager reflected that Rome was not built in a day.

> (It is preferable to avoid clichés by replacing them: *The machine slowly stopped, . . . those responsible had disappeared, . . . the manager reflected that there would be other opportunities.*)

## 5. Quotation Marks with Other Marks of Punctuation

In the United States a comma and a period are placed inside quotation marks. If a sentence ends with both single and double quotation marks, the period and comma are inside both. (Originally this style was for the convenience of printers.) Works printed in England and some publications in English printed abroad in places like Holland and Italy do not follow this rule, nor do writings in a few disciplines, such as philosophy and theology (for terms enclosed in single quotation marks to indicate special meaning) and philology. Writers quoting from such works should retain the punctuation of the original texts.

The question mark and exclamation point go inside or outside the quotation marks according to whether they are part of the quotation or of the rest of the sentence; the colon and semicolon appear outside the quotation marks.

Have you read Smith's latest article, "The Congressional Stampede"? (The question mark is not part of the title of the article.)

Have you read his most recent review, "What Price Glory?" (The question mark is part of the title and so is inside the quotation marks. Only one mark of punctuation in addition to appropriate quotation marks, parentheses, and brackets appears at the end of a sentence.)

Agreeing with the chairman's statement that the problem was "incontrovertible and incriminatory," the committee decided to meet the next morning.

The charge is thoroughly examined in Smith's "The Congressional Stampede"; in his latest book Black also devotes some space to it in Chapter 9, "Pathways and Pitfalls."

## 6. Single Quotation Marks for a Quotation within a Quotation

A quotation within a quotation takes single quotation marks. With other marks of punctuation the rules for double quotation marks apply.

The letter stated, "All three of us confirm the opinion of Dr. Jones, who says, 'This traumatism cannot have been caused by the injury in question.'"

The investment counselor cautioned, "Read the annual report carefully, especially any paragraphs that contain the words 'unassigned funds,' before you accept the offering."

The student asked, "Can you understand the instructions that state, 'Use the D-algorithm'?"

He does not agree with the article "The Convoluted Motivation of 'Who Killed Hilda Marsh?'" which he read last night. (Although the final clause is nonrestrictive, there may be only one other mark of punctuation with the quotation mark.)

NOTE: In British usage single quotation marks generally appear where Americans use double.

## 7. Quotation Marks around Words Following *Known as, Called,* and Similar Expressions

Many writers place quotation marks around the word or phrase following *known as, called,* and such. (See also Underlining, Section 3.)

Computers use a method called the "octal number system."

If the results cannot be predicted, an experiment is known as a "random" experiment.

## THE SLASH (SOLIDUS, VIRGULE, DIAGONAL, SHILLING MARK)

Known by a number of names (slash, solidus, virgule, diagonal, shilling mark), the mark that resembles a diagonal line has been increasingly used in the past few decades. Unfortunately, the meaning of the slash is sometimes misunderstood; writers, therefore, should beware of ambiguity.

## 1. The Slash with Alternatives

The slash is used to separate alternatives.

> The animal would then walk/run to its cage.
> All employees who have worked for the company for five years or more will receive this bonus and/or a three-week vacation.

> NOTE: The *and/or* in the preceding sentence means *a bonus or a vacation or both*. It is misread so often that careful writers avoid *and/or* and spell out the meaning. (See Appendix B.)

## 2. The Slash with Time Divisions

The slash is used for successive divisions of extended periods of time.

> 950/949 B.C.     Academic year 1897/1898

## 3. The Slash to Represent *per* in Measurements

> 50 km/hr     75 feet/second

## 4. The Slash with Fractions

Fractions written on one line in the text use the slash instead of the built-up form. (See The Punctuation of Equations, page 252.)

> 12/25     a = 18b/7

# THE HYPHEN

## 1. The Hyphen with Compound Words

There are no rules for forming compound words. An unabridged dictionary is the best guide. In consulting a dictionary, writers should look up both parts of the compound. If the desired compound does not appear in the dictionary, it usually is best to write it as two words or to hyphenate it, rather than to write it as one word; however, the modern tendency is to eliminate hyphens in compounds.

> vice president *but* president-elect
> marketplace *but* stock market

old-fashioned *but* newfangled
postwar *but* post-office (adjective)
redheaded *but* blue-eyed
downpour *but* down under

## 2. The Hyphen with Words Modifying a Following Noun

*a.* Adjectives and adverbs, and sometimes modifying nouns functioning as adjectives before a noun, may be connected by hyphens in order to clarify which words are modified.

All prospective employee interviews are scheduled for Tuesday. (all employee interviews that are anticipated)
All prospective-employee interviews have been scheduled. (interviews with prospective employees)
Her twenty-nine-year-old son is still at the university.
Use the bluish-purple solution.
The light-colored paper is used for reports.
The light colored paper is for airmail sent by the sales department.
Watch those misinformation-today, denial-tomorrow press meetings.
He ordered twelve-foot boards.
They reported on employer-employee relationships.

NOTE: The hyphen is usually omitted after an adverb that ends in *ly*:

skillfully drawn figures
unexpectedly high temperatures

*b.* The hyphen followed by a space suspends the first half of a hyphenated compound which precedes another hyphenated compound with the same final word:

three- and four-syllable words
four-, five-, and six-hour meetings

(If commas are needed, the comma follows the hyphen.)

*c.* Many common compound modifiers do not require a hyphen.

heat transfer projects
high school students
osteopathic arthritis symptoms
intensive care units
fundamental research grants
bull market reports
labor relations studies

### 3. The Hyphen with Prefixes and Suffixes

*a.* Placing a hyphen between a word and a prefix is becoming obsolete unless the absence of a hyphen would be misleading or awkward.

predigest
unusual
nonconformist
reformation (the act of amending or improving)     *but*   re-formation (the act of forming again)
resign (to disassociate oneself from)     *but*   re-sign (to sign again)
coworkers (but some prefer co-workers)
iso-octane     *but*   isobutylene
anti-icteric     *but*   antiperistalsis
contraindicant     *but*   contra-anthropomorphic

(Many writers prefer to hyphenate words in which the initial vowel of the word repeats the final letter of the prefix.)

NOTE: Before a capital letter the prefix is always hyphenated:

pre-Reformation
anti-American
ante-Roman
post-World War II

(With common compounds that are not hyphenated, a hyphen appears only after a prefix.)

*b.* Suffixes are usually attached to the preceding word, but there are exceptions. If adding a suffix results in the same letter occurring three times or in the doubling of a letter that is not normally doubled in English, a hyphen is placed before the suffix. Some writers prefer to hyphenate suffixes following proper names.

worker
jobless     *but*   thrill-less
nationwide
praiseworthy     *but*   crew-worthy
godlike     *but*   doll-like
Nero-like

### 4. The Hyphen with Compound Capitalized Names

Compound names modifying a noun take a hyphen between the names but none between the parts of the names. In print this hyphen, known as an "en dash," is slightly longer than the regulation hyphen.

the New York–Washington shuttle
the United States–Great Britain treaties

## 5. The Hyphen with Numbers

*a.* The use of a hyphen to clarify what word a number modifies is particularly important.

> He ordered two pound packages. (two packages)
> He ordered two-pound packages. (packages weighing two pounds each)
> The pipe was cut in four foot lengths.    (four lengths
> *or*                                     of a foot each)
> The pipe was cut in 4 foot lengths.
> The pipe was cut in four-foot lengths.    (Each length was
> *or*                                     four feet.)
> The pipe was cut in 4-foot lengths.

*b.* The hyphen is used with compound numbers from twenty-one to ninety-nine.

> twenty-four     sixty-fifth     one hundred one

NOTE: Numbers above ninety-nine do not need additional hyphens:

> one hundred twenty-seven     three thousand forty-two

*c.* A hyphen is required in fractions only if the fraction immediately precedes the word it modifies.

> The jug was one-third full.
> He had lost two thirds of the liquid.
> Her reserves were depleted by three fourths.
> The remaining one quarter was barely adequate.
> Even a one-half–percent increase is significant. (In print the second hyphen is an en dash. Omitting the second hyphen would not cause a misreading of this sentence.)

*d.* When used between numbers or dates, the hyphen means *up to and including.*

> pages 123–197
> the decade 1991–2000

## 6. The Hyphen at the End of a Line to Divide a Word

A hyphen is placed at the end of a line of type to indicate that the last word has been divided and continues onto the next line. Formerly there were strict rules for the hyphenation of words according to specific syllables. The advent of computerized typesetting, unfortunately, seems to have abrogated these rules, and a word is often broken without regard for syllabification, particularly in newspapers. The hyphen, however, remains. When at all possible, proper names should not be hyphenated.

> People who think quickly are sometimes guilty of lipog-
>     raphy.
> They gladly acclaimed their illus-
>     trious colleague.
> He promised to finish the report for John
>     Anderson, his section head. (not *for John Ander-*
>     *son*)
> He promised to finish the report for John Anderson, his section head.
>     (preferable)
> She offered a Spanish-
>     English course.
> The registration was a so-
>     called triumph.

## THE APOSTROPHE

### 1. The Apostrophe to Show Possession

*a.* The apostrophe and *s* indicate the possessive case of singular nouns.

the manager's office     the boss's orders     Bruce's report
Dickens's novels     ship's keel (*The keel of the ship* is preferable, as inanimate objects do not possess. [Ships, however, traditionally are considered feminine in literature: *Her sails were furled.*] Exceptions are such expressions as *a day's work, an hour's pleasure, money's worth.*)

NOTE 1: If a long name or word ending in an *s* or *z* sound seems awkward with another *s*, the possessive may be formed with the apostrophe alone:

Thucydides' history     Socrates' dialogues
for goodness' sake

NOTE 2: Many Biblical names ordinarily take only an apostrophe, but adding an *s* also is acceptable.

Jesus' *or* Jesus's     Moses' *or* Moses's

*b.* Plural nouns ending in *s* show possession by adding only an apostrophe.

employees' lounge     managers' cars
the Joneses' house     customers' complaints

249

*c.* Plural nouns not ending in *s* add an apostrophe and *s* to form the possessive.

men's clothing      women's movement
alumni's bequests    geese's feed

*d.* When an object or concept relates to several people, only the last mentioned person is in the possessive form. If more than one object or concept separately concerns several people, the possessive form is used for all the people mentioned.

The executives and employees' benefit plan (one plan for all) needs improvement.
The executives' and employees' dining rooms (one room for executives, one room for employees) will be air-conditioned.
Forbes and Carey's agreement was dated May second.
Forbes's and Carey's contracts were mailed the same day.
Smith, Jones, and Brown's law office is in this building.
Smith's, Jones's, and Brown's offices are in the same building.
The Joneses and the Smiths' lawyer handles all their real estate.
The Joneses' and the Smiths' brokers seldom agree.

*e.* Compound words form the possessive by adding an apostrophe to the last unit of the compound.

Her sister-in-law's estate was large.
The brothers-in-law's land will be sold.
The senators-elect's places will be assigned tomorrow.

*f.* Indefinite pronouns form the posssessive by adding an apostrophe and *s*.

another's    anybody's    everyone's    one's

*g.* Personal pronouns have a possessive (genitive) case and do not use an apostrophe.

| | | | |
|---|---|---|---|
| my | mine | our | ours |
| your | yours | their | theirs |
| his | his | its | its |
| her | hers | | |

*h.* Reflexive pronouns (pronouns ending in *self*) do not form the possessive.

## 2. The Apostrophe in Contractions

Missing letters in words or missing numbers are represented by an apostrophe.

| | | |
|---|---|---|
| can't (cannot) | doesn't (does not) | it's (it is) |
| who's (who is) | five o'clock (of the clock) | in '83 (1983) |

Except for *o'clock*, these contractions are avoided in formal writing.

NOTE: Formerly some abbreviations substituted apostrophes for missing letters, but this style is no longer used: *nat'l. (national), m'f'r. (manufacturer), 1'st., 3'rd.* If abbreviations must be used, the preferred forms are *natl., mfr., 1st, 3d, 5th* (also *2d*).

## 3. The Apostrophe in Plurals of Some Words, Letters, and Numbers

Words out of context, lower-case (small) letters, some capital letters, and sometimes numbers form the plural by adding an apostrophe and *s*.

There are three *as*'s in that sentence.

He uses so many *questionable*'s throughout his report that one is tempted to consider the entire work questionable.

Without an apostrophe the *x*'s and *y*'s might be misread, but the *X*s and *Y*s are quite clear.

Many would misread momentarily a*s, is, A*s, *I*s, and so *a*'s, *i*'s, *A*'s, *I*'s are preferred.

How many M.B.A.'s are there in the company?

The number of PAC's has increased amazingly.

People born in the 1980's will really be considered part of the twenty-first century.

NOTE: Some authorities favor using an apostrophe and *s* only when abbreviations contain periods. Some use only *s* for the plural form of all abbreviations:

At least 500 M.D.s attended the conference.

Our inventory of TVs is badly depleted.

NOTE: The tendency now is to drop the apostrophe before the *s* for the plural of numbers except 1. (For consistency many writers use the apostrophe with all numbers if a sentence contains 1's.)

Her 3s are often mistaken for 8s.

The 1980s might be considered the roller-coaster years for interest rates.

All her 1's look like 7s *or* all her 1's look like 7's.

## THE PUNCTUATION OF EQUATIONS

If an equation is part of a sentence, it is punctuated as any other sentence element would be unless the equation is displayed (centered).

> If you write $a = (2/b)^2$, your result will be correct. If you make a seemingly minor change, such as $a = 2/b^2$, however, your result willl be incorrect.

Both variables can be calculated from

$$\frac{mc^3 + \tan c}{2Y^2} = 5ab \qquad (4)$$

and

$$ab = d^3$$

If the displayed formula is not numbered, some publishers prefer to run it into the text unless it would be unwieldy:

> Both variables can be calculated from $(mc^3 + \tan c)/2Y^2 = 5ab$ and $ab = d^3$.

## UNDERLINING, OR ITALICS

Italic type, used to distinguish particular material from the bulk of the text, is indicated in typescript by underlining.

### 1. Italics for Titles

The titles of books; magazines; newspapers; dramatic works; art works; very long reports, poems, and musical works are underlined, or italicized. Titles of articles; of short poems, stories, and musical works; and of sections of a book or long report appear within quotation marks. (See also Quotation Marks, Section 3.)

> *Contamination of the Local River* is the first sizable report from the new committee.
> Many people have heard of *Paradise Lost*; fewer have read it.
> Michelangelo's *David* alone is worth a trip to Florence.
> John Smith's new book, *Vesting, Investing, Divesting,* is likely to win an award.
> Section 7, "Problems of Evaporation," in Jones's *Manual of Cryogenics* should be rewritten.

### 2. Italics for Words Out of Context

Words discussed as words rather than as concepts are underlined.

252

*Operation,* a commonly used term in many fields, has a frightening connotation when uttered by a surgeon.
Have you looked up the origin of *high-muckety-muck?*


### 3. Italics for Words Introduced by *Called, Known as, Labeled,* and Similar Terms


Many writers underline words following *called, known as,* etc., but occasionally they are enclosed in quotation marks or not distinguished at
all. (If italics are used for the text, the material to be set off is not italicized.) A good guide is the emphasis desired.

Known as *Marielitos,* the Cubans who came to the United States in 1980
embarked from the coastal town of Mariel.
The tops of supporting columns are called *capitals.*
The curve labeled *profit* on this chart is of special interest.


### 4. Italics for Foreign Words

Foreign words in an English text are usually underlined. Familiar
words that have been adopted into English, however, are not set off.

Although the French *rouge* has been adopted into English, it does not always
mean the same in both languages.
The Spanish consul agreed to shorten his *licencia.*
*Bitte* and *danke schön* are all some people seem to need when they tour
Germany.
The hacienda style of her home blended well with its surroundings.
He speaks in clichés, which he seems to consider original.


### 5. Italics for Letters Used Alone

In mathematics letters are italicized, or underlined. When referred to
as letters in text, they also are italicized.

$x + y = a^2$
(In typescripts containing a good deal of mathematics, underlining is not
usually necessary unless ambiguity is likely. In printed material italics
are used.)
It is difficult to distinguish the *l*'s and *e*'s in her handwriting. (See also
Apostrophe, Section 3.)

NOTE: The *s* that forms the plural of letters or numbers is not
underlined.

## 6. Italics in Specializations

In many fields italics are used for particular material, such as genuses and species in science and cases in law.

> *Homo sapiens* was once considered a humorous way of referring to the general populace.
> *Roe v. Wade* is one of the important precedents of the late twentieth century.

## 7. Italics for the Names of Vehicles

Names of ships, boats, airplanes, and such are italicized.

> The French liner *France,* which made her last transatlantic voyage in 1974, eventually became the *Norway.*
> He named his boat *The Make Waves.*
> The missing plane *The Flub Dub* was never found.
> Countless landlubbers have sailed on HMS *Pinafore.*

## PROOFREADERS' MARKS

For writers who are editing or revising their work and for supervisors who are editing or reviewing the writing of their staffs, a consistent method of editing expedites the completion of reports and other papers. The use of conventional proofreading marks is one way of gaining consistency.

The proofreaders' marks, or symbols, need not be inserted in the margin of papers; they may be incorporated in the typescript. (See the list below.) Rewriting, if not too extensive, may be done between the lines of the typescript, with the original crossed out. Long additions or changes should be written on a separate page and their insertion in the text located by numbers or letters identifying the new material: "Insert A attached."

When proofreaders' symbols are used, writers and typists—or those using word processors—should have a list of the marks. The most common marks, or symbols, appear below.

PROOFREADERS' MARKS, OR SYMBOLS

(The words *in the margin* in the following list apply to printers' proofs or to papers that are to be corrected on a word processor.)

|  |  |
|---|---|
| ℐ | This symbol in the margin on a line with a word or words that are crossed out indicates that material is to be deleted. |
| ̲ | This symbol under a letter means that it should be capitalized. The term *cap* appears in the margin. |
| / | A slash through a capital letter means that it should be a |

254

small, or lower case, letter. The term *lc* is placed in the margin.

—— A line under a word or letter has the value of italics. The term *ital* is written in the margin. If the typist does not use italic type where indicated, the underlining should be typed. If an entire report is typed in italics (which many people find distracting to read), underlining should be used or the material underlined should be typed in regular, or Roman, type.

〰 A wavy line beneath a word or letter indicates heavy, or boldface, type. Again, many typists do not change type in a report. Writers do not have much use for this indication except, perhaps, to distinguish vectors and some other mathematical expressions. The symbol in the margin that indicates boldface is *bf*.

⁋ Placed in the margin with a line drawn to the desired position, this symbol indicates that a new paragraph is needed at that position. In typescript the symbol may be placed where the new paragraph is wanted.

∽ This mark, or symbol, is drawn around words or letters that should be transposed: "The table only shows the estimated increase." The symbol or the term *tr* appears in the margin.

*LOAN*  A caret indicates an insertion: "County Savings and Association." In proofs the insertion appears in the margin; in typescript it may be above the caret.

‿ Connecting two words or letters, this symbol means that the space between them should be closed: "text book."

# This symbol placed above words that are run together indicates that a space is needed: "bankteller."

- - - A broken line beneath material that has been crossed out means that it should be retained. The term *stet* in the margin calls attention to the change.

[ or ]  A single bracket at the beginning or end of a group of words means that the material should be moved to the left or right. The bracket should be in the margin also.

⌐ This symbol shows that material should be moved up: "The symbol also appears in the margin."

∟ This symbol indicates that material should be lowered: "This symbol too should appear in the margin."

(center)  This instruction and similar ones should be circled and written in the margin. The symbol for *center* is two square brackets ] [ placed in reverse before and after the material to be centered.

255

This symbol joins material on separate lines; that is, the material on the second line is to be "run in" with the preceding material. "The symbol appears in the margin also."

 Material that is to be moved should be bracketed or circled and an arrow should be drawn to the new position. The margin of the proof should show *tr*. If the material is to be moved to another page, this information, circled, should appear in the margin of the typescript as well as of the printed proof. The place to which the material is to be moved should be indicated by a caret, and in the margin of the proof or of the typescript the words *insert from page*-- should be circled.

Numbers or abbreviations that are circled should be spelled out. *Spell*, circled, should appear in the margin.

This mark in the margin indicates that a comma is to be inserted where a caret appears in the text.*

This symbol in the margin shows the need for a period where a caret appears.*

This mark indicates the need for a semicolon.*

This mark requires the insertion of a colon.*

This symbol shows that quotation marks are to be inserted.*

This symbol indicates that an opening single quotation mark is wanted.*

This symbol stands for a single closing quotation mark or an apostrophe to be inserted where indicated.*

Two short vertical lines enclosing a horizontal line indicate a dash.*

This symbol indicates that a hyphen should be inserted. (Manuscripts edited for a printer show the hyphen as two short horizontal lines, but unless the mark is labeled, this form should not be used where it could be mistaken for an equal sign.)*

*? or !* These punctuation marks in the margin indicate that they should be inserted where indicated.*

This symbol indicates that a prime should be inserted.*

This symbol in the margin calls for the insertion of a superscript.* In typescript the inverted caret, (∨), or the roof ,

* In marking a typescript an editor may place these punctuation marks where they belong in the text. The circles, carets, or inverted carets (roofs) should be used for clarity.

should be used if the superscript is not raised sufficiently above the line: $y \sim a+b$ .

$\hat{\underset{\sim}{2}}$ This symbol in the margin calls for the insertion of a subscript.* The caret above the subscript is necessary in the typescript if there could be any doubt that a subscript is wanted: $CO_2$.

 In equations the equal sign is not likely to be misread. If it is used in the text, however, it should be labeled; $a = c$ could mean $a$ hyphen $c$ or $a$ equals $c$.

---

* In marking a typescript an editor may place these punctuation marks where they belong in the text. The circles, carets, or inverted carets (roofs) should be used for clarity.

Where more than one correction appears in a line, the symbols in the margin are separated by slash marks (virgules).

Editor's comments which are not meant to be incorporated in the text should be written in the margin and circled.

### A SAMPLE OF AN EDITED PROOF

The following are corrections that an editor would make to a printers' proof of an excerpt from the Charter of the United Nations. On a typescript an editor would not necessarily repeat the corrections in the margin. They are added here as illustration of what would be needed on a printers' proof or by a typist using a word processor.

Chapter V ⁀(bf)  ⌉  The Security Council [

COMPOSITION

Article 23  (ital)  (Center)

1. The Security Council shall consist of ⑪ members of the (spell)/cap
United Nations. The Republic of China, France, The Union of  lc/∧
Soviet Socialist Republics, the United Kingdom of Great Britain
and Northern Ireland, and the United States of America shall be  lc
permanent members of the Security Council, due regard being specially
paid, The General Assembly shall elect six other members of the  (cap)
United Nations to be non-permanent members of the Security
Council, due regard being specially paid, in the first instance,  §
to the contribution of Members of the United Nations to the  #
maintenance of international peace and security and to the other
purposes of the organization, and also to equitable geographic
distribution.

2. The non-permanent members of the Security Council shall
be elected for a term of two years. In the first electyon of  #/
the nonpermanent members, however, three shall be chosen for a term
(spell) of ① year. A retiring member shall not be eligible for immediate
re-election.

3. Each member of the Council shall have one representative.
— Security

257

# Part Three
# STYLE

# CHAPTER 9

# Style: The Personality and Character of Writing

*In science the credit goes to the man who convinces the world, not to the man to whom the idea first occurs.*

*Sir William Osler*

Style is the personality and character of writing. Like the personality and character of a person, style makes a single impact on first acquaintance. When later analysis reveals that many qualities were combined in that first impression, they often prove difficult to trace and understand. The characteristics of style blend just as intricately and subtly as the characteristics of a person.

Style and the writer are so closely related that critics frequently describe style with the adjectives that are used for people: weak, strong, muscular, flabby, graceful, clumsy, plain, ostentatious, cold, warm, coarse, elegant. The sports column is breezy, the book review is keen and witty, and the begging letter is too obsequious. The style of the politician we dislike is pompous and hypocritical; of the one we like, natural and honest. Shakespeare is profound, Milton is grand and lofty, and the King James Bible is earnest and sincere.

"The style is the man himself," said Georges Louis Leclerc de Buffon to the French Academy. And in the two hundred and fifty years since he spoke, hundreds of less penetrating thinkers than the members of the French Academy have repeated, "Style is the man." Many who quote these words apply them superficially: they expect a man with a sloppy appearance to express himself carelessly, an earnest-looking writer to produce serious and ponderous works, and a prim-looking person to write neatly and properly. When they read a book written by someone they know or when they meet the author of a book that they have read, they may be shocked. The woman with egg stains on her blouse has written precisely and carefully, the doctor who is dour and unapproachable has

261

written love letters in baby talk, and the dry-as-chalkdust schoolteacher has written bawdy stories in an earthy style.

Georges de Buffon was addressing a group of keen and subtle thinkers who did not expect human beings to be easy to understand. His statement is no less true because it does not necessarily apply to the outer person as perceived by an insensitive or casual observer. People present many images to the world and to themselves; the true person may be a medley of all or some of these. The egg stain on a writer's blouse may have escaped her notice because all her attention was concentrated on writing accurately. The stern and dignified doctor may have needed desperately the relaxation of playful affection. And the schoolteacher may have escaped the restrictions of his prim life through inner rebellion and revels, or he even may have been guiltily concealing a bawdy nature under a prim exterior. The style is the person—in all his or her complexity, elusiveness, and mystery.

One would expect these human qualities to be generally recognized in this Freud-ridden age. But so many people see only the obvious that writers have difficulty overcoming an impression left by poor style. This may be true even though the style has been forced upon them by teachers or supervisors and is not their own. Their annoyed readers (and who is not annoyed by poor style?) assign the defects of the style to the person. And when writing pleases readers by its good style, they assume that the writer has the virtues of the writing. Therefore improving style enhances reputation.

The first step in achieving a good style is organizing effectively (Chapter 3). De Buffon states:

> This plan is not indeed the style, but it is the foundation; it supports the style, directs it, governs its movements, and subjects it to law; without it, the best writer will lose his way, and his pen will run on unguided and by hazard will make uncertain strokes and incongruous figures.[1]

An effective style for the professions is a clear, concise style (Chapters 10 to 13). A well-organized, clear, and brief paper is adequate for many purposes and certainly better than much of the writing that circulates in industry and appears in learned journals. But such a style often leaves an intelligent writer vaguely dissatisfied and readers apathetic or bored. Avoidance of error does not of itself constitute good writing; indeed, pedantically correct writing may be dull and colorless. Many writers, supervisors, and instructors sense this and wonder what to do.

A most helpful answer is to begin to improve by eliminating the characteristics of weak and awkward style. Most writers find that as they remove these characteristics from their writing a good style emerges and develops. One most exciting teaching experience is to free persons of superior intelligence from the clutter of weaknesses common in functional prose and to observe them acquire not merely the rudiments of style but an individual style that transforms their writing.

The first time this happened to one instructor she was astonished. The class ranged from a distinguished director of research to the newest B.Ch.E. The new employee and the director were average writers in the group of twelve, the first displaying such a spectrum of typical freshman errors and weaknesses and the second writing in such choppy, incoherent sentences that reading their reports was painful.

When she conferred with the neophyte at the end of the course, she found that he had finally made the progress that he should have made in his freshman writing course—an accomplishment to satisfy but not to dazzle his corporation. When the director, whose conference followed the young engineer's, began by giving her an article that he was writing for a journal, she was not optimistic. From the very first sentence, however, she was delighted, but not just because the choppy sentences had disappeared. That much improvement she had expected of him. What surprised her was that his article was written better than any technical article she had ever read—written so well, in fact, that she soon forgot her amazement and became absorbed in the reading.

"All that I did was stop counting words and syllables and start applying the principles of good writing that you gave us," he explained. With intelligence, imagination, and good taste, he had applied the principles; and for the first time she saw how a superior mind trained in science and technology can effect an improvement in writing, how it can root out errors and leap to mastery of the style that the writer has always needed and wanted.

That day the thought of the director's wasted years of poor writing made the instructor sad, but he was so happy with his success that she had to share his pleasure. They shared it again when his article, published without change, brought him many enthusiastic letters from strangers who had not expected to enjoy an article in a journal of engineering and from friends and acquaintances who congratulated him on what they called "the marvelous change in his writing."

There was nothing marvelous in the basic principles he had applied so well. His flexibility in discarding bad habits of long standing, his superior intelligence, and his diligence were unusual. But the principles that improved his writing are those that any person with intelligence enough for a professional career can understand easily and apply effectively. The speed and quality of the improvement vary with each individual, but every professional person who sincerely tries to improve can acquire a style to be proud of.

The first step—eliminating the bad habits that are impediments to the development of a good style—is simplified by the fact that the same bad habits are shared by nearly all competent writers in a particular occupation: writers on business, education, engineering, law, medicine, politics, science, sports. Each group has its common weaknesses and errors. The repetition of errors makes it possible to correct them in a book of reasonable length. If there were more variety in the errors, several volumes of instruction would be necessary, but in our experience writers in the same occupation seldom exhibit errors or weaknesses of style that are not common in what they hear and read. Many, perhaps most, writers in a profession come to write and speak—and even to think—in the same style. An improvement in writing often becomes, therefore, an improvement in thinking. Avoiding the unnecessary use of the passive voice, for example, can lead to more decisive thinking. Thus the usual order of good thinking leading to good writing becomes good writing demanding good thinking and good thinking well expressed leading to better decisions and better support for the decisions.

## THE PERVASIVE PASSIVE

"Always use the passive voice" is a prescription so frequently pressed on writers of informational prose that it has proved to be one of the most harmful fallacies, if not the most harmful. Frequently enunciated by a person in a position superior to a writer's, the fallacy bears the heavy weight of law. This erroneous advice may confront an engineer or scientist first in graduate school. There professors may insist that students write as the professors do, in the passive voice, in order to appear scholarly, to show objectivity, to acquire a style like that of journal articles, or—more brutally—to make papers acceptable. In business and industry some supervisors have discovered the use of the passive to evade responsibility and therefore use it as ruthlessly as politicians do.

Fortunately some scientists and engineers so instructed break away later to return to normal use of the active and passive voices; and luckily in the past few years more and more professors, supervisors, and managers have been avoiding or at least toning down the fallacy of always using the passive voice. Writing in business, government, industry, and journals, however, still shows so much unnecessary use of the passive that writers and editors need to become familiar with the correct and effective uses of the active and passive voices.

The differences in the forms of the active and passive voices are not difficult to understand. The active voice is clear and complete, and the order of the active voice is familiar to readers. The subject acts, the object receives the action, and the verb expresses the action clearly and directly.

Group A: Subject (Actor)–Active Verb–Object (Recipient of Action)
1. He *struck* his wife.
2. The Tropical Belt Coal Company *made* a profit of ten million dollars.
3. God *helps* those who *help* themselves.
4. A stitch in time *saves* nine.

But in the passive voice the subject receives the action, the agent is omitted or is named in an awkward phrase beginning with *by*, and the verb needs the assistance of an auxiliary verb:

Group B: Subject (Recipient of Action)–Verb
1. His wife *was struck.*
   His wife *was struck by* him.
2. A profit of ten million dollars *was made.*
   A profit of ten million dollars *was made by* the Tropical Belt Coal Company.
3. Those who *are helped by* themselves *are helped by* God.
4. Nine stitches *are saved by* a stitch in time.

Sometimes the passive voice serves a purpose. George's mother might say. "His wife was struck," to avoid admitting that George did it. A variation of Sentence 2 in Group B might be useful in a discussion of profits if the name of the company were of no significance: "Profits of three million, seven million, and ten million dollars were made on the drug this year." But who would use the passive voice in all four sentences of Group B? Writers on science, technology, business, and computers would and do, as in the following examples:

Group C: Additional Examples of the Passive Voice
1. His secretary was fired.
   His secretary was fired by him.
2. Mistakes were made in the stability tests.
   Mistakes were made in the stability tests by Reliable Testing Company.
3. The members of his team who are helped by their own ability and diligence are assisted in every way by this supervisor.
4. One hundred work hours a week are saved by this method.

Many writers in the professions would never use the active voice in these sentences or in any sentences like them. When these writers were students, their instructors taught them to use the passive voice, and later their supervisors required it. When the writers prepared articles for journals in science and technology, they found that the journals also used the passive voice. At conventions of professional societies and at business conferences, their colleagues talked in the passive voice. The passive voice became so familiar that the active voice sounded strange and wrong to these writers. It even alarmed them.

The prevalence of the passive voice in government and industry and in science and technology amazes those who meet it for the first time. They find the following experience unnerving. A director of training asked a consultant to confer with a foreign-born engineer who had learned English while working for two years in the United States. When the consultant met the engineer, she apologized for her lateness.

"It is nothing," he replied courteously. "A cigarette was smoked and a book was read while waiting."

He was learning engineering English fast—not only the passive voice but the incorrect ellipsis that often accompanies it.

Later she asked a native engineer and a native chemist, "Can you improve the sentence 'A cigarette was smoked and a book was read while waiting'?"

"No error can be seen," decided the engineer.

"No improvement can be made," said the chemist, studied her expression, and added, "by me."

The passive voice weakens style when it is used, consciously and unconsciously, to evade responsibility. A popular passive construction is "It is thought that. . . ." Used anywhere but in science or technology, this indicates that a general opinion or truth follows. But when writers on business, science, or technology use it, they may mean, "I think that . . . ," "we think that . . . ," "the Committee thinks that . . . ," or even "I hope that somebody reading this report thinks that. . . ." By close attention some readers may learn to interpret "it is thought that." Readers of a committee report may mentally substitute, "The committee thinks that." But after the readers have interpreted two paragraphs successfully this way, they come to "it is thought that" followed by a common misconception, which is corrected in the next sentence. Do the writers mean that the committee held that misconception? Indeed, no. Now they are using "it is thought that" to introduce a general opinion that they wish to correct.

Readers of minutes written in the evasive passive often find themselves sounding like hoot owls as they scream, "Who? Who? Whooooo?" to statements like

> The cost of the TV program was estimated incorrectly.
> A letter to the FDA will be written.
> A report on these dyes will be submitted.
> It was said that the ruling will be ignored.
> A suggestion was made that marketing be postponed two months.
> It was reported that the new package had been well received.

In such writing all agents are anonymous and many statements are ambiguous. Helpful and useful information is omitted. Nobody is responsible for anything.

For a whole year an apocryphal head of a college department evaded responsibility by writing sentences like the following:

> The new curriculum has been approved.
> The two suggested promotions from assistant to associate professor have not been approved.
> Budget allotments for travel have been reduced by twenty percent.

266

An annual increase of $2,500 for the head of the department has been
recommended.

Nobody knew that it was this worm who had approved and not approved,
had reduced and recommended. But at the end of a year his department
discovered the evader behind the passive voice, and in the unmistakable
active voice the members demanded his resignation.

When writers using the passive voice name an agent, they have to
employ a clumsy, wordy construction like "The recommendation was
made by the Safety Committee that testing be continued." The active
voice is much neater and briefer: "The Safety Committee recommended
that testing continue." The sentence "Papers were presented by John
Jones and John Smith of our company" buries the credit due Jones and
Smith. The sentence "John Jones and John Smith of our company presented papers" gives them due credit. If a writer wishes to emphasize the
company, he may also do that in the active voice: "Two Polymer Company
chemists John Jones and John Smith presented papers."

The most common mixed-up sentence in the passive voice appears in
reports of visits to factories and plants. The most inexperienced member
of the visiting team writes the report. He begins with this gem, "On May
4, 1985, at the Polymer Plant Dr. A. J. Anderson was visited by the author
accompanied by Drs. Smith, Brown, and Jones."

When asked why he has used the passive voice and that flattering
phrase *the author*, he is sure to reply, "It's more modest. You see, I was
less important than the others and I didn't want to seem to be making
myself important. I've always been told to use *the author*." It is difficult
for him to see that *accompanied by* discourteously and immodestly subordinates his learned colleagues to himself, *the author*. But if someone
asks him to put the sentence in the active voice, he writes, "On May 4,
1985, at the Polymer Plant Drs. Brown, Jones, and Smith and I visited
Dr. A. J. Anderson," and he may even add, "to discuss the use of computers
in our polymer research." He knows that common courtesy requires that
*I* be last in the list, and if he does not use the passive, he reports clearly
and informatively, as well as courteously and modestly.

Passive verbs do have their proper place in style and are effective when
used correctly. They may be used if (1) the actor is obvious, is unimportant, is not known, or is not to be mentioned; (2) the receiver of the action
should be stressed, or the actor should appear in a subordinate position
at the end of the construction; (3) the thought comes too rapidly in the
active voice and needs a more deliberate presentation; (4) variety is
needed in a passage expressed in the active voice; and (5) a weak substitute for the imperative is desired:

   (1) He was injured in action.
      Commencement will be held on May tenth this year.
      Awards will be presented at the annual dinner.

The equipment was placed in Laboratory B, which has adequate space and proper lighting.

(2) His brother has been elected.

A gold Cadillac was presented to the star by her producer.

A bonus of a thousand dollars for every year of service was voted by the board.

He was removed from office by a unanimous vote.

(3) and (4) The active voice occurs far more often than the passive voice in exposition and narration, more than 90 percent of verbs being active according to one authority. The passive voice requires more words, blurs meaning, and uses the verb *to be*, which occurs in so many other constructions in English that it becomes monotonous. Inexperienced writers who employ the passive to avoid *I* or *we* create awkward sentences and shift voices without reason. They should use the active voice whenever possible and the passive voice only in those few constructions where the active voice is ineffective. Scientists and technologists must make special efforts to use the active voice because much of their writing requires the passive voice, and, being accustomed to it, they tend to use it where others would use the active. Often poor scientific style can be improved by changing the passive to the active voice to gain brevity, clarity, force, and liveliness.

(The use of the passive voice in the final sentence of the preceding paragraph adds variety, changes the pace, and by contrast stresses the final sentence.)

(5) The letter should be typed today.

(This polite command differs in tone from *Type the letter today* and *Please type the letter today.*)

This stationery should not be used for my business letters. Use it for my social and personal correspondence.

(The imperative in the second sentence is more forceful than the disguised command in the first sentence. Two direct commands might seem peremptory.)

The inventory should be completed this afternoon.

(Addressed to an executive, this disguised command implies that he should order things so that the inventory will be completed. It does not suggest that he should complete it himself as the direct command does: *Complete the inventory this afternoon*; and it does not put him in the position of merely receiving commands to pass along as does *Have the inventory completed this afternoon.*)

Customers are not to be addressed this way.

(This sentence implies that it is not correct or desirable or suitable to address customers this way. It seems to leave to the employee the decision to cooperate.)

An adroit use of the passive voice where it is suitable benefits style by permitting variations in meaning, stress, pace, and rhythm; but excessive use of the passive limits meaning, stress, pace, and rhythm. To write entirely in the passive would seem not just unwise but impossible; yet some misled scientists and engineers attempt it.

To write impersonally does not require frequent use of the passive voice; the active voice can be just as impersonal. Boring, Langfeld, and Weld, who use both *we* and *you* in *Foundations of Psychology*, also write impersonally in the active voice when their subject matter dictates an objective point of view. In the following paragraph the passive voice appears only in the last sentence:

> Although these tests point to the possibility of detecting accident-prone individuals before they have had serious accidents, the past accident history of the individual is still the most reliable indicator we have of accident tendencies. For one thing the tests measure only some of the factors which make an individual accident-prone, and for this reason a person can be accident-prone even though his test performance is good. On the other hand, a person can very often compensate for his deficiencies and consequently can have a good safety record although his test scores are not at all satisfactory. At present, therefore, best results are obtained when the diagnosis is based both upon test scores and upon previous accident history.[2]

In *International Science and Technology* R. G. Neswald begins an article with a lively, humorous paragraph that uses the active voice where many scientists and technologists would consider the passive necessary:

> In days of old, before space became so headily and incestuously intertwined with time that it tangled back on its own warped self, its various "continua" shot through with wormholes and chopped up into quanta, Nature was a pretty unsophisticated anthropomorphic simpleton. She abhorred vacua, favored simple symmetries, and, when pinched, punched, or otherwise perturbed, she reacted in kind with restoring forces, which (to a first approximation) were proportional to the magnitude of the original disturbances. The LeChatelier-Braun principle in chemical solutions, Lenz's law in electrical circuits, Hooke's law in mechanical systems reigned supreme.[3]

## THE EVASIVE *I*

Another harmful fallacy about style is the rule "Avoid all personal pronouns. Never use *I* or *we*." The pervasive passive is sometimes due to the evasive *I*. First-person pronouns have long been absent from technical writing. They disappeared in the United States about 1920, when impersonal style began to dominate in science and technology. In the writing of many divisions of the government of the United States, and particularly in the writing of the Pentagon, *I* and *we* and any other indication that a human being is writing were and in some divisions still are taboo. During the early twentieth century business and industry avoided pronouns in the first person by using business jargon like "Yours of the 10th inst. received and in reply beg leave to state . . ." and "Herewith are forwarded the reports under active consideration. Please be advised of general concurrence in desire to expedite matters." Since the 1940's business and industry, being closely associated with the government and with science and technology, have been influenced by the impersonal, inflexible style of federal and technical writing.

An attempt to achieve objectivity by avoiding personal pronouns is a mistake, and the idea that using the third person instead of the first person achieves modesty is equally wrong. Discarding necessary common words like *I* and *we* merely leads to awkward writing marked by excessive use of the passive and by reliance on weak indirect constructions. Writers deprived of *I* and *we* turn to unnatural and objectionable substitutes like *the author, one, the present writer, this reporter, your correspondent,* and *the undersigned* or even to titles—*the vice president, the chairman:*

1. "The writer cannot accept these findings." The woman who wrote this sentence would seem just as modest if she wrote simply and naturally, "I cannot accept these findings."
2. "The national secretary of the society initiated the following improvements in the management of the central office." Without his title, the national secretary of the society sounds more modest: "I initiated the following improvements in the management of the central office."
3. "When one observed that the floors were made too slippery by the new wax, one reported one's observations to the superintendent of maintenance." The new employee who wrote this sentence would sound less pretentious and pompous if he were to write, "I reported to the superintendent of maintenance that the new wax made the floors too slippery for safety."
4. "The present man of letters is much indebted to the aforementioned authors." The fact that the "aforementioned authors" and the self-titled "present man of letters" produce some occasional technical articles does not make them authors. The word *author* and the title *man of letters* are reserved for writers and scholars of distinction. A writer should not call himself a *man of letters* even if his mother and his admirers are kind enough to do so. The writer of Sentence 4 has misused these words and has dragged in the unnecessary *aforementioned* because he did not use *I* and write his sentence naturally as, "I am much indebted to these writers."
5. "The minutes will be sent to you every month by the undersigned." The passive voice and the legal word *undersigned* are both unnecessary if *I* is used: "I will send you the minutes every month."

Some writers consider *we* less personal than *I.* In imitation of an editorial writer, who is expressing the opinions of a board, a writer may mistakenly use the plural *we.* The obvious result is confusion. In the first paragraph of the following letter such a writer uses *our* to mean himself and his correspondent and *us* to mean himself; in the second paragraph he uses *we* for his company, for himself, and for his division of the company; and in the third and fourth paragraphs he uses *we* and *our* for all or none of these.

> Dear Professor Keppelbacker:
> Your letter with its references to our meeting at the Chicago convention brought back to us memories of a very pleasant evening—and a profitable one too.
> We have revised our formula as you suggested, and we are now testing the results. We will send you a copy of our report as soon as it is ready. In the meantime we are sending you the samples you requested.
> We hope to see you at the New Orleans meeting and would like to discuss having you consult with us at regular intervals. We can offer you a fee of four hundred dollars a day plus expenses.
> Please give our regards and the regards of our wife to Mrs. Keppelbacker.

The substitution of *we* for *I* in this letter is awkward. The use of *I* for the writer, of *we* for the Research Division (*we of the Research Divison*), and of *the company* or *our company* when the company is meant would at least be clear. Although the quoted letter probably caused no serious misunderstanding, the substitution of *we* for *I* can cause dangerous or embarrassing confusion. When writers using company stationery state, "we will do," they obligate their company unless in the first sentence in which *we* appears they clearly define the *we* to mean something else. The possible embarrassment is illustrated well by an experience of the Reverend John A. O'Brien when he proposed a conference on population growth. The last paragraph of a newspaper account states: "Although Father O'Brien used the phrase 'we propose' in connection with his proposals, a spokesman for Notre Dame said the professor was speaking for himself." Writers who uses *we* should be sure that they have authority to speak for others. And the only safe authorization is written.

Most scientific and technological journals now permit authors to use *I* for a single writer and *we* for more than one writer, especially when the material is personal, as in interpretation of results and in predictions. Indeed, many editors urge this use wherever appropriate. The *American National Standard for the Preparation of Scientific Papers for Written or Oral Presentation*,[4] which includes an impressive listing of the many organizations represented in its views, states, "When a verb concerns action by the author, the first person should be used, especially in matters of experimental design ('to eliminate this possibility, I did the following experiment')." It then warns against "constant use" of the first person, a warning that teachers of freshman composition present so frequently that few students who pass a good course should have trouble. Obviously, however, avoiding for years the use of pronouns in the first person necessitates some adjustment when writers begin to use them in scholarly or business writing. It is well to bear in mind the following aids to good use:

Minimize the use of the first person as the first word and subject of sentences. It easily becomes monotonous.

Be consistent; provide an antecedent for *we* or otherwise identify *we* unmistakably.

Avoid using *we* for the writer and the reader or readers if this implies agreement that you may only be assuming or if the *we* is not clearly stated to mean the writers or the writer and the reader or the readers.

Be especially careful to avoid an overdose of the personal in persuasive prose where it may increase the heat of argument.

Use the possessive and objective cases of the first person pronoun more freely than the nominative.

Oddly, some editors never objected to the use of *my, mine, our, ours, me, us* although they broke their pencils in wrath when they found an *I* or *we*. One senior manager of a major corporation told us, "Nobody in my division is allowed to use the first person for any reason whatsoever. See, this is how *I* write." The example he exhibited had *me* in the first sentence and *us* in the third, the *us* being, of course, without an antecedent or other identification. This manager was appalled by his error and thereafter tried to audit as many writing courses as he could. Writers should be sure to check any advice about grammar and style given by supervisors who have so little knowledge of the fundamentals of English.

Our advice is to use personal pronouns judiciously after dissertations have been approved and degrees received. The impersonal passive prose insisted upon by some graduate school advisers, even to the extent of rewriting theses themselves, is not the best style for publication (except in a few old-fashioned journals) or for use in the professions or business. Writers should discard it as soon as possible before it becomes a bad habit painfully difficult to change.

In our experience writers who read only what is essential for their business or profession and who dislike reading and writing have the most difficulty learning a new style. When the shackles are removed, they overwork the first-person pronouns and the active voice. Lacking a background in reading and writing, they have no literary judgment, no feeling for the appropriate use. For example, the manager who recognized only the forms *I* and *we* as personal pronouns had to work hard to learn how to use personal pronouns and the active voice gracefully, but the fifty scientists and engineers in his division who liked to read improved their writing as soon as they were liberated from dissertation style. He struggled for many months; they succeeded in their next papers.

Today in business and industry and even in academia there is less pressure than there was twenty years ago to avoid personal pronouns and the active voice. Indeed, many major businesses and industries are pressing hard for readable prose. The government is, as it so often is, divided: some states have moved toward legislation for understandable contracts of various kinds, while some military bodies still use and demand the dull, virtually unreadable prose of the thirties and earlier. Recent trends in writing for business and industry have been toward clear, concise, effective style. To achieve it, good writers and editors in business and industry, in government, and in colleges and universities have been freeing themselves from unnecessary rules and regulations. Instead of droning *never use the active voice* and *never use personal pronouns*, they have been concentrating on the functions of the active and passive voices, on the functions of personal and impersonal pronouns, and on the avoidance of usage and style not suited to the idiom of the English language. It will be interesting to watch the changes that occur.

## DULL STYLE

When the active voice and personal pronouns were reduced to a minimum or eliminated, much writing on science and technology became lifeless and dull. This led to the fallacy that writing on professional subjects has to be dull; therefore there is no use trying to do anything about it. Such writing is by its nature dull and heavy, the fallacy-maker insisted, and cited scholarly journals in proof. True, journals are often hard reading, even for scholars who are used to them; and some articles come as close to being unreadable as anything published today with the exception of a few textbooks and many doctoral dissertations. The writers of turgid prose either do not know how to write better or, like the residents of Laputa, they scorn to cast light on the world and would rather cast a shadow. Some of them condescendingly assume that their readers are so stupid as to admire only what cannot be read and understood easily. But few readers are duped by unintelligibility. Even the greenest undergraduate views skeptically an incomprehensible professor: "Well, maybe he knows his subject, but you'd never guess it from the way he explains it."

The lively lucidity of many scholars proves that poor style is not inevitable in the professions. In our experience there is a marked correlation between the excellence of writers' understanding of their subjects and the clarity and grace of their expression of their written thoughts on those subjects. Those wishing to write a livelier style should apply the sections on emphasis in sentences and paragraphs (Chapters 11 and 12) and the advice about choice of words (Chapter 10), as well as the recommendations in this chapter.

## STRINGS OF MODIFIERS

One of the characteristics that mar the style of the prose of information is placing a string of modifiers before a noun. In some Germanic languages one may write such a string as the following without arousing comment: *the-meeting-at-two-o'clock-on-alternate-Thursdays-for-social-and-philanthropic-purposes-ladies-of-the-Church-of-St.-Peter-and-St.-Paul-cooking-and-sewing society*. In English, however, modifiers may be placed before and after the words they modify. For example, "the blue-eyed, curly haired asthmatic patient" is clumsy. Usually some modifiers are placed after the noun: "the blue-eyed, curly haired patient with asthma" or "the asthmatic patient with blue eyes and curly hair." The American National Standard advises that nouns be modified by only one other noun but permits two and restricts the number of modifiers preceding a noun to three. Difficulty in understanding and ambiguity, it suggests, result from the use of expressions like "heavy beef heart mitochondria protein" and "the constant pressure heat capacity temperature maxima."[5] A mountain of modifiers before a noun is unidiomatic and graceless in English, but many writers accustomed to such strings of modifiers in their own specializations use them freely although they object to those from other specializations. The following examples illustrate the awkward style and ambiguity that results: *the culturally disadvantaged child conference; specific diesel fuel system design variations; high level Armed Forces manuals speaker; environment electronic systems specification sponsor information; chemist and engineer advance placement information; satiated, 6 hour trained Strawberry Hill Farm rats; overall information exchange complex; a complicated kind of producer-to-middleman-to-consumer sequence; difficulty obtained objectives; biodegradable-detergent feedstocks; kerosene-range (C10–C16) normal paraffins; 1606-type earthquakes; raw materials and fuel cost differentials*. Too often nouns have been used to modify nouns. Instead of *chemist and engineer advance placement information*, for example, why not *information for the advanced placement of chemists and engineers* or *advance information about the placement of chemists and engineers*?

These strings of modifiers are lumpy, bumpy reading. The argument that they save a few words is specious. The words saved are prepositions, which writers may as well use to improve the quality of their writing. Indeed, many writers who omit prepositions in these phrases with the excuse of seeking brevity are the same writers who insert unnecessary prepositions and other words in phrases like *head up, as to the nature of* (for *about*), *by means of* (for *by*), and *in order to* (for *to*). An occasional pile of modifiers, even of nouns modifying nouns, may do little harm to style; but few who use them can write chains of modifiers only occasionally. Moreover, so many terms in the sciences and technologies and pseudosciences contain chains of modifiers that writers forced to use these terms should avoid adding more. A writer who must use technical terms like *subject-action, verb-target of action; subject-linking verb-description of subject; liquid-liquid-liquid extraction; approach-avoidance conflicts and avoidance conflicts; zero-sum, two-player games; 1500-MeV electron synchrotron;* and *Van de Graaf accelerator and electrostatic deflection sys-*

*tem* should avoid other chains of modifiers. Such unidiomatic constructions make any prose sound like a poor translation from a foreign language. The foreign language might be called Insensitive-to-Style Science-Technology Pseudo-Science-Pseudo-Technology Anti-English. And if that does not hurt their ears, writers are in danger.

Besides being cumbersome, chains of modifiers without hyphens may be confusing. If a writer uses such poor style, one cannot tell whether *sulfur containing additives* means additives that contain sulfur or sulfur that contains additives. And what does *synthetic plant construction* mean? or *superior test monkeys?* or *economically more attractive pure quality chemicals?* or *productive time estimates?*


## TELEGRAPHIC STYLE

The telegraphic style found in the publications of government and industry should be avoided even in telegrams. Omitting articles, some pronouns, and conjunctions may save money in telegrams and cables if it does not lead to other telegrams requesting clarification. But it is difficult to determine what the writers of telegraphic style are saving. They are not saving money, for the confusion they cause is expensive. They are not saving time, for their readers have to waste time puzzling out the meaning. They are not saving typing or paper, for they often prefer long words to short. They may be trying to save face by attempting to conceal that they cannot write English, or they may be evading responsibility by clouding their meaning so that no one can tell who is responsible for anything or even whether the writer understands what is happening.

Some writers of minutes are addicted to telegraphic style, and because minutes should be useful records of decisions and of responsibility for action, the smudged meanings of telegraphic style are particularly dangerous in such records. The following sentences are typical of the opaque expression of some writers of minutes:

> Drug 109 discussed re government approvals recommended further animal experiments. To be investigated.
> Limited marketing Drug 111 Hospital Service suggestion. No action.
> Suggested approved MD complete Drug 141 May pamphlet circulation. Pamphlet proofs pending action. Postponed March 6.

Even someone who attended the meeting has trouble understanding such reporting.

Sometimes a telegraphic writer does not even attempt to save words but just moves them from one position to another. Here is a sample from a journal on writing: "This symposium was held for the purpose of exploring problems and the promise of 'More Effective Communication of Scientific and Engineering Information.'" The writer might begin his exploring by searching for the misplaced *the*; he meant "for exploring the problems and the promise of."

Telegraphic writers also scorn conjunctions and carelessly produce sentences like this: "For a molecular weight of 500 this would be 1.5 mg/ml for practical purposes we would like to have at least 15 mg in this case." The writer should have placed a *but* after *ml*. If her purpose was to save words, she might have omitted *in this case*.

Writers who use telegraphic style in a misguided search for brevity can achieve it more effectively through other means. If by using a telegraphic style they are concealing carelessness or attempting to evade responsibility, they are self-convicted nonwriters.

## REPEATING AND REPEATING AND REPEATING

Another fallacy that misleads writers is the advice to repeat ideas three times. In an anthology on technical writing one contributor assigns to a "Navy publication friend" the recommendation that "first you tell them what you are going to tell them; then you tell them; and then you tell them what you have told them." Another contributor to the same volume attributes this advice to a preacher who developed his sermons by such repetition. And according to our students, some high school teachers strongly recommend that paragraphs be developed this way.

Yet few writers can repeat ideas three times effectively. A paragraph that begins by telling readers that a drug has three side effects and cites convincing proof certainly does not benefit from two or more statements that the drug has three side effects. Perhaps poor students benefit from repeating the main idea three times in each paragraph; at least they may achieve some paragraph unity by reminding themselves of their main subject. But experienced readers and writers do not need three reminders of the topic of each paragraph.

## WRITING THE WAY YOU TALK

A common cause of misunderstanding that leads to poor writing is the advice "Write the way you talk." Writing and speech are different in vocabulary, in grammar, in sentence structure, and in organization. Sometimes a few sentences can be retrieved from the taped discussion of a committee, but who would have the patience to read the repetitive and discursive whole? There is a vast difference between actual conversation, dialogue in a book or play, and other writing. As Chapter 2 illustrates, the professional writing that sounds easiest and most natural has been worked over the most, for writing that is easy to read is a long way from talk. Some few novelists have good enough ears to catch the nuances of talk and to use them skillfully in writing dialogue that reads like talk. But unedited talk is confusing and boring reading.

Some years ago a professional society received complaints about the recording of discussion at its meetings. Irate speakers accused the stenographers of inaccuracy, the editor of incompetence, and the publishers of dishonesty. The society had been recording discussion stenographically, editing it into some semblance of meaning, and sending typed copies to speakers for correction. After many complaints the society recorded the discussion on tape, transcribed it in the pure form in which it had been uttered, and sent the speakers their verbatim comments. Screaming in dismay or denial, they pleaded pitifully for rewriting before publication. The talk of government officials at press conferences and at televised proceedings also illustrates the unsuitability of talk for print—the rambling verbosity, the tendency to backtrack, the inadequate vocabularies, and the failure to state clearly what is meant. Such poor talk does not make good writing.

This is not to say that writers cannot benefit from thinking of their readers in order to direct writing to them. Considering how one would explain a process to a particular mechanic, to a known professor, or to a certain administrator may help in selecting material, in planning the paper, and in expressing the ideas, as the preceding chapters show.

The reason behind the admonition "Write the way you talk" is often sound; the adviser wants writers to adjust their material and their expression to specific readers. And the question "How would you explain this to Jones?" is sometimes useful to writers who have failed repeatedly to explain a thought and are weary of revising a paper: it may free them from their typed sentences and enable them to start fresh. And this may be what they need when a particularly knotty passage will not untangle.

But if writers think that "Write the way you talk" is an invitation to use the vernacular of the shop and the cafeteria, to employ sentence fragments and incorrect grammar, and to wander and repeat as they might in conversation, what they produce is not writing or talk or dialogue. Writing that gives the impression of being talk is not achieved that easily, even by the best novelists and dramatists.

## WORDS, SENTENCES, AND PARAGRAPHS

Readers of this book who avoid the errors and weaknesses described in this chapter are ready to consider the more complex and more satisfying elements of style presented in the following chapters. Any order is suitable. Some may follow their own interests and choose to begin with diction, for example, because words have always aroused their curiosity (Chapters 7 and 10). Most writers benefit from studying Appendix B. Others may be guided by their own weaknesses and study the style of paragraphs first because they have trouble developing paragraphs (Chapter 12). They may review their writing and start with sentences (Chapter 11) because sentence structure presents difficulties for them. Or browsers may read here and there in these chapters to learn a little about sentences, study some words that they use often, and find the answer to a paragraphing problem. Whether writers burrow through the chapters like determined moles or flit back and forth like butterflies makes little difference in their improvement. It is just as easy to improve words, sentences, and paragraphs together as to work on one problem at a time.

Our experience shows, however, that certain instruction quickly helps seekers after better style and pleases them. Those who have not chosen plans for their writing benefit markedly from Chapters 3 and 4. Writers who have not consciously striven for coherence between sentences and paragraphs can gain obvious improvement and satisfied readers from the application of the principles of coherence. Correct but dull writers can apply to obvious advantage the principles of emphasis in phrases, sentences, paragraphs, and the entire paper. Experienced supervisors who have been editing for a long time often improve by reviewing brevity, correct diction, and changing rules. Even young managers may find old-fashioned diction and insistence on out-of-date punctuation and grammar dulling their style and causing problems in editing the work of others. Indeed, all of us have to guard against vogue words, jargon, and gobbledygook in our writing if we meet these contagious faults often in our writing or reading.

Writers should never forget that the purpose of study and review is not to memorize rules and recommendations but to apply them. Writers can always look up a rule that they have half forgotten, but what good is a correct rule in the head and an error or weakness from the same rule broken in the published paper? Merchants can testify that customers buy expensive cosmetic creams and scents that they use once or twice and then throw out. Doctors object to similar conduct of patients who pay well for their doctor's time, buy the prescribed drugs, take a few tablets—and then never take them again. Unfortunately, they seldom throw out the drugs but press them on others who at best may not need them and at worst may be made ill. Many writers have suffered from changes in their reports made by supervisors unloading the drugs they did not take themselves. Our advice to all writers is to concentrate on the rules and recommendations that they need and to measure their progress not by the amount of instruction they memorize but by the improvement in their writing.

## NOTES

1. Georges de Buffon, "Discourse on Style," trans. Rollo W. Brown, *The Writer's Art* (Cambridge: Harvard UP, 1932) 280.
2. Edwin Garrigues Boring, Herbert Sidney Langfeld, and Harry Porter Weld, *Foundations of Psychology* (New York: Wiley, 1948) 485.

3. R. G. Neswald, "Ultrasound in Industry," *International Science and Technology* (Feb. 1964) 40.

4. ANSI Z39.16-1979, *American National Standard for the Preparation of Scientific Papers for Written or Oral Presentation* (New York: American National Standards Institute) 12.

5. ANSI Z39.16-1979, 12.

# CHAPTER 10

# Style and Diction

*. . . and his word burned like a lamp.*
*Ecclesiasticus 48:1*

Once writers have satisfied professional standards for diction by avoiding errors in the choice and use of words, they face the more subtle task of selecting from all the correct words those that fit their thoughts and style most exactly. Finding and choosing the word that is precisely the right one is both an exasperating problem and an enticing pleasure. The fun and the difficulty coexist because English is rich in synonyms. It is so rich that some writers do not take time to view the supply fully.

If writers want to convey, for example, that a committee praised their report, they may choose from many synonyms for praise:

> acclaim, admire, advocate, applaud, appreciate, approve, ascribe perfection to, bow to, boost, butter up for, celebrate, cheer, clap, commend, compliment, defend, defer to, do homage to, endorse, entertain respect for, esteem, eulogize, express approbation of, extol, flatter, give credit to, give a favorable opinion of, hail, hail with satisfaction, have a good word for, have respect for, hold in high esteem, honor, look up to, overpraise, pat on the back for, pay deference, pay homage, pay respect, pay tribute, peal hosannas, praise to the skies, prize, puff, recommend, render a panegyric, render homage, render honor, render plaudits, respect, revere, salute, say a good word for, say it is OK, set great store by, shout applause for, slaver over, sound the praises of, speak encomiastically of, speak well of, speak highly of, stand up for, stick up for, support, think good of, think highly of, think much of, think well of, thunder plaudits, uphold, value highly, venerate, worship.

If none of these expresses the precise meaning of the writers or if they are discussing their own report and wish to sound modest, they may deny the opposite of praise. They may say that the committee did not disapprove their report, did not cast a slur upon, cavil at, criticize, dislike, disparage, disregard, frown upon, fulminate against, make light of, overlook, poke a hole in, rebuff, scorn, slight, speak ill of, sneer at, take exception to, turn up their noses at, undervalue it, view it in a bad light, damn it with faint praise, pick it to pieces, push it aside, run it down, or shrug it off.

They also may say of the report that it accomplished what they wanted, answered their purpose, came off well, came off with flying colors, conquered the committee, gained praise, had good fortune, hit the mark, scored a success, took effect, turned up trumps, triumphed, was go all the way, was well received, went up like a missile, won over the committee, won the day (crown, cup, medal, palm, prize). Or they may express this negatively by stating that the report did not come to nothing, fail, fall short of expectations in its reception, fare badly, flunk, mess things up, misfire, or miss the mark.

## THE WRITERS' INFLUENCE ON CHOICE

What influences their choice? Many characteristics of the speakers or writers—their training, knowledge, intelligence, taste, judgment, personality, circumstances—all influence the decision. If speakers with small vocabularies are conversing, they choose from the few words that come to their mind, however lamely those words may express the meaning. If writers are revising, they may consult dictionaries, books of synonyms and antonyms, or thesauri and choose from a large number of words. If they have sufficient taste to reject a cliché, they will avoid "They praised it to the skies." Of their own work, they will not write, "The president lauded the report," but may say, "It did not go badly." Writing the minutes of a meeting, they may record that the negotiators acclaimed the report that ended a long strike. Speaking at a memorial service, one of them may say of the reports of a deceased president, "The committee venerated his advice."

## THE READERS' INFLUENCE ON CHOICE

Writers naturally and often unconsciously adapt their language to their listeners or readers. They may tell a professor, "The committee spoke well of the report." But when their young sons ask how the report was received, they say, "It was go, go, go all the way." One of them writes to her brother that she hates to hear her colleagues slaver over the president's report. To her colleagues she says only that they are flattering the president. When her son brings home an F on his report card, the writer warns her husband, "Now, when you go to school, don't stick up for him." But she admonishes her assistant, "If you don't check those figures, we won't be able to defend this report." Speaking or writing to professors, brothers, colleagues, husband, son, and assistant, she has adapted the language to the reader or listener. Thus the characteristics of the readers or listeners—their training, knowledge, intelligence, taste, judgment, personality, circumstances, and connection with the writer— also influence the choice. Whether writers are conveying information or

282

sensations, influencing judgment, moving to action, arousing emotion, or evoking a mood, they must consider their readers when they choose their words.

## DENOTATION

More keenly than other communicators, writers on professional subjects must strive for extreme precision in their choice. They are likely to consider denotation, the exact dictionary definition of a word, more important than connotation—the associations of a word, its overtones, its aura. The fact that words like *idiot, moron, neurosis,* and *sublimate* have acquired connotations in general writing does not prevent psychologists from using them without these connotations in the technical writing they direct to fellow specialists. But when psychologists communicate with other readers, the connotations may cause them difficulty; then it may be necessary for the writers to avoid confusion by explaining away the general connotation and emphasizing the definition or by substituting another word or phrase.

Medical doctors face a similar problem daily, because they communicate with their colleagues in technical terminology that might confuse or frighten patients and relatives of patients. Successful communication with colleagues and laymen requires that doctors shift points of view frequently. One who neglects to shift may mislead listeners, especially when a term has a different meaning for a lay person. When an ophthalmologist says, "Good afternoon, Mr. Jones. I just saw your wife; she has a foreign body," he may startle Jones. Specialists become so used to the terminology of their fields that they forget what the terms mean to others. The first paper in chemical engineering assigned to one editor was "Heat Transfer in Mixed Packed Beds." It took the chemical engineers in her office fifteen minutes to understand why she thought the title was amusing.

## CONNOTATION

It is also easy to forget that a technical term has different connotations. No matter how much writers of informational prose are involved with careful selection of terms for precise meaning, they must not forget that there are many kinds of meaning and that one of the most important kinds is the atmosphere that a word conveys to readers. A dog, for example, may be a *bitch, canine, cur, hound, mongrel,* or *mutt.* The word *dog* may have pleasant associations, as in a *dog's life* and *watchdog,* or uncomplimentary connotations, as in *dog in the manger, to go to the dogs, to put on the dog.* And what *bitch* and *dog* mean at the American Kennel Club is not necessarily what they mean at a cocktail party when someone says that Jones has married a bitch or a dog.

Many engineers and scientists would like each word to have a fixed meaning and no connotations except those that they establish. But this is not likely to come about. As soon as people read or hear a word, the word has some overtones for them. Even the dictionary definition—"a domesticated, carnivorous mammal of many varieties"—has bookish or scientific flavor as compared with simple *d o g*. Sensitive awareness of the connotations of words is necessary for effective selection. Whether writers want colorful or colorless words, the problem is the same. They must understand connotations well enough to make a knowing selection. There is a marked difference between a *prize-winning poodle* and a *mop of a lapdog* and between a *visionary egghead* and a *far-sighted thinker*, even if the difference is only in the eye of the writer. Listening and reading help writers to develop the sensitivity necessary for distinguishing between words charged with associations and words that are more neutral and for selecting those appropriate to their writing.

## LENGTH

Many writers are confused by the advice that they should use only short words or by the statement that long words will make their work less readable. It is true that if a short word will function as well as a long one, a writer should generally choose the short one. But there is no point in replacing a polysyllabic word by three sentences of short words. Yet the fallacy makers insist on short words and even insist on counting syllables and checking averages against some magical number or numbers. Although the defenders of this fallacy are fewer than they were at the height of the craze, we still meet writers who think that the only important difference between synonyms is length.

Writers in the professions need not become counters of syllables. Length of words is not important. If readers understand the long words and do not understand the short ones, the long words are obviously the appropriate choice. The pertinent question is Will readers understand the words? Writers must also consider that variety in word length is desirable. Papers composed entirely of short words tend to read like primers, and papers composed entirely of long words look forbiddingly difficult. While revising, writers should strive for variety and should not forget the value of the happy medium.

## EMPHASIS BY CONTRAST

Writers should know one important principle about the length of words. A word that is markedly different—in tone, in quality, in length—from the surrounding words receives emphasis. Writers of expository prose in the professions need not use this device of emphasis, but it is important that they understand it well enough to be able to keep it from operating against them.

EXAMPLES:

The proposed changes and modifications will increase the economy of operation, simplify semiannual budget planning, prevent competitors from bugging our meetings, and improve administrative efficiency. (The slang word *bugging* attracts too much attention in this sentence of routine business phrases.)

He is a conscientious and painstaking assistant but a schmo about people. (The supervisor who wrote this satisfied his conscience by stating two good characteristics, but then he minimized them by using *schmo*. It receives undue emphasis.)

Proofread carefully your typing of formulas. Formulas that are inadvertently erroneous may ruin tons of chemicals. (The long and pompous *inadvertently erroneous* distracts attention from the main point. *Errors in formulas* would be more appropriate.)

This catalogue exemplifies the administrative confusion, unnecessary expense, and injury to the reputation of the institution occasioned by the incorrect positioning of one device of punctuation. (The short specific phrase *one comma misplaced* instead of *the incorrect positioning of one device of punctuation* would contrast effectively.)

## NEVER REPEATING A WORD

It is usually wise to avoid repeating unimportant words. But important words may, and even should, be repeated. Skillful repetition is useful and effective: it provides coherence and emphasis. The naysayers would have writers avoid repetition by using synonyms and synonymous phrases. This technique, known as *elegant variation*, is, in spite of its pleasant name, undesirable. It soon leads a writer to high-sounding terms. A *cat* in the first sentence becomes in later sentences a *tom cat* or a *grimalkin*, a *mouser*, a *feline being*, and *felis catus*, not to overlook *puss, pussy, kitty,* and *kitten*. Thus elegant variation encourages pseudoscholarly, flowery, and extravagant diction. If a writer wishes to avoid repeating a word or phrase, he should know that pronouns serve better than roundabout expressions and inexact synonyms.

The preceding objections to awkward avoidance of the repetition of words and phrases concern mainly taste and judgment, but writers of functional prose have another objection to elegant variation—the shortage of exact synonyms. Using an inexact synonym may mislead readers. Even a more general or a more specific term is undesirable when, as often happens, it confuses. If engineers write about Experiment A taking place in a tank and Experiment B in a unit, readers assume that they are using different equipment, not elegant variation.

## NEVER ENDING WITH A PREPOSITION

The taboo against ending with a preposition is a dimly perceived point of emphasis incorrectly applied. Near the end of an English sentence a major stress falls, sometimes on the last word, sometimes on a word just before the last word, sometimes on the final phrase. For effective emphasis, the word or phrase stressed should be important:

> He authorized us to spend ten thousand dollars. (The stress is on *ten thousand dollars*.)
> He emphasized the final point. (The stress is on *final point* or *final*.)
> She said that she would complete the work on Monday. (The stress is on *Monday*.)

Careful writers avoid stressing an unimportant word, like a preposition or adverb, particularly an unnecessary word:

> These pipes are difficult to connect up. (*Up* is unnecessary and should be deleted.)
> Everyone thought it a poor way for the speech to end up. (The stress falls on the unnecessary *up*.)
> This is the drawer to put cancelled checks in. (The stress falls on *in. This is the drawer for cancelled checks* stresses *cancelled checks*, and *Put cancelled chekcs in this drawer* stresses *this drawer*.)
> He said that after he returned from California he would contribute two hundred dollars however. (The stress falls on *however*. If *however* is placed after *he said*, it performs its proper function of connecting sentences—which it cannot perform well at the end of a sentence—and the final stress falls on *two hundred dollars. He said, however, that after he returned from California he would contribute two hundred dollars*.)
> Baton Rouge is the division of the company that he is coming from and Seattle is the division he is going to. (This might be sensible emphasis if there had been a disagreement as to whether he was coming from or going to Baton Rouge or Seattle. The *to* would be stressed, and *to* would be important. Otherwise the sentence has better emphasis if it is rephrased: *He is coming from the Baton Rouge division of the company and going to the division in Seattle*.)

But in many a sentence that ends with a preposition, the stress falls on the word before the preposition. If that word is important, there is no need to avoid ending with a necessary preposition or adverb:

> He is a difficult person to disagree with. (The stress falls on *disagree*, and *disagree* is an important word in the sentence.)
> Children should have bright objects to play with. (The stress falls on *play*, an important word.)

## NEVER SPLITTING AN INFINITIVE

Another shibboleth—never splitting an infinitive—arises from a similar misunderstanding. The natural place for the sign of the infinitive is next to the infinitive, as in *to come, to understand, to demonstrate*. A word or phrase between the sign of the infinitive and the infinitive disturbs normal word order, and a word out of its usual place in the sentence attracts stress: *to complain surreptitiously* is the normal order, and *to surreptitiously complain stresses surreptitiously*. Obviously there are occasions when a writer might want such stress. A supervisor discussing two employees—one who had warned the supervisor that he was taking a complaint to a vice president and another who complained secretly to the vice president—might write that he regretted that the second employee found it necessary to surreptitiously complain to the vice president. Thus he would emphasize the distinction between complaining and surreptitiously complaining.

If writers do not realize that splitting an infinitive emphasizes the word or phrase after *to*, they may accidentally stress a word that they do not want emphasized and thus confuse or mislead readers. Good writers split infinitives when splitting is desirable for emphasis and when avoiding it would be awkward. But splitting an infinitive can also be awkward, particularly when several words come between *to* and the infinitive:

> He attempted to by every means gain promotion.
> *Improved:* He attempted to gain promotion by every means.
> By every means he attempted to gain promotion.

> His motion to carefully for three months examine these findings was passed immediately.
> *Improved:* His motion to examine these findings carefully for three months was passed immediately.

Some split infinitives are redundant. Careless writers tend to insert trite intensifiers or qualifiers between *to* and the infinitive, as in the following examples: *to actually realize, to better know, to clearly understand, to definitely believe, to really comprehend*, and *to virtually have*. Unnecessary adverbs, like other unnecessary words, should be deleted during revision (Chapter 13).

## NEVER BEGINNING WITH . . .

To avoid beginning sentences with certain words is another common taboo. Writers may begin sentences with any words that they like. The first word or words in a sentence are usually stressed, and they should be important words. But occasionally even the much maligned *however* may be important because a writer wishes to emphasize for a reader or listener that an unexpected shift in thought follows. *But* and *and*, which are also listed as forbidden first words by some teachers who teach *don't's* instead of *do's*, seldom are stressed when they introduce a sentence; and they are, therefore, useful, unobtrusive initial conjunctions.

Writers who have been avoiding them should notice how often and how effectively *and, but, for*, and other short conjunctions introduce good sentences in modern prose. Writers in the professions might use them more often to begin sentences. They should take care, however, to avoid placing the same unimportant word at the beginning of so many sentences that it becomes monotonous. Sentence variety, essential for an interesting style, may be defeated by monotonous sentence beginnings ("Variety in Beginnings," Chapter 11).

## ABBREVIATIONS, ACRONYMS, AND CLIPPED WORDS

While revising, writers should remove clipped words, like *fridge, hypo, mike, perm*. At best such words impede the smooth flow of reading; at worst they are ambiguous, like *hypo* (*hypochondria? hypodermic? hyposulfite?*). All are inappropriate except in informal style: for example, in a note on the kitchen chalkboard.

Some abbreviations are acceptable: *Dr., Mr., Ms, Jr., A.M., Ph.D.* Within each specialization of science or technology, abbreviations peculiar to that division are used. The American National Standards Institute issues lists of many of these abbreviations and advises on using or omitting periods. For clarity and for the goodwill of readers, writers should be careful not to carry the abbreviations of one specialization to another. In the rare instance that a carryover is unavoidable, the term should be spelled out the first time it is used.

For readers unfamiliar with the initials alone, acronyms should also be spelled out at first use. Some acronyms are known to most readers (NASA, snafu, OPEC) and require no explanation; nor do acronyms of a special field when they appear in writing specifically designed for that field.

Unfortunately, the use of abbreviations and acronyms common to a particular specialization results frequently in papers that are difficult or impossible for readers in other fields. Thus specialists are discouraged from reading outside their fields. It is a pity, sometimes a tragedy, to discourage such reading and thus lose the benefits often resulting from cross-fertilization.

In any case writers should avoid overuse of shortened forms. It is confusing and irritating to attempt to read pages sprinkled with initials. When the forms are used, however, they should be used consistently.

## FANCY WORDS

Sometimes fancy words are confused with long words. Fancy words, short or long, are undesirable. They are used by writers who display language. Such writers *initiate* work rather than *begin* it; they *activate* a project rather than *start* it; they *proceed* to the administration building rather than *go* there; their work *comprises*, not *is*, research and development; and they earn *compensation*, not *pay*. In these examples of writers trying to be elegant the falsely elegant words are long, and the natural ones are short. This is often true, but there are also fancy short words, like *doff, dwell, prior.* The fancy word is the pretentious word, the pseudoscientific, would-be-learned word. Sometimes a word is fancy because it is out of place. It is ridiculous for a worker to report, "They commenced waxing the floors at ten o'clock." It is acceptable for him to state, "The procession commenced at ten o'clock when my daughter graduated from college." Many misguided writers, conscious of their inadequate vocabularies, cultivate pompous language; and some professors in graduate schools and some supervisors in industry and government deplore simple, natural language and train writers to use the stuffy language of pedants, the clichés of business, the gobbledygook of government, or even incongruous combinations of all three.

> EXAMPLE: Adumbrating that the complaints of the siblings of exceptional children would commence with concomitant discontinuance of overall parental manumission of affections, the speaker asseverated the diurnal requisite of minimizing the antipathy of the latter while potentiating the amelioration of the former by gratification at the advance and development of accomplishment.

Writers who use words to fill a page or to attempt to impress readers instead of to convey meaning are in danger of writing sentences like "Unfortunately the kind of case that causes trouble in practice is that in which the nature of the use made of language is not of a transparently clear type." If writers dot their work with *case, character, condition, degree, instance, line, nature,* and *type,* they should remove them when they revise. As pointed out in Chapter 13, these words are characteristic of a weak, diffuse style.

Writers who choose fancy words instead of simple words may find that their fancy overblown style is inaccurate. Even a professional writer's attempts to use elegant language often end in malapropisms. The *New York Times* writer of a signed special article from East Hampton, Long Island, blundered thus: "The houses of East Hampton generally dissuage visitations, but guests on today's tour will see. . . ."

## FOREIGN WORDS AND PHRASES

Some writers err in considering it elegant or learned to decorate their writing with unnecessary foreign words and phrases. Foreign words are appropriate when they add associations to meaning or when there is no good English equivalent. But the need for foreign words occurs less often in informational prose than in more atmospheric writing. And writers often misuse and readers often misunderstand foreign words. *Versus,* or *vs.,* is frequently misused for *and, or,* and *compared with.* And many professional people misread *i.e.* and *e.g.;* therefore when writers use these abbreviations to save time, they may waste it. Even *etc.* is misused; the items that precede *etc.* should make the category clear, but many writers place before *etc.* two items that belong to more than one category. "Such candidates are found at Harvard, Dartmouth, etc." does not clearly identify the category.

## CLICHÉS

Stereotyped phrases, which appear in informational prose as commonly as fancy words, cover an absence of thought. The user has borrowed not merely some worn-out words but an often-expressed thought. Perhaps the wording or the thought was striking and effective when it was fresh. The first lover who told a woman that her lips were like cherries and her cheeks like roses is a different thinker from the man who borrows the phrasing millions of users later. The first lover was probably honest, imaginative, and perceptive; the borrower is probably dishonest, lazy, and unperceptive (the lipsticks and rouges of today provide more subtle shades). Some clichés are old—*acid test, at a loss for words, believe it or not, better half, better late than never, beyond the shadow of a doubt, bitter end, conspicuous by his absence, equal to the occasion, few and far between, field of endeavor, filthy lucre, first and foremost, goes without saying, in the last analysis, the irony of fate, it stands to reason, last but not least, no sooner said than done, on the ball, pinnacle of success, sadder but wiser, sneaking suspicion, words fail to express.* Some are newer (a word may be overworked in a few years or even in a few months)—*at this point in time, the bottom line, drastic action, elder statesman, Iron Curtain, pending merger, put teeth into.*

Whether old or new, a hackneyed phrase betrays a user's willing acceptance of a vague approximation of his thought, if indeed he has one. Cliché writers use the phrase *the foreseeable future* without a thought to how unforeseeable the future is. Frank Sullivan's cliché expert. Dr. Arbuthnot, uses old and new worn-out phrases when he testifies, and his ideas are as hackneyed as his words. Given a question, he supplies the answer automatically, whatever the subject. In talking about nuclear energy, for example, Arbuthnot uses "usher in the atomic age," "prove a boon to mankind," "spell the doom of civilization as we know it," "boggles the imagination."[1]

Some authorities find clichés less objectionable in conversation than in writing, but the advantages of keeping one's mouth shut when one has nothing new to say and no fresh way to say it should not be ignored. We find clichés particularly objectionable in political oratory, where they flourish. And the after-dinner speech stuffed with familiar phrases is always painful. Beginning with "unaccustomed as I am" and running through "which reminds me of a story," the after-dinner bore finally reaches "and so in conclusion" half an hour before he delivers the final trite phrase.

However, one can give new life to a worn-out phrase by twisting it wittily or comically. In a formal speech such turns of familiar phrases are best when they are unobtrusive. In extemporaneous talks the effects of such twists are heightened by the spontaneity; for example, a speaker at a dairy convention replied to a question about methods of successful dairy farming, "Well, I like to start with a few well-chosen herds." A girl listening to her family argue at length in favor of a Sunday picnic remarked, "It takes you a long time to agree to agree." A banker who declined to risk his bank's money on a world's fair observed, "None but the brave preserves the Fair." And when an office weeper was given a raise in salary, his coworker quipped, "There was method in his sadness." This device is effective in the title of a book or article and particularly useful if most of the suitable titles for the subject have been used. For example, in July of 1951 *Chemical and Engineering News* published the first of a series of articles by Robert L. Dean on writing. Nearly every possible title for that subject had been used more than once, but Dean thought of a twist. He called his series "Watch Your Language."

## EUPHEMISMS

Euphemisms, which soften unpleasant or offensive concepts and attempt to raise the lowly, are, like other lies, undesirable except when they are indisputably necessary. The word *die*, for example, is simple and dignified. It is far better than many of the substitutes for it: *to cash in one's chips, to pass away, to pass over, to pass to one's reward, to go beyond, to depart this world, to expire, to enter the valley of the shadow, to come to an untimely end, to perish, to be taken, to resign one's breath, to give up the ghost, to end one's days, to breathe one's last, to depart this life, to join the great majority, to kick the bucket.* The number of such euphemisms illustrates that words designed to soften reality quickly lose their veiling power and are replaced rapidly. The poor become the *underprivileged, those in want,* the *deprived,* the *disadvantaged*; the aged are the *elderly,* the *mature,* the *old folks, those who have had their day, our seniors, the patriarchs, the senior citizens.* To those who are sensitive to language as well as to death and age, a euphemism often seems worse than what it is trying to soften. Besides, it soon becomes trite.

## VOGUE WORDS

A writer should not think that odd words or odd forms of words that are in vogue lend freshness. "Praisewise my report did all right" contains a faddish form that is just as trite as "praised to the skies." Overuse of a form like the *-wise* suffix is common in government and industry, and we object to it most strenuously there because it is weak from abuse by the time it arrives.

An editor first heard about the fad of *-wise* suffixes when she was abroad in 1951. "Wait until you get home and read, 'Parkingwise I had trouble,'" teased a history professor who had just come from Washington. On returning to the United States, she found her writers talking of "improving *wordwise*," and in books on industry she read, "*Specificationwise* we wish to state . . ." and "*Psychologicallywise*, it is clear. . . ." Soon afterward the author of a satirical article expressed the feelings of many readers by concluding that "wordwise" he was "fed-up-wise." Years later at a performance of *How to Succeed in Business Without Really Trying*, the query "How is he doing successwise?" drew a laugh from the entire audience. Nevertheless much later when we removed *medianwise, budgetwise, batchwise, vatwise,* and *responsibilitywise* from a technical report and told the writer, "In addition to being ugly and unnecessary that -*wise* suffix has been overworked for at least thirty-five years," he responded sheepishly, "I thought it was a new way to put it."

Writers in the professions should be sparing in their use of vogue words. They should be especially wary of those that have one scientific meaning and another popular meaning, like *allergy, exhibitionism, neurotic, sadism, subliminal.* Employing faddish words reveals, as the use of hackneyed phrases does, a failure to think, laziness, evasiveness, or a pathetic and belated attempt to seem up to date.

The vogue words *overall, type* as an adjective, and *hopefully* are overpopular in technical writing. *Hopefully* with the meaning *it is hoped that* is used to avoid stating who is hoping, but *hopefully* also means *in an optimistic manner* and *with agreeable expectations.* And so when a user of vogue words writes, "Hopefully management will meet with the union leaders again tonight," he means *it is hoped that*, and he is expressing some doubt that the meeting will occur. But some of his readers will think that the sentence means that management has agreeable expectations, that it expects a good result from the meeting. In German *hoffentlich* does not confuse readers, but *hopefully* used like *hoffentlich* may confuse readers of English. A precise writer avoids the careless use of *hopefully*, and a careful reader is supicious of it.

We once heard an administrator say to a project head, "But you wrote to me that you hoped that the results would be complete this month."

"No, I said I doubted that the results would be complete this month."

"Here's your memo. Look at your last sentence: 'Hopefully the results will be complete before December.' Now why did you write that if you didn't expect to be ready?"

"That's exactly why I wrote it. It means, 'It is only hoped that the results will be complete.'"

When we left they were still arguing. The project leader, who had tried to avoid trouble by using the ambiguous *hopefully*, was in trouble anyway; and the supervisor had learned to query every *hopefully*. He tells us that he has heard ten different meanings assigned to it by ten members of his division, and hopefully, or perhaps despairingly, he expects to hear more.

*Type*, another offender, should be used as an adjective only in technical expressions, like *A-type blood.* In other expressions the adjective *type* suggests the speech of gangsters, as it does in *marriage-type love* and *delicatessen-type sandwiches*, or telegraphic style, as in *an IBM-type employee*. Sometimes it is redundant: *a future-type bonus, detergent-type cleansers, a wise-type decision*. We have even seen *an electric-type typewriter*.

*Overall*, which means *from one end to another*, is popular as a synonym for *total* and *general*. We recommend that writers of functional prose use it only for the first meaning and use other words, like *total* and *general*, for the other two meanings. *Overall* is likely to appear five or six times on one page of industrial writing and to have one meaning in the first sentence, another in the second, and sometimes no meaning at all. A mechanic wrote, for instance, that an engine needed "a general overall overhaul of all its parts." Hopefully it will be totally overhauled overallwise.

## VAGUE WORDS

Closely related to the vogue is the vague. Words that convey too many meanings convey none at all. Some girls use *nice* to describe their fiancés, their aerobics class, their hairdressers, and their new shades of nail polish. And a chemist may say that the apparatus for her experiment is fine, her supervisor is fine, her vacation was fine, and it will be fine to send out her letters tonight. Adjectives like *real, good, bad, horrible* convey little meaning when they are used loosely: *horrible cooking*, a *horrible movie*, a *horrible accident, horrible language*. A word that has fewer meanings says more, but if the search for a suitable word requires effort, there is a temptation to revivify a stock adjective by adding an intensifier— *very nice, real fine, definitely horrible, actually real*. Such overworked intensifiers not only lose their strength but become weakeners. Writers in industry and government must try harder than writers in other fields to avoid vague words and overworked intensifiers because they hear and read them often. Danger words are *actually, definitely, exceedingly, exceptionally, extremely, really*, and *very*. These are often used carelessly and thoughtlessly, as in the memorial resolutions that referred to an executive's "very fatal final illness."

## TECHNICAL LANGUAGE

Technical language is indispensable in conveying ideas to readers in the same specialization. It is the simplest, clearest, quickest road to precision. Much technical language has come into English in this century, and many writers use it where it is unnecessary and even confusing. A careful reviser replaces unnecessary technical language with nontechnical language when it is possible to do so and retain brevity and clarity.

One of the characteristics of poor writing in all the professions is a tendency to employ terms from other specializations. Chemical engineers, for example, should consider their readers and refrain from adding to the technical language of chemical engineering legal expressions like *aforementioned, hereinafter, versus,* and *abovementioned.* Business administrators should not use *in the order of magnitude of* for *about;* it means something else. The language of one specialization is enough—often too much. Readers can hope that the time will come when leaders in the professions will take an interest in removing some jargon from their specializations to improve technical language. There must be better ways of saying *liquid–liquid extraction, situational variables in interpersonal verbal interactions, maximization and minimization of linear functions subject to linear constraints, postdepositional and detrital origin,* and *cloud-track pictures obtained in uniform field spark breakdown experiments with timed square-wave impulse potentials at near atmospheric pressures.* Until the learned societies assume their aesthetic, and possibly their ethical, responsibility for improving language, the best that writers can do is use the technical terms of their specializations sparingly and avoid the needless use of technical language from other specializations.

## GENERAL AND SPECIFIC WORDS

A problem in style that troubles many writers—the use of general and specific words, seldom causes difficulties for experienced writers on professional subjects. They choose the word that is appropriate to their meaning—not the too-general word which blurs meaning and not the overly specific word that is pedantic or fussy. But beginners may be lazy about seeking a specific word that does not come quickly to mind and substitute an ineffective general term. In the *Handbook of English in Engineering Usage* A. C. Howell warns writers not to say *contrivance* for *motor-generator, machine* for *apparatus, bottle* for *Erlenmeyer flask, a thing to measure angles with* for *protractor.*[2] And even experienced administrators, scientists, and technologists do not always realize that specific words can lend vividness and precision to their writing on subjects outside their profession. "An assistant who investigates thoroughly, anticipates needs, and follows through" is a clearer description than "a competent, efficient assistant." A request for "a report on the training division's use of visual aids, such as samples, pictures, videocassettes, slides, filmstrips, and teaching machines," will elicit a better report than will a request for "a report on what they're doing in training with some of these new teaching things that help you to see what you're learning." An executive who is inclined to use slovenly diction and to rely on meaningless general words like *thing* and *gadget* should take the time to replace vague words by specific terms. The incompetent spoken requests of many executives suggest that if they were surgeons they would say not, "Scalpel, Nurse," but "Will somebody give me that thing over there."

294

## ABSTRACT AND CONCRETE WORDS

T. A. Rickard ably illustrates another weakness of technical writers—the unnecessary use of abstract language:

> The hankering for the abstract is exemplified by the vogue attained by "value" and "values" in mining reports. In a stope or in a mill the use of "value" in this way may cause no confusion even if it be an objectionable colloquialism, but in technical writing it should be taboo, as the very type of all that is nondescriptive and unscientific. "This mill is intended to catch the *values* in the ore" is a vague way of stating that the mill is designed to extract the gold or silver, the copper or the zinc—in short the valuable metals in the ore. In one mill the zinc, for example, may be not only valueless but a deleterious impurity; in another the copper may be insufficient in quantity to be extracted profitably, but sufficient to interfere with the saving of the gold by cyanidation. "Value" is the worth or desirability of a thing; it is an attribute, not a substance. A man who designs a mill "to catch the values" might as well build a railroad to pursue a quadratic equation.[3]

Just as specific words are often more effective and precise than general words, concrete are more effective and vivid than abstract. Superfluous abstract words make writing fuzzy or meaningless. An appropriate use of concrete terms, even in writing on an abstract subject, enlivens and clarifies. George Holbrook uses concrete, specific diction and concrete, specific examples to develop his abstract thought in the following passage. He begins with an abstract idea, the importance of chemistry, enhances it with concrete illustrations, and concludes with another well-founded abstraction:

> Of all the sciences, none penetrates so deeply into the structure of the human environment as chemistry. Chemical processes and principles are basic to life in all its phases. They appear in such familiar activities as building a fire, raising crops, preparing food, relieving illness, and in producing virtually every commodity of the commerce which has marked civilization's ascent from the cave.
>
> For the better part of six thousand years of recorded history, however, men established their living conditions very much on the products of nature as they found them. They built with stone, wood or clay; they wove cloth from cotton, wool or silk; they made shoes, harnesses, and saddles from animal skins.
>
> Gradually, they began to make changes in the products of nature to satisfy their desire for a better way of life. They won metals such as copper and iron from natural ores, made concrete, paper, and gunpowder. In the course of the past half century or so, this innovating process has been most radically accelerated. The natural directions of chemical reaction have been replaced or augmented by a series of induced reactions which rearrange the limited raw materials of nature into endless patterns of usefulness. Now we build with metal alloys, glass, hundreds of plastics; weave nylon and

other synthetic fibers; manufacture countless dyes and pharmaceuticals; make fuels, rubber, automobiles, airplanes, and spacecraft largely from synthetic materials.

The process through which these opportunities have been seized and the benefits enlisted to the service of mankind is now recognized as chemical engineering, one of the newest of the professions.[4]

## CLARIFICATION OF WORDS AND INFORMATION

### Brief Interpretative Comments

Words that require interpretation, such as *appreciable, appreciably, considerable, considerably, relative, relatively*, may confuse readers. An appreciable difference to a microscopist may not be appreciable to an artist, a considerable amount in chromatography may be inconsiderable at a refinery, and a relatively mature student may be an immature employee. To convey thoughts accurately, it is necessary to specify how large a difference, how great an amount, how mature a person. If one is writing on specialized subjects for nonspecialists, explanations of the significance of the amounts are essential. An effective technique to use for all readers is describing the amount meaningfully: *an unexpectedly large yield of two grams; a difference of 1.4, surprisingly small; two of the ten candidates— the usual proportion; even a new employee; three exceptions in the thousand instances, not enough to affect the conclusions; ten complaints from customers, and we usually investigate when there are as few as two*. Professor Reginald O. Kapp illustrates this well in *The Presentation of Technical Information:*

> You can point to the conclusion that is to be drawn from the information by the addition of words like "fortunately" or "unfortunately", "surprisingly" or "as is to be expected". Other, more specific, more thoroughly informative words can be even more helpful. I can imagine a situation in which it would be good to say, "and yet the points all lie on a straight line", while it would be even better to say, "and yet the points all lie, *deceptively*, on a straight line". Similarly it might be good to say, "the temperature reaches the high figure of 750°F", and better to say, "the temperature reaches the *dangerously* high figure of 750°F".[5]

No doubt some scientists and technologists think that these comments are not so impersonal and objective as they have been taught that their writing should be. But there is a difference between a comment based on an expert's thoughtful interpretation of evidence and a personal, subjective comment. Moreover, wishy-washy, evasive statements, however impersonally they are expressed, betray timid, overcautious, self-protective attitudes. A chairman of safety who writes to a committee, "Eight accidents seem due to faulty inspection; it has been suggested that there be an investigation," has tried to evade responsibility. A less equivocal statement is more objective: "The insurance investigators attributed eight accidents to faulty inspection. This large number clearly warrants an investigation by the Safety Committee." The comments *large number* and *clearly warrants* are not personal, subjective statements. They are the clarification of a responsible expert interpreting evidence for others. But to introduce a suggestion feebly with "it has been suggested" when an investigation is the normal procedure that a chairman of safety should demand is so misleading as to be dishonest.

A reader seeing "it has been suggested" may justifiably conclude that the safety chairman is not backing an investigation. Thus this objectively worded statement is intentionally or unintentionally highly personal and subjective; it is obviously the work of a fence sitter who should not be trusted with any decision less objective than how many pencils are in the storeroom. Unfortunately some experts are deterred from conveying their knowledge and experience to readers because of a mistaken belief that all comments are personal and subjective. Others write timid statements with premeditated self-protection. "I'm not sticking my neck out," such writers will state unofficially, although scientific objectivity is their official excuse for depriving their readers of judgments that the writers are being paid to present, judgments that might save time, money, and lives.

## Longer Explanations

A brief comment leading the reader to a correct interpretation of numbers and other technical information will often spare both writer and reader lengthy, inconvenient explanations. But sometimes longer explanations are necessary, and they are necessary early. For example, the terminology of a technical subject presents problems to writers of popular articles, who must write clearly and simply for readers who are not experts. The best writers for the general public use nontechnical language whenever they can; and when they cannot, they define or explain as they first present the technical term or concept. They do not use a term three or four times and then explain it. By then the reader may have stopped reading. They place the explanations where readers need them, make the information unmistakably clear, and interpret when necessary. The exposition must be as long as needed to make readers understand. An explanatory passage shortened so much that readers do not grasp the points wastes the time of writers and readers. Writers also must be careful to polish explanations as well as they polish other sections. Some writers become bored explaining points they know well themselves and skimp on the writing and revising. Usually this results in readers being unable to understand the paper and rejecting it.

## Comparisons and Definitions

Good writers for the general public never forget the value of analogies with familiar concepts for explaining the unfamiliar. Glancing comparisons are also often helpful—a factory machine compared to an eggbeater, company benefits compared to an insurance policy, a book of laboratory instructions compared to a cookbook. The comparisons both clarify and vivify the concepts. The important point in comparisons and analogies is that the familiar concepts should be familiar not merely to the writer but also to the reader.

Necessary definitions and other explanations should be expressed clearly. Some specialists write merry-go-round definitions that leave readers confused: "An appositive is a word in apposition"; "nonrepresentational—not representational"; "revocation, or the act of revoking." Specialists who are prone to fine writing (unnecessarily ornate writing or pretentious writing) define words by using more difficult words as though in emulation of Samuel Johnson's definition of *network*: "anything reticulated or decussated at equal distances with interstices between the intersections."

### Achieving the Desired Emphasis

The comparisons and definitions necessary for a reader's understanding create problems in emphasis when the concepts clarified and the terms defined are not important. Then the amount of space devoted to explanation or the vividness of comparison may inappropriately emphasize a minor idea. What can writers do to restore proper emphasis? Usually one of the following techniques works:

1. Eliminate a word or concept that is only partially relevant to a main idea. Close examination of words and concepts that require lengthy explanations often shows the words and concepts to be unnecessary. Removing them restores emphasis to main points and improves paragraph unity. In many paragraphs the topics become clearer and sharper when material that is not completely germane is removed. The thought or the word may have captured a writer's interest, but consideration may convince him that the reader is better off without it—and so is the writer's paragraph.

2. Many necessary definitions may be slipped unobtrusively into dependent clauses and phrases if writers do not wish to stress them. Instead of using a paragraph or a sentence as they would for material they wish to stress, writers may place the definition in a phrase, a clause, or a parenthetical expression:

> The company, including the New York Office, the Trenton Plant, and all foreign subsidiaries, would be a desirable purchase only if ten million dollars were available for construction and repairs.
> The company, that is the New York Office, the Trenton Plant, and all foreign subsidiaries, . . .
> The company (the New York Office, the Trenton Plant, and all foreign subsidiaries). . . .

A book review in *Science* subordinates a term in an *or* phrase instead of giving it a sentence to itself:

298

> ... is an excellent selection of important articles in the general and not too well-defined area of anthropological linguistics, or linguistic anthropology as Hynes prefers to call it.[6]

In an article in the same issue of *Science* an appositive and its modifier convey information without overemphasizing it:

> It is remarkable that carbon monoxide, a known toxicant constituting 4.2 percent of tobacco smoke, has not been considered as an explanation of the diverse effects of tobacco smoke on the circulation.[7]

3. Sometimes definitions in a separate section—an introductory list or a list in an appendix—may be necessary. If a paper requires a number of definitions or explanations for words or concepts that are essential but not important and if writers can find no way of introducing them unobtrusively into the paper without shifting emphasis away from the main thought, then a glossary may be desirable. The need for one seldom occurs, and writers planning such a section should be convinced that it is necessary.

For example, if we were writing on standards of diction to undergraduates who have little knowledge of grammatical terms, we would consider appending an alphabetical list of definitions of grammatical terms. But because our readers have been educated in professions, we assume that some will immediately understand a term like *appositive*, others can recall it, and some few who do not know and cannot recall it can use a dictionary. For a phrase like *fine writing*, however, we provide a definition because the phrase might mislead those who do not recognize it as a technical term. The decision should be based on our judgment of whether readers need the definition, not on our wish to teach or not to teach grammar, to expand or shorten our book, or to impress our readers.

### Necessary and Unnecessary Definitions

In industry and government scientists, technologists, accountants, lawyers, and consultants of all kinds writing for management have problems very close to those of a writer of popular articles. The difference is mainly that they are writing for more experienced readers who are familiar with some terms. This means that a decision as to whether a technical term should be defined or a process explained becomes more difficult. Sometimes personal knowledge helps. Writers may remember from committee meetings that their readers are familiar with some terms, or they may know from past reports that the terms have been explained so often that their readers understand them. When they lack this knowledge, their best decision is to explain. Technical reports in industry and government often contain unnecessary details, but they seldom offer too much explanation of technical matters for readers without technical training. Few readers in industry or government object to explanations and definitions.

If readers find one unnecessary, they need the next one; and anyway, they are inclined to skim the ones they do not need. If they give the unnecessary ones a thought, they feel superior because other readers apparently lack their knowledge.

A writer preparing her first financial report for a company consulted us about supplying the definitions of some terms. "I did it where I worked before this, but I am not sure whether this management needs the definitions," she explained. "Many of these terms do not appear in reports previously submitted to management here." We advised her to include whatever definitions and explanations seemed essential and to note the reactions. She reported that most of the readers did not react at all. But one division manager said, "I'm glad to see I know some of the financial terms that others don't know." And a vice president told her, "This is the first time I've been able to understand an annual report in this company. I always had to get someone to translate financial reports for me until yours came along."

## The Problem of Condescension

Writers sometimes worry too much about seeming to condescend when they explain, and some worriers insist on apologizing for offering explanations or on detailing their reasons for explaining. Expressions like *as you know, as you no doubt remember, as you may recall*, and *of course you are aware* irritate some readers.

One executive wanted to know, "Why do they write, 'as you no doubt recall,' and then spend a page and a half recalling it to me? Either they think I recall it, or they think I've forgotten it and need to be reminded. But they can't have it both ways."

"Do you usually recall it?" I asked.

"Usually it's something I never heard of, and for one report in five it's details I don't need in order to understand the report and will never need."

Writers in industry and government should courageously include the explanations necessary to their readers at the point where the readers need them. The authors should phrase the information in the briefest manner consistent with clarity and refrain from apologizing. About unnecessary details there is only one word of advice: omit.

## FIGURES OF SPEECH

By using figures of speech John Donne presents the abstract subject of brotherhood in memorable concrete terms (See "Allusion" below.) Many writers in the professions who are interested in other qualities of language never give much thought to figurative language. They consider it the province of poets and never notice how often the best writers in the professions find it useful. By omitting consideration of figures of speech, books on business writing and technical writing encourage writers on business, science, and technology to ignore the techniques of using rhetorical language, and so uninstructed writers use poor figures of speech for inappropriate material, omit figures where they would be of value, and even misread them in the works of others.

Everyone uses some figures of speech—*break the ice, cold shoulder, family tree, red tape, skeleton in the closet.* Every writer should know enough about them to avoid common errors, to read them correctly, and to use them effectively. Figurative language properly employed can clarify, strengthen, and vitalize at the same time that it lends a welcome touch of imagination. The principal figures of speech as well as the advantages and problems of using them in the prose of information follow.

## Allusion

An allusion is a brief, passing, casual reference to a well-known person, organization, place, event, or literary passage. An allusion is indirect; a reference is direct. Writing, as they usually do, for educated readers, professionals may use many allusions. But their writing should not seem studded with references, for readers consider such writing pretentious. Successful allusions are not hackneyed. Even the best literature does not stand up well under incessant quotation, as witness "to be or not to be." A reference to "Pavlov's dog" to explain habit is equally trite; so are allusions to Waterloo and to St. Patrick driving the snakes out of Ireland, quotations like "a plague on both your houses," and proverbs that are too familiar, like "a stitch in time." A successful allusion is woven neatly into the writing and introduced without fanfare.

Timothy Ferris uses allusion effectively in two passages: "Theoretical particle physicists ponder questions that, by comparison, make Alice in Wonderland seem as mundane as a baseball game in Kansas" and "Such tests must of necessity be highly derivative, like hunting for planks from Noah's ark."[8]

For an allusion to be effective, it must be familiar to the readers to whom it is directed, it must be accurate, and it must enable readers to see "a world of meaning in a grain of sand." It does not matter whether the allusion offers penetrating thought, evokes emotion, or creates an atmosphere. In informational prose allusion serves best when it is terse and pregnant. Sometimes the allusion to well-known words, often from Shakespeare or the Bible, may capture attention by playing on a word of the quotation, as David Brower, the famous defender of the wilderness, does in the title of his article "The Meek Shall Lose the Earth."[9]

Allusions within the family to family jokes—the company joke at a dinner for a retiring officer, the college joke at a commencement, the engineering joke at a convention of engineers—succeed because an audience feels comfortable when it recognizes the familiar; an audience in Columbus, Ohio, for example, is pleased by a speaker's allusion to the latest Ohio State football victory. But the mention should be a passing one; attempts to build a successful family allusion into a successful paragraph usually fail. And speakers who are using an allusion lightly to win their audience or to bring themselves into harmony with their audience should choose a pleasant allusion; families are not delighted by references to their mistakes or their black sheep. But allusions, unlike puns, are not restricted to the aptly pleasing or entertaining; they are more likely to be serious.

Some allusions are echoes of phrases, characters, titles of literature. Titles may spring from the writing of another; John Donne's words, for example, have been suggestive:

> No man is an island, entire of itself; every man is a piece of the continent, a part of the main: if a clod be washed away by the sea, Europe is the less, as well as if a promontory were, as well as if a one of thy friends or of thine own were; any man's death diminishes me, because I am involved in mankind. And therefore never send to know for whom the bell tolls; it tolls for thee.

The titles *No Man Is an Island* and *For Whom the Bell Tolls* are much richer in meaning when one knows their origin. And this is the advantage of an allusion; it clarifies, enlarges, and heightens meaning and feeling; and it does so with economy. If an allusion is the best kind, readers may not feel cheated if they do not recognize it and may feel pleasantly enriched if they do. This requires that when an allusion is woven into a sentence, the meaning should be clear for those who do not recognize the allusion—less rich, less full, less vivid, perhaps, but still clear.

### Analogy

One of the most useful figures of speech for writers in the professions is analogy: extended treatment of similarities (often of properties, relations, behavior, or function) in things unlike in kind, form, or appearance, for example, the heart and a pump.

Whole paragraphs and even whole short works may be developed by analogy (Chapter 12). Writers of informational prose often find it effective to compare difficult technical concepts new to readers with simpler concepts familiar to them. This puts the readers in a proper frame of mind to understand the technical concepts because the familiar by mitigating fear of the unfamiliar puts readers at ease. Such a comparison may clarify better than a formal definition and may even extend the meaning of a term beyond its definition and thus enlarge a reader's understanding of the definition. An analogy of three or four sentences is not difficult to write, but a long analogy requires skill in organization and expression.

An analogy explains and clarifies; in general argument it does not prove. It may be used to enforce or enhance proof, but in itself analogy is not evidence. Although comparison of similar subjects may be useful, things alike in some respects are not necessarily alike in others. In the absence of other proof, they may be thought to be alike, but the analogy, although it may lead to this supposition, does not prove anything about the points that are not compared.

## Irony

Irony, expressing a thought by stating its opposite, is more common in persuasion than in exposition. It is not unusual in conversation and in speeches, for example, *my great success* (a project that failed), *his usual short, pithy remarks* (two hours of rambling discussion), *my mink* (an inexpensive cloth coat). Irony may be as brief as these examples or as long as a whole work, like Jonathan Swift's *A Modest Proposal*. It is the opposite of the cliché in that it presents the unexpected juxtaposition, the imaginative contrast of the incongruous. But it is easily misunderstood, even by sympathetic readers if they are insensitive to irony. To stress the inhumanity of the English, Swift presented a mock proposal that the Irish avoid starvation under absentee landlords by eating Irish infants; but some English readers, hardened in their hearts toward the Irish, reproached him for having gone too far—for a clergyman. And some book reviewers missed the obvious irony of Hannah Arendt's *New Yorker* series on Eichmann. Irony is not for the literal-minded reader, nor the insensitive, nor the unsubtle.

Irony is often confused with sarcasm, particularly by those who dislike ridicule, whether it is intended to correct or to hurt. Irony may be sarcastic, and sarcasm may be ironic; but each may exist without the other. Sarcasm is intended to hurt or injure a person or a group; irony is designed to correct follies and evils or to express, wryly perhaps, the oddness of the world. When a supervisor says to a good writer, "Another poor report, I suppose," his words are ironic though not witty. But when he says the same thing to a poor writer, the words are sarcastic and still not witty. When a woman speaks of another's "charming pink-candy hair," the expression is sarcastic if the pink-haired one will be injured by it but otherwise the "charming" is ironic. After unsuccessful labor negotiations, disgruntled representatives may report with sarcastic irony. The company representative states, "Showing its usual concern that our company should not lose money, the union proposed increases of three dollars an hour"; the union representative reports, "With its usual concern for the welfare of the workers, the company proposed wage cuts of one dollar an hour."

## Litotes

Litotes is negation of the opposite of what is meant: *The danger is not insignificant; His bonus was by no means a small one; The bikini was no Mother Hubbard.* This facet of English offers variety and nuances to expression. It is common in Anglo-Saxon writings and suits the style of English well. It is effective without being obtrusive and is appropriate for formal and informal style.

### Metaphor and Simile

Metaphors and similes, which present a similarity in objects otherwise unlike, are used frequently in technical prose, but not always effectively. One editorial in *Science*, "The Research and Development Pork Barrel," has a metaphor in its title and a sprinkling of figurative language throughout: "on the road to becoming new Appalachias," "a Middle West 'brain drain,'" "face-to-face with a most serious kind of over-concentration," "road to manpower chaos," "whatever midwestern site appears to be most in the running after initial screenings," "the region exerts its maximum potential pressure."[10]

Similes and metaphors are particularly useful for making abstract general ideas concrete and specific. In Robert Oppenheimer's talk to the Princeton Graduate Alumni, "The Scientist in Society," appropriate figures of speech abound:

> There is something inherently comforting about a panel of experts. One knows that the partial and inadequate and slanted and personal views that he expresses will be corrected by the less partial, less personal views of everyone else on the panel; it is not unlike the experience of the professor who always is glad that he has to meet his class again because he can correct the mistakes that he made the last time. . . .
>
> This is a vast terrain—one full of strange precipices, chasms and terrors. . . .
>
> I know that it is a very happy occasion at the Institute [for Advanced Study] when some piece of work turns up which is of interest to both the mathematicians and the physicists. It is a very rare occasion and we tend to ring bells when a small bit of cement can be found between their interests. . . .
>
> The experience of science—to stub your toe hard and then notice that it was really a rock on which you stubbed it—this experience is something that is hard to communicate by popularization, by education, or by talk. It is almost as hard to tell a man what it is like to find out something new about the world as it is to describe a mystical experience to a chap who has never had any hint of such an experience.[11]

Similes state a comparison—*a backbone like jelly, no more sense of value than a magpie, a report as interesting as a mud flat and just about as clear*. Some writers think that only *like* or *as* may express the comparison, but other words may be used. Metaphors imply a comparison instead of expressing it—*rubberneck, his coltish manners, expecting stocks to zoom*.

A simile or metaphor should be appropriate to the context in which it appears. If a comparison in a formal text is expressed in slang, if a comparison in a sober work is too extravagant for the meaning, or if a comparison in a paper of workaday prose is too decorative, the writer should discard it. *The Government refused to play ball on this issue* is unsuitable for a formal report to stockholders. *Research and Development rocketed into space with this new tranquilizer* has a metaphor inappropriate to the meaning. *After the fire the company resurrected itself like a phoenix* contains a figure too far-fetched and decorative for a financial report.

Trite comparisons should also be deleted; removing them usually strengthens a sentence. A writer should avoid clichés such as *at one fell swoop, beggars description, bolt from the blue, the Buckeye State, busy as a bee, chalk up sales, cheap as dirt, clear as crystal, fall by the wayside, ironclad agreement, memory like a sieve, method in his madness, spice of life, sweat of his brow.*

Many comparisons are dead or dying—*the head of a pin, the tail light of a car, the foot of a bed, the heart of the matter, toe the mark, point a finger of scorn, show guts, shoulder a suggestion aside, elbow one's way to promotion, knuckle under to authority, scratch for ideas, a thin* (or *thick*)-*skinned person.* If a comparison is dead, the literal meaning does not come to life and confuse readers: *He stepped on the foot of the ladder; He engraved the Declaration of Independence on the head of a pin; The hurricane lashed a branch of a tree against the tail light.* Here *foot, head,* and *tail* mean nothing but *bottom, top* and *rear.* And this is true even though *lashed* suggests a literal meaning for *tail.* Therefore a writer need not avoid every dead metaphor; it might not be trite, and a substitute might be awkward. But a few metaphors still have some evocative qualities, not strong enough to make the figure effective but strong enough to make it ridiculous in certain contexts: *He could see the long arm of coincidence in her broken leg. Because of all the red tape, they could not bandage his finger. Showing guts* says nothing more than *showing courage* yet has vigor enough to prevent a sensitive person from using it. And the literal meaning may revive awkwardly: *In the middle of the operation the surgeon showed guts.* Between the dead and the dying metaphors is a metaphor that seems dead but revives in some contexts. *Rubberneck,* for example, would seem to be a dead metaphor, but it shows life in *bouncing rubberneck.*

This literal meaning may seem so obtrusive to readers that they cannot understand why the writer did not see it and remove the figure. But many writers in the professions are unconscious of the need for care in the use of figurative language or are unaware that they are using figures. We have read poor examples like, "As president of the American Goiter Society, he said that he intends to 'stick his neck out' at the April meeting and stump for less conservative use of thyroid." When a figure revives its literal meaning in a sentence concerning the ill, the disabled, or the dead, the result may be unpleasant enough to open a writer to a charge of poor taste.

305

## Metonymy and Synecdoche

Metonymy and synecdoche are not so common in prose as are metaphor and simile, but they do occur. *Synecdoche* is a substitution of the name of an important part for the whole or of the name of a whole for a part, as in *factory hands, hungry mouths, using his head, a strike against a plant.* Somewhat similar is *metonymy*, the substitution of an attribute or an association for the word meant, as in *playing to the gallery* (for *the audience in the gallery*), *sweets* (for *candies and desserts*), *Blue Points* (for *oysters dredged near the village of Blue Point*). A company may use the name of a town or city for the plant located there—*Baton Rouge, Yonkers* (for *the divisions at these locations*). Synecdoche and metonymy are useful when writers have to repeat a word often, and they are especially useful in avoiding the repetition of long names or titles. The only requirement is that the substitute be clear. Metonymy sometimes seems informal. When it does, writers of formal papers should use the original term or a pronoun rather than confuse the tone of their writing.

## Overstatement and Understatement

Akin to irony are overstatement, or hyperbole, and understatement, or meiosis. Overstatement is extravagant exaggeration used for emphasis or for comic effect: *He acts as though an error in typing will blow up the plant; I expect to be worrying about that point in my grave; He has worked here a million years at least; This report has more fancy words than the unabridged dictionary.* Understatement, deliberate restraint or playing down, is most effective when a subject is overpowering, as in John Hersey's controlled, almost pent-up presentation of the shocking material in *Hiroshima.* When it is difficult for words to match the strength of the subject matter, a writer may well use understatement. And sometimes minimizing one's angry convictions is humorous. Samuel Johnson defined a Tory as "one who adheres to the ancient constitution of the state, and the apostolical hierarchy of the Church of England, opposed to a Whig," and then coolly expressed his scorn in an understated definition of a Whig as "the name of a faction."

## Paradox and Oxymoron

A paradox is a seemingly self-contradictory or absurd statement that, properly understood, is reasonable, a statement opposed to common sense but true in a larger sense. "His bankruptcy in 1930 was so profitable that it became the foundation of his success in the company he started in 1932." Sometimes a paradox is based on a pun. A patient explained his visits to a psychiatrist as necessary "because his office is the only place where I can lie down and relax undisturbed."

Paradox is closely related to oxymoron, the juxtaposition of words that are apparently contradictory. A writer discussing the population explosion may call it the death from too many births. The Christian martyrs were said to die to live eternally. A worker who performed poorly on the first few days in a new job and thereafter performed well said that he failed only to succeed.

## Personification

Personification, endowing something not human, for example, objects or abstract qualities, with human attributes, is not unusual in prose. It may sound old-fashioned or strained or fancy, as in "The subtle villain Envy tricked him into that wrong decision." A personification such as *Liberty added her voice* is not likely to be used today except in Fourth-of-July oratory, where old-fashioned devices flourish. But *alma mater* means only *college*; it has no more life than any other dead metaphor. Contemporary prose also uses personification: *And Number 10 Downing Street looked melancholy; The IRS computer had a nervous breakdown; The evidence did not have even a speaking acquaintance with the conclusions.* Personifications lend emphasis and vividness because of our ability to understand our own reactions and experiences: *Machines grow old and have to be retired.*

## Puns and Other Plays on Words

A pun is the use of a word (or of two words that sound alike) in more than a single sense at once. Today puns in prose are used mainly for humor, but puns have long been used for wit. In the second paragraph of *Victory* Conrad writes,

> The Tropical Belt Coal Company went into liquidation. The world of finance is a mysterious world in which, incredible as the fact may appear, evaporation precedes liquidation. First the capital evaporates, and then the company goes into liquidation. These are very unnatural physics. . . .

Called the lowest form of wit, puns often convey the highest meaning. Poets use puns as vehicles for their profoundest insights, as does the Bible. That punster Mercutio conveys more than obvious meaning in *Romeo and Juliet* by his plays on words, not the least of all by his pun after he has received his mortal wound, "Ask for me to-morrow, and you shall find me a grave man." In the work of a witty writer puns may be effective even in a serious context, and in speeches they may be welcome light touches. The humorous pun should not be strained because when one of the meanings does not quite fit or when the use of the pun lacks point, listeners flinch. But listeners who wince and groan at every pun lack originality and perception.

### Selection of Figures of Speech

Skillful use of figurative language can clarify and enliven thoughts at the same time that it entertains and pleases readers. A figure of speech is an effective way to emphasize a main point. It may be persuasive and mnemonic, for its vividness makes it powerful; therefore an imaginative writer who can provide fresh, appropriate figures of speech can convey ideas precisely and forcefully. But trite figures of speech advertise that they are worn out. It is better to state that she is smart than that she is smart as a whip. And expressions like Mother Nature and Father Time and their progeny should be shunned.

It is dangerous to shift so rapidly from one figure of speech to another that they blend in a reader's mind. One figure may be incompatible with the other in tone or meaning: for example, *Whenever he feels dog tired he behaves like a bull in a china shop, multiplying blunders like rabbits,* or again, *The blushing bride was green with envy.* Mixed figures may be delightful when reprinted in *The New Yorker,* but they are an embarrassment in a serious work. Writers should avoid mixed figures like the following: *Faced with concrete criticism, he buries his head like an ostrich and hopes that the egg he laid will go away; The company newspaper flooded us with dead news that was dry as dust; These high school dropouts will land with a thud in the face of present employment conditions.*

An inappropriate figure of speech—one that does not suit the tone of the writing, one that overemphasizes a minor point, one that confuses rather than clarifies—is an obtrusive error because of the vividness of figurative language. If writers feel unsure of figures, they are wise to omit them. And writers should not so crowd their work with figures that the writing becomes tiresome and therefore ineffective, as this published sentence does: "This is where your proposals will either hit pay dirt and lead to a contract or strike out." Figurative language should be used with taste and judgment. A writer may develop these through reading and through studying effective examples.

## THE SOUND AND RHYTHM OF PROSE

Rhetorical figures—climax, anticlimax, antithesis—concern structure and arrangement. They are discussed and illustrated in Chapter 11. Some rhetorical constructions and sound patterns are seldom used or are avoided in informational prose—alliteration, onomatopoeia, and rhyme. A careful writer avoids ineffective repetition of a word used in two senses: *A timer timed out; They could not understand his stand on the budget; They took no stock in the stockbroker's advice; As soon as we meet her, let us make sure that she meets our qualifications; The letter entered the case because it did not satisfy the letter of the law.* Such awkward uses are easy to correct: *They could not understand his position on the budget; They did not trust the stockbroker's advice.* Repetition of an unimportant word gives poor emphasis: *He said that the regulations said that a doctor should be present.* Accidental rhyme or alliteration is also undesirable: *His attitude would be less cold if the plan involved more gold; He will iron out the cation problem; Neither his calculation nor the secretary's negotiation improved the situation; In space flight the spacecraft speeds sufficiently to supply necessary centrifugal force; With the weight of this warhead why would we wish for wider spread weapons?* Sentences like these may appear in first drafts; a reviser improves them by a judicious use of pronouns and synonyms, by omitting, and by recasting.

The sound and rhythm of informational prose should not attract unfavorable or unwanted attention. Disturbing sounds distract a reader from the sense; therefore harsh combinations of sounds should be avoided. Because English has a large number of consonants, writers must strive consciously for euphony. Reading aloud helps them to avoid ugly sounds like these: "In this display all textbooks exceeding the approximate size are without prejudice disqualified, but registration and admission fees are returnable by the administration on continuation or retention of registration for succeeding equivalent exhibits." Jogging rhythms are also unattractive and distracting in prose. The rhythms of parallel construction are acceptable, but contemporary readers prefer even these to be subtly modified to avoid regularity. The drumbeats of a list of words with similar rhythm should be avoided. The following example unhappily combines rhyme with monotonous rhythms, as many jogging lists do: *Introductions, presentations, graduations, celebrations, grand occasions demand special services.*

Writers sometimes use harsh sounds, rhymes, and regular rhythms for special effects in expository prose, but such effects are more common in fiction and essays than in reports and memorandums. Advertisers use these devices to capture and hold attention because they consider them catchy. But constant dinning makes such phrases irritating. A phrase like *from womb to tomb* captures so much attention the first time one sees it that it does not bear much repetition. Professionals are more likely to use special effects in speeches than in writing. If a catchy phrase is not injudiciously repeated, it may be a success in a speech or an interview.


## AVOIDING CONFUSION IN MEANING

### Technical and Nontechnical Meanings

When a word has a technical meaning in a specialization and a different meaning in general use, it should not be used with first one meaning and then the other. It is better to use it with the technical meaning in writing about the specialization and with the general meaning in other writing. *Cohesion*, for example, has, besides its general meaning, one special meaning in physics and another in botany. The physicist should use *cohesion* not in its general sense, not with the meaning that it conveys

309

in botany, but with the special meaning it has in physics. In an unusual paper a physicist might use *cohesion* with the meaning it conveys in botany but should indicate that he is doing so. He should also carefully review such a paper to be sure that a reader is not likely at any point to assign *cohesion* one of its other meanings. In some specializations this problem arises more often than in others. Economics, for example, has a terminology based largely on assigning special meanings to words in the general vocabulary—*bond, distribution, stock, market, run, panic, depression, recession* (how did the economists miss *obsession?*), *production, consumption*. Chemistry on the other hand has taken few words from the general vocabulary to give them special meanings.

Whenever writers use technical terms for readers who do not have technical training in the field, they must define the words. Whenever a technical term is also a word in the general vocabulary, writers must be certain that their readers continue to think of the word in its technical sense throughout their reading of the work even though this requires reminding readers of the special sense. Writers complicate their tasks unnecessarily if, having established a special definition for a word, they then employ the word in its general sense. No reader should be expected to follow such switches in meaning.

## Many Meanings

A word that has two or more dissimilar meanings may cause difficulties:

> FORMAL: The report is not quite ready.
> His recommendations are quite acceptable.
> INFORMAL: She is quite pretty.
> He is quite successful.

*Quite*, which means *totally* or *entirely*, also means, in informal English, *rather*. When writing is not distinguishably formal or informal but lies on the border, writers should be certain that their use of a word like *quite* cannot be misunderstood. If there is any question, they should replace *quite* by a word that is not ambiguous.

## Changing Meanings

Changing meanings cause difficulties. The word *lady*, for example, has changed drastically in meaning. In some professional societies older members will not change the old-fashioned title Ladies' Program although *lady* has been denigrated by terms like *scrublady* and *lady of the evening*. This usage provides some problems for convention chairmen, one of whom wrote to the wife of a member, "I know that the women in your group will enjoy the Ladies' Program." Changing times—in this example the disappearance of the lady of leisure and an increase in the number of career women—create problems by changing the meanings of words. But changing times may eventually solve the problems; the title *Ladies' Program* will disappear when more husbands of engineers enjoy the programs for guests while their wives listen to papers on engineering.

*Disinterested* and *uninterested* also indicate the difficulties due to changing meanings. Handbooks of English used to state flatly that *disinterested* meant *impartial, unselfish, without selfish interest* and that *uninterested* meant *indifferent* or *not interested*. Some handbooks still differentiate between the two. But writers have ignored the difference so stubbornly that *disinterested* has come to mean *lacking in interest*. We have noticed it used for that meaning more frequently than *uninterested*. If a writer of minutes states, "The committee was disinterested in voting on the proposal," a few readers will assume that the committee was impartial but most, that it was not interested. Unless the context guides readers to a clear understanding of words with changing meanings, a writer should avoid such words. *Not interested, uninterested,* or *impartial* would not have been ambiguous in the minutes.

Clinging to outmoded language is just as foolish as adopting the latest fad in language, but there are those who do both and speak a language all their own: "Hopefully the girls will find the Ladies' Program successful moneywise even if they are disinterested in the choice of goodies."

## FINDING THE RIGHT WORD

Writers in the professions are interested in the qualities of diction and the techniques of using language mainly because they hope to learn to express their meaning precisely and accurately. They may think that this separates them from other writers, but it does not. Accuracy is the foundation of all good prose style. To avoid blurred expression writers select from synonyms the exact shade of meaning that they want. In spite of the number of synonyms in English, there is often just one word that expresses the sense precisely, one pat to the meaning, the rest being makeshifts.

General dictionaries list some words of similar meaning. But writers searching for an elusive word may not find it in such a brief list. They should turn to *Roget's International Thesaurus* or to a dictionary of synonyms and antonyms. Roget lists not just synonyms that are the same part of speech but also other parts of speech that have similar meanings. Thus a person who looks up a noun will find handy not only nouns with similar meanings, but verbs, adjectives, adverbs, phrases, and sometimes interjections. These lists are particularly helpful because many writers habitually use nouns where verbs would be more effective. The inclusion of all parts of speech in one listing offers freedom to writers; they do not have to use an adjective simply because the word that came to mind and the synonyms in their dictionary are adjectives.

A thesaurus is helpful too because it refers by number to other sections that list words of somewhat similar meaning. When writers are searching, the word that they start with is sometimes not near enough in meaning to enable them to find the word they want among the close synonyms a dictionary provides. But the thesaurus with its longer listing and its many references to related words is likely to supply the word or to enable them to express their ideas by denying the opposite, for it lists antonyms. So much to choose from may be too much for a beginner, but it is just what other writers need when the word they want will not come to mind.

The right word presents an idea to readers with such startling precision that they rejoice in grasping the thought as though a screen between them and the writer had been removed. Readers of carefully selected words need not struggle through a haze of inexact words and a clutter of useless modifiers; they do not have to translate the idea from jargon to understand it; they do not experience the sensation of almost but not quite perceiving the idea. They grasp it as surely as though it were their own thought; only they have, too, the gratification of knowing that they are understanding another person's speech and thought with ease. But they must never be deluded into thinking that the writing was effortless. A reader's perception of an exact shade of subtle meaning comes from a writer's untiring search for words that are precisely the right ones.

## NOTES

1. Frank Sullivan, "The Cliché Expert Testifies on the Atom," *The New Yorker* (17 Nov. 1945): 27.

2. A. C. Howell, *Handbook of English in Engineering Usage*, 2 ed. (New York: Wiley, 1940) 26.

3. T. A. Rickard, *Technical Writing* (New York: Wiley, 1931) 42.

4. George Holbrook, in *Listen to Leaders in Engineering*, Albert Love and James Saxon Childers, eds. (Atlanta: Tupper; New York: McKay, 1965) 63.

5. Reginald O. Kapp, *The Presentation of Technical Information* (New York: Macmillan, 1957) 75.

6. Osward Werner, "Anthropology and Linguistics," *Science* (9 July 1965): 168.

7. S. M. Ayres, Stanley Giannelli, Jr., and R. G. Armstrong, "Carboxyhemoglobin: Hemodynamic and Respiratory Responses to Small Concentrations," *Science* (9 July 1965): 193.

8. Timothy Ferris, "Physics' Newest Frontier," *The New York Times Magazine* 26 Sept. 1982: 38, 44.

9. David Brower, "The Meek Shall Lose the Earth," *Greenpeace Examiner* (Jan./March, 1985): 18.

10. "The Research and Development Pork Barrel," *Science* 149 (2 July 1965): 11.

11. J. Robert Oppenheimer, *The Open Mind* (New York: Simon, 1955) 119, 123, 126.

# CHAPTER 11

# Style and Sentences

*Pregnant in matter, in expression brief,*
*Let every sentence stand with bold relief.*

*Joseph Story*

## WRITING ONLY SHORT SENTENCES

No other advice has been pressed upon modern writers so relentlessly during the past fifty years as "Write short sentences." Because papers on science, technology, and business management are not always easy reading, the advocates of short sentences attack most vigorously writings on these subjects. One adviser comments on a forty-three-word sentence in an article on engineering: "If this sentence is read aloud the reader will be out of breath by the time he reaches the end." A speaker at a technological convention summarizes: "We have been hearing much from the readability experts about the need to write simply. They say, 'Use short words and sentences.'"

Panaceas are misleading. Offering the short sentence as a cure-all demonstrates the dangers of half knowledge. When a short-sentence addict complains that reading a sentence of forty-three words aloud leaves him out of breath, he has reached the extreme of pneumatic testing. The idea that a sentence must be read in one breath was buried long ago in a family plot next to the idea that a comma always represents a short pause for breath (or is it a pause for a short breath?); a semicolon, a longer pause; and a period, a full stop. Lincoln's *Gettysburg Address* was written to be read aloud. Could even a trained singer read its eighty-two word sentence in one breath? Picture a breathless husband and wife at the breakfast table panting to each other the fifty- to seventy-word sentences in their morning newspaper. And what of the Elizabethans, who wrote even longer sentences? Were they longer-winded?

The final irony of the sentence choppers is that they often urge scientists and technologists to read the classics to improve their style. But they do not account for the exceedingly long sentences in the classics—even in the least likely modern fiction. In one Hemingway story a critic

313

did find sentences averaging eight and one-half words in length, but in another Hemingway story in the same volume, forty-nine.[1] Businessmen, scientists, and engineers who are ignorant of the rhetorical principles involved may well and truly conclude that *no* particular sentence length is correct or that *any* length is correct.

## THE READER AND SENTENCE LENGTH

*Short* and *readable* are relative terms. An intelligent person's concept of sentence length changes with age, education, and reading experience. A sentence that looks long to a fourth-grade child may look short to a college junior. A man whose only reading is a few letters and newspaper headlines will consider too long a sentence that looks short to a reader of current literature. It seems reasonable, therefore, to adjust sentence lengths to readers. Appropriately a story for children has sentences averaging fifteen words, and newspapers and magazines directed to those of little education also use short sentences; but *The New York Times*, learned journals, and technical reports just as appropriately have longer sentences. In general it is safe to say that sentences in much material for educated adult readers average between twenty and thirty-five words. A higher or lower average indicates that the writer should inspect the sentences critically, but it does not necessarily mean that anything is wrong. Sufficient variety in type, construction, and length is more important than the average number of words.

To a certain extent writers on science and engineering, like other writers, adjust sentence lengths to readers as material permits. An engineer at a meeting wrote relatively short sentences in his letters to his children about the local zoo. His letter to his wife, telling her about the trip, explaining some banking arrangements, and relating the news of the convention, had longer sentences. The sentences in his discussion of a paper presented were similar in length to those in his letter to his wife. When his discussion, taken from tape, was sent to him for editing, he put several related thoughts into one sentence and compressed the whole into slightly longer sentences. He made this adjustment of sentence to reader on the basis of common sense without any consideration of rhetorical principles.

## THE INFLUENCE OF THE DEMAND FOR PRIMERS

Those who prescribe the short sentence for all writing ills would have writers go against this natural adjustment of sentence lengths to their reader and their subject matter. Some of the most demanding of these choppers cannot read with ease above the high school level. They find the short sentence easy, just as the school child does, and they mistake their needs for universal needs. If professional people must write for them, these writers must adapt not just their style but also their material. Elimination of difficult and subtle material and simplification of sentences and of vocabulary can place their writing at the level of these poor readers. Very likely only a page of simple summary of a fifty-page report will suffice, for that is probably all that the readers can master.

Those few readers who insist that complex scientific and technological material be brought to the level of a dull high school student would probably demand the article in comic-book form if they dared. They need special writers and cartoonists to simplify adult material for immature readers. Such readers cannot grasp complex ideas, fine distinctions in meaning, adult thinking. Even when scientists and engineers write with consummate skill, such readers still complain. It is not the style that makes it difficult for them to grasp the ideas; it may be the ideas themselves that are too difficult for them.

Writers in the professions who try to suit the whims of readers or of "experts" who demand primer style will soon find themselves in the position of the professors in *Gulliver's Travels* who engage in compressing polysyllables into single syllables or who strive to abolish words altogether. Swift presents comically the inevitable result:

> However, many of the most learned and wise adhere to the new scheme of expressing themselves by things; which hath only this inconvenience attending it, that if a man's business be very great, and of various kinds, he must be obliged in proportion to carry a greater bundle of things upon his back, unless he can afford one or two strong servants to attend him. I have often beheld two of those sages almost sinking under the weight of their packs, like pedlars among us; who, when they met in the streets, would lay down their loads, open their sacks, and hold conversation for an hour together; then put up their implements, help each other to resume their burthens, and take their leave.

A writer burdened with advice that will gradually eliminate language is in danger of sinking under the difficulties, as Swift predicted. Short sentences and short words spread ideas so thin that it may take more than twice as long to express an idea. Subtle distinctions disappear, and only the most obvious ideas and the most elementary vocabulary and style remain; in effect, writer and reader might just as well communicate by displaying objects. That might satisfy the executive who no doubt thought his expression brilliantly economical when he advertised for a secretary able to write "short basic English." He, of course, would object to Swift's long sentences as wasteful.

## OTHER INFLUENCES ON SENTENCE LENGTH

A writer who wishes to use sentences of various lengths should examine the other influences on sentence length. The reader has already been considered. The subject matter, type of writing, and demands of variety and emphasis are also important.

The demands of subject matter vary. A plea for better salaries does not require sentences as long as those in a journal article on ultrasonics. A discussion of a new name for a division of a company may well have shorter sentences than are required for stating the research objectives of a highly specialized project.

The nature of the presentation also influences the sentence length. A good speaker, for example, will tell an anecdote in relatively short sentences and use longer ones in the argument. And the description of apparatus in a technical article may be expressed in long sentences that group related details. Informal speaking has the shortest sentences; formal writing, the longest: an explanation to a fellow worker during a coffee break calls for much shorter sentences than those in a published article.

A writer's natural style also influences length. Some people think in terms of groups of related details; others naturally separate small divisions of thought. There is nothing wrong with either style provided that it is not allowed to become monotonous. Writers with a natural tendency to aggregate should practice the short terse sentence, which they may use most effectively for emphasis by contrasting it with their longer sentences. Writers who naturally place every division of a thought into a separate sentence should try showing relationships by reducing some unimportant short sentences to words, phrases, or dependent clauses.

## USES OF THE LONG SENTENCE

The long sentence is suited to grouping a number of related details clearly, neatly, and economically. Journalists illustrate this when they answer in their first sentence the most important of the six questions Who? What? Where? When? Why? and How? Causes and reasons, lists, results, characteristics, minor details may all be expressed tersely, clearly, and effectively in long sentences.

To illustrate the advantages of grouping related details, the following four sentences have been compressed into one. In order that the illustration be fair, the four sentences chosen are those of an average writer on chemical engineering and the changes made are only those necessary for the alteration in sentence structure.

> The seventeen materials used for study were examined by several methods other than the compression-permeability method, in order to obtain as much information as possible concerning their physical properties. The materials in dry form were examined under the electron microscope, and the photomicrographs of Figures 4 to 7 were taken to establish size and shape of ultimate particles. Nitrogen-adsorption measurements were also made, and specific surface values were calculated for each material, employing the Brunauer-Emmett-Teller method. Air-permeability measurements were made on the dry materials in most cases, using the Fisher subsieve sizer, and the corresponding specific surface was calculated for each by use

of the method and slip-flow correction presented by Arnell, and by Carman and Arnell, employing a slip factor Z 3.5.

To obtain as much information as possible about the seventeen materials used for study, they were examined by several methods other than the compression-permeability method: (1) examining the materials in dry form under the electron microscope and taking photomicrographs (Figures 4 to 7) to establish the size and shape of ultimate particles, (2) measuring the adsorption of nitrogen and calculating specific surface values for each material by the Brunauer-Emmett-Teller method, (3) measuring the air permeability of most of the dry materials by the Fisher subsieve sizer and calculating the specific surface by the method and slip-flow correction of Arnell and of Carman and Arnell, employing a slip factor Z 3.5.

The difference in effect and the compression result solely from the changes in sentence structure.

Because of its usefulness in bringing a number of related thoughts together, the long sentence is often effective for a summary conclusion. Whenever they are expressing several conclusions in one summary sentence, writers should try to place them in parallel construction and to arrange them in logical, chronological, or climactic order. A long summary sentence may be emphasized by contrast with several short sentences. Then the sentence stands out by the bulk of its length, the weight of the important ideas, the careful construction, and the contrast with surrounding sentences. It is a strong ending.

## CHOPPING SENTENCES

The following sentence, which was used as an example in an article advocating short sentences, provides a striking illustration of the economy of the long sentence:

Although it is possible for the refiner to control the selectivity deterioration by continually discarding a fraction of the catalyst in the unit and replacing it with fresh material of good selectivity, this expedient increases operating costs because of the additional catalyst consumption.

The comment of the article follows:

If this sentence is read aloud the reader will be out of breath by the time he reaches the end. This is a fine example of trying to do too much in one sentence. The author has three thoughts here and each one should have its own sentence.

The refiner has a way to keep the selectivity up. He can keep adding fresh catalyst and throwing away part of the old. But this costs him money because he uses more catalyst.

Once more, fewer words, with no sacrifice of meaning spell a gain in simplicity.

317

"No sacrifice of meaning"? On two scores objection must be raised—to changes in meaning and to using three sentences to do the work of one.

The original sentence stated, "this expedient increases operating costs"; the revision reads, "this costs him [the refiner] money." There is a decided difference in meaning. The chemical engineer who wrote the original quite properly did not go into the problem in economics of who would pay the increased operating costs—refiner, middleman, consumer. The revision changes the meaning by stating that the expedient costs the refiner money. If such carelessness were shown in revising a legal paper, the change in meaning might be serious.

The original speaks of "fresh material of good selectivity"; the simplification drops the modification "of good selectivity." Discarding these necessary words would invalidate a whole experiment if the use of replacement material of average or poor selectivity were to give a different result.

Reducing length by discarding necessary words is not to be recommended. By failing to convey accurately the exact thought of the original long sentence, these three short sentences illustrate the disadvantages of simplifying too much. Oversimplification not only loses distinctions in thought but soon makes writers insensitive to fairly obvious differences in meaning. Simplifying at the expense of meaning is dangerous.

The reviser states that the original sentence has three thoughts requiring three sentences, but the thoughts in the original sentence are so closely related that separating them makes them harder rather than easier to read. Only one sentence is needed:

> The refiner may maintain selectivity by continually replacing part of the catalyst with fresh material of good selectivity, but he thus increases production costs.

The original sentence has forty-three words; the revision in three short sentences has thirty-three; the revision in one sentence has twenty-four. The longer revised sentence is more economical than the three short ones. It has fewer words even though it includes the modifications in meaning omitted in the three short sentences.

Writers should never be urged to use short sentences where longer sentences present ideas to their readers more effectively and economically. Writers for professional journals have intelligent adult readers. For their fellow engineers and for intelligent readers not in the engineering profession, a well-constructed long sentence is clearer than a jumble of short sentences. Many writers have as much trouble constructing a series of good short sentences as one good long sentence; therefore they gain nothing by using the short sentence where it is inappropriate. By selecting the most effective sentence length, writers make their own task and the task of their readers easier.

## USES OF THE SHORT SENTENCE

Short sentences do have their uses. A crisp, clear sentence is a good opening, a polished short sentence gives an effect of wit and sparkle, and a terse concluding sentence may be used for sharp emphasis. A short sentence may also be used for effective contrast after a series of long sentences if an important idea appears in the short sentence. An editorial beginning

> In almost every talk we give, and in many conversations we have with our members on the problems of publishing an engineering magazine, the question is invariably asked, "Why don't you sell more advertising?"

might have started with the short question

> "Why don't you sell more advertising?"

A book review beginning

> As the authors tell in the preface, this book is not a manual of engineering drawing. Rather it is a summary of the revolution that took place when the General Electric Company decided to simplify its drafting practices.

is an attempt to use a short opening sentence. But a really short one would be more effective—

> Here is a revolution in drafting practices.

Once writers have their rough draft of a letter or an article completed, they will often find in the first or second paragraph a terse sentence that with a little polishing will make a good short opening. Not many writers automatically begin a first draft with an epigrammatic sentence, but many writers bury among the first few sentences one that may be reduced in words and sharpened in expression to provide sparkle.

An effect of force and wit is sometimes achieved by a brisk closing sentence. Brevity is rapierlike in the last sentence of many anecdotes. A chemist discoursing on the eccentricities of her first boss may end this way, "And so sometimes before he left a conference the boss would turn to us and say, 'I suppose you all think I'm crazy,' and you know we did think that he was insane." Ending with a terser statement increases the effect: "And so sometimes before he left a conference, the boss would turn to us and say, 'I suppose you all think I'm crazy.' We did." This two-word closing gains effectiveness from contrast with the longer preceding sentence. The shorter the contrasting sentence, the greater the emphasis.

319

## THE HAPPY MEDIUM

Sentences of medium length are frequent in the writing of professional people. The material expressed in sentences of intermediate length receives less emphasis than the material expressed in sentences of contrasting lengths. And sentences of medium length are likely to become monotonous. Therefore when they are revising, writers should add interest to sentences of similar length by varying beginnings, constructions, and rhetorical types. And by removing unnecessary words, they may shorten some sentences. They should construct their sentences of medium length carefully, because, being dominant in their papers, these sentences may determine the general style and tone.

## VARIETY, THE FOUNDATION OF SENTENCE STYLE

Variety in sentence length is much more important than average length. Too much worrying about reducing the average length may lead to short sentences of similar extent and construction and thus result in monotony. Any continued repetition of sentence length is dull, but a passage of short sentences is unbearably monotonous when there is little variety in structure and rhythm. Dull paragrahs of sentences formed in one mold are as depressing as the vista presented by a typical cheap housing development—hundreds of houses all the same size, all with the same architectural plan—deadening duplication. The short-sentence enthusiasts should remember that although one short sentence may have punch, a paper of short sentences leaves one punch drunk.

Knowing the uses of sentences of all lengths helps a writer achieve pleasing variety in length. And infinite variety contributes to interesting sentence style just as it does to attractiveness in people. Other variations in sentence structure that contribute interest to style are variety in sentence beginnings, in grammatical structure (simple, complex, compound, compound-complex), in rhetorical structure (loose, periodic, balanced), in the elements (words, phrases, clauses), and in sentence rhythms. A tendency to use one kind of beginning, one grammatical structure, or even one order should not be allowed to escape a writer's notice and lull readers into drowsy inattention.

We do not believe that it is necessary to check everything that one writes for all these kinds of variety, but we have noticed that the monotony found in one sample of a writer's work also usually deadens other examples of the writing. We recommend therefore that writers check the variety of their sentences in two or three typical examples of their writing. The findings will enable them to improve their future papers by removing the weaknesses found in the sample.

Anyone with very little time for revision should devote this time to varying the most important sentences—the first and last in the paper and the first and last in each paragraph. These sentences ought to contain

320

important ideas. They deserve a reviser's labor because they attract a reader's attention.

## Variety in Beginnings

Much writing for business, government, science, and technology has monotonous sentence beginnings. In published engineering papers, for example, we found a strong need for variety in sentence beginnings. Too many engineering writers begin sentences similar in length with subjects or with conjunctions like *however* and *hence* followed by subjects. There are many ways of beginning any sentence, as the following illustrations show.

ORIGINAL ORDER

This gadget, marketed under the name of Glamorglow, was first offered to industry in the fall of 1980 after a period of development and laboratory tests.

VARIED BEGINNINGS

Marketed under the name of Glamorglow, it was first offered to industry in the fall of 1980 after a period of development and laboratory tests.

In the fall of 1980 after a period of development and laboratory tests, this gadget, marketed under the name of Glamorglow, was first offered to industry.

After the gadget had been developed and tested in the laboratory, it was offered to industry under the name of Glamorglow in the fall of 1980.

First offered to industry in the fall of 1980 after a period of development and laboratory tests, this gadget was marketed under the name of Glamorglow.

ORIGINAL ORDER

Accurate prediction of either bubble-cap or perforated-plate behavior is complicated by the large number of variables that must be considered.

VARIED BEGINNINGS

Because of the large number of variables that must be considered, accurate prediction of either bubble-cap or perforated-plate behavior is complicated.

The large number of variables that must be considered complicates accurate prediction of either bubble-cap or perforated-plate behavior.

Because a large number of variables must be considered, accurate prediction of either bubble-cap or perforated-plate behavior is complicated.

Writers may vary their sentence beginnings by placing first an adverb, an adjective, a participle, a prepositional phrase, a participial phrase, or even the simple conjunctions too often spurned by writers on science and technology—*and, but, or, nor, yet.* A little effort can avoid the deadly droning that occurs when every sentence begins with its subject:

There were six designs chosen to receive Aluminum Company of America's Student Design Merit Awards. All six were conceived by industrial design students. Each winning design was selected by the school's design faculty as the outstanding student project employing aluminum. Three of the award-winning designs are described below: . . . .

## Variety in Grammatical Types

Writers should use all grammatical types of sentences as they need them. Any long work should have examples of the following types.

> SIMPLE: Engineers wishing instruction in statistics may wish to consider the two courses offered next month: (1) a course in the fundamentals for beginners and (2) a course in the latest developments for engineers with a mastery of the fundamentals.

A simple sentence is not, as some writers think, necessarily a short or uncomplicated sentence. It is a sentence with one main clause and no dependent clauses. Such a sentence may have complicated appositives and phrases and may be just as long as any other type; or it may, of course, be short and simple, like these three examples: (1) "Go." (2) "He increased the pressure." (3) "It is hot."

> COMPLEX: When she tried to register for the advanced course, she found that it conflicted with the meetings of her division.
> Because he had taken two elementary courses, he did not register for the first course.

A complex sentence contains one main clause and one or more subordinate clauses. In spite of its name, it may look like a simple, easy sentence: "Give him the report as soon as he comes."

> COMPOUND: The first course concerns the fundamentals of statistics, and the second course concerns new developments in statistics.
> The students in the first course found it too fast, and the students in the second class found it too slow, but the instructor progressed at the same speed and covered the same amount of material in each course.

A compound sentence contains two or more main clauses and no subordinate clauses. The two or more main actions are coordinate: "Then the attention of the class faltered, and the instructor worried." This is quite different from the unequal emphasis of a complex sentence: "When the attention of the class faltered, the instructor worried."

> COMPOUND-COMPLEX SENTENCE: Although the students in the first class found it too fast and the students in the second class found it too slow, the instructor progressed at the same speed, and she covered the same amount of material in each class.
> In March, when our director, the head of training, said that he would

322

take the course, we believed him, and we continued to believe him—right up to registration day in June.

A compound-complex sentence contains two or more main clauses and at least one subordinate clause. It is usually long or at least of medium length. Its length and complexity permit some juggling of ideas. A number of main clauses, a number of subordinate clauses, and of course other minor constructions for ideas less important than those of the subordinate clauses give a writer the opportunity to try various arrangements.

Sometimes writers overwork one grammatical sentence structure, particularly the simple or compound. Excessive use of simple or compound sentences equates every thought with every other and shows only the most elementary thought relationships, as a child does. Such style needs the subordinate constructions for minor ideas that are provided by complex and compound-complex sentences; by compound subjects, predicates, and objects; and by participles, appositives, phrases, and other modifiers.

> EXAMPLE 1: This executive trains new employees of his division and he directs research and he supervises writing.
>
> EXAMPLE 1 WITH COMPOUND PREDICATE: This executive trains new employees of his division, directs research, and supervises writing.
>
> EXAMPLE 2: Harriet Jones is the director of research and she supervises writing and she recommended this course.
>
> EXAMPLE 2 WITH APPOSITIVES: Harriet Jones, the director of research and supervisor of writing, recommended this course.
>
> EXAMPLE 3: The schedule of the secretarial pool was crowded and no one could type the report immediately.
>
> EXAMPLE 3 WITH PHRASES: Because of a crowded schedule for the secretarial pool, no one could type the report immediately.
>
> EXAMPLE 4: He is a well-trained assistant and he is conscientious and he always shows tact.
>
> EXAMPLE 4 WITH ADJECTIVES: He is a tactful, well-trained, conscientious assistant.
>
> EXAMPLE 5: This company is one of the largest publishers of instruction manuals in the world, and it has a schedule of hourly work, and that schedule should be helpful to us.
>
> EXAMPLE 5 AS A COMPLEX SENTENCE: This company, which is one of the largest publishers of instruction manuals in the world, has a schedule of hourly work that should be helpful to us.
>
> EXAMPLE 5 AS A COMPLEX SENTENCE: Because this company is one of the largest publishers of instruction manuals in the world, its schedule of hourly work should be helpful to us.
>
> EXAMPLE 6: He fired his secretary and he felt guilty and so he gave her a good recommendation.
>
> EXAMPLE 6 AS A COMPOUND–COMPLEX SENTENCE: He fired his secretary, but because he felt guilty he gave her a good recommendation.

These examples illustrate the looseness of the compound sentence and its stringy effect. Subordination provides the necessary tightening and indication of thought relationships.

## Variety in Rhetorical Structure

Writers of English may use three rhetorical structures for the sentence—(1) the loose sentence, which moves from subject and its modifiers, to verb and its modifiers, to object or predicate nominative and its modifiers; (2) the periodic sentence, which suspends the completion of grammatical structure until the last word or nearly the last word; and (3) the balanced sentence, which contains clauses similar in length and in movement in parallel constructions. The loose sentence has a more natural, conversational word order: *The applicant was interested in coffee breaks, holidays, fringe benefits, job security, and quick advancement.* The periodic sentence has a more unusual order, sometimes almost bookish, but useful for emphasis and suspense: *Coffee breaks, holidays, fringe benefits, job security, and quick advancement—only these interested the applicant.* The loose sentence may have a suspense of its own—that of climax:

> The instructor began in a leisurely manner with the simple principles, moved faster over more complicated advanced material, and at the end left us confused as he rushed through the most difficult points.

Although readers of this loose sentence may stop many times before the end, the arrangement of material in the order of increasing interest is designed to keep them reading. Periodic order, however, demands that they read to the end for a complete thought:

> Beginning in a leisurely manner with the simple principles, moving faster over the more complicated advanced material, and at the end rushing through the most difficult points, the instructor left us confused.
> The instructor, beginning in a leisurely manner with the simple principles, moving faster over the more complicated advanced material, and at the end rushing through the most difficult points, left us confused.

The balanced sentence is useful for emphasis on an idea, for comparison of ideas, and for contrast of ideas. When the balanced ideas contrast, the device is called *antithesis*, the classic example of antithesis being Alexander Pope's "To err is human, to forgive divine."

Too much writing on business, government, science, and technology uses the loose sentence almost exclusively. It might with good effect include periodic and balanced sentences for variety, emphasis, and rhythm.

## Variety in Sentence Elements

Writers should always bear in mind the many ways of subordinating ideas. Careful choice provides variety, enables them to express subtle relationships, and makes possible a desired rhythm. A description, re-strictive or nonrestrictive, may be expressed in different kinds of dependent constructions with varied effects:

1. A CLAUSE: Caesar King, *who is the best customer of our perfume department.* . . .
2. AN APPOSITIVE: Caesar King, *the best customer of our perfume department.* . . .
3. AN ADJECTIVE PHRASE: Cleopatra, *sweet with musk.* . . .
4. A PRESENT PARTICIPLE: Cleopatra, *wearing Naughty Nile.* . . .
5. A PAST PARTICIPLE: Caesar King, *scented with Roman Pleasure for Manly Men.* . . .
6. PHRASE CONTAINING A GERUND: *After selling the perfume to them,* the clerk felt he needed deodorizing.
7. AN ABSOLUTE CONSTRUCTION: He watched her approaching on the gaudy barge, *his chance of resisting her growing fainter as she drew nearer. Her journey completed,* she smiled slightly.

The clause (1) is more leisurely and less taut than the appositive (2). The appositive verges on the telegraphic. The adjective phrase (3) is neatly attached to the noun it modifies. Such close positioning tightens a sentence like

My secretary, who is always careful and accurate about details, found the error.

This becomes

My secretary, always careful and accurate about details, found the error.

Participles (4 and 5), having much more flexibility, may replace strings of verbs or short clauses for a smoother, sophisticated effect. Gerunds and absolute phrases (6 and 7) also permit subtle expression.

VERBS: We finally *talked* with Dr. Jones, *asked* him many questions, and *realized* that frequent conferences with him would be necessary.
CLAUSES: *We talked with Dr. Jones,* and *we asked him many questions;* then *we realized that frequent conferences with him would be necessary.*
PARTICIPLES: *Talking* with Dr. Jones and *asking* him many questions, we realized that frequent conferences with him would be necessary.
GERUNDS: After *talking* with Dr. Jones and *asking* him many questions, we realized that frequent conferences with him would be necessary.
ABSOLUTE CONSTRUCTIONS: *Our conferences completed,* we reported that the problem was still unsolved.
*Its morale lowered, its plans confused, its relationships with other divisions tangled in ill feeling,* the research division needs immediate reorganizing, *Smith being the person to do it.*

These varied effects tempt writers. Experimenting with them is interesting and profitable if writers are not rank beginners addicted to dangling participles and misplaced modifiers. Effectively using a variety of forms pleases readers and gives a writer a well-earned sense of accomplishment.

## SENTENCE EMPHASIS

### Contrast

As the discussion of subordination illustrates, sentence variety and sentence emphasis are Siamese twins. Revision to achieve variety aids emphasis, and many of the techniques of emphasis provide variety, none more than the principle of contrast: Any sentence markedly different from the preceding sentences receives stress—a short sentence after several long ones; a periodic sentence after loose sentences; a simple sentence after a series of complex, compound, or compound-complex sentences. To make contrast effective, a writer must be sure that the sentence that is different contains a major thought. If the contrasting sentence states an unimportant point, that point receives stress regardless of the writer's intentions.

Like other principles of emphasis, contrast functions relentlessly, and the writer ignorant of the principle may find it his master rather than his tool. All writing has emphasis, whether the author knows it or not; and if the stress is on the wrong idea, the results may be just as disastrous as the results of illogic. That they are often far worse makes it essential for a writer to understand and use the principles of emphasis correctly.

It is unfortunate that many textbooks and instructors consider emphasis advanced study that is optional, just icing on the cake. A teacher who leaves a writer unaware of the principles leaves him at their mercy. Many writers in the professions spurn emphasis as "fancy technique all right for fiction and poetry and such literary works," but not necessary in the prose of information. One might just as well spurn the laws of gravity. Gravity does not cease to function because someone refuses to acknowledge it. To keep falling on one's face thorugh defiance of gravity is idiotically dangerous. Sometimes one falls far and hard.

We once met a supervisor who was meticulous about correct grammar and diction, particularly in the writing of his subordinates. Emphasis he considered a decoration suitable only for the work of those who had nothing better to do with their time. Every time he sent a memorandum, he stated furiously that his colleagues could not read. Everyone, he complained, always got the wrong idea from his writing. If he wrote about the side effects of a drug, they remembered only some trifling remark about the label. If he wrote about vacation schedules, readers thought he was writing about coffee breaks. And every time that he directed an assistant to do something important, the assistant performed two less important tasks. Why, even vice presidents could not read: they never seemed to get the meat out of his memorandums.

This supervisor had a dull, turgid style that tempted readers to skim as much as possible. Such style not being uncommon in industry and government, this alone would not have caused misunderstanding. But the supervisor defied the laws of emphasis. At the end of a memorandum written in long prosy sentences, he would append an afterthought expressed briefly and colloquially, sometimes as a question or exclamation. These brief postscripts, besides being in the position of principal emphasis, contrasted sharply in language, length, and structure with the rest of his writing. Naturally the principle of contrast operated, and his readers noted and remembered the afterthoughts.

## Position

A sentence containing an important thought sometimes lacks emphasis because instead of being in an important position—the beginning or the end—the main idea is buried in the middle. Placing important words in the emphatic positions is natural to English idiom. An undergraduate says, "If I don't pass that English test today, I'll die," not "If I don't pass that English test, I'll die today." Her father says, "Tonight I want you home by ten o'clock," thus stressing both *tonight* and *ten o'clock*. An engineer says, "Another incorrect pressure chart from that technician, and I'll fire him," with the stress on *fire*. But when they write, they neglect the idiomatic stress of English. "In my opinion," writes the undergraduate, emphasizing her opinion as though she were an expert on the subject, "Chaucer is serious when he describes the prioress and in many other instances too." She could not have found less important words to stress than those at the beginning and end of her sentence. A chemist writes, "In accordance with your urgent request, I am replying as soon as possible," and not only throws away the positions of stress in his sentence but opens his letter with an unnecessary sentence.

An improvement in emphasis may reduce wordiness. Placing important words first and last leads to more compact expression in the following sentences:

Prior to the start of this experiment
Before this experiment

When one proceeds to this further step of written communication
When one writes

In experiments with water the air bubbles were gotten rid of by thoroughly boiling the liquid.
The water was boiled thoroughly to eliminate air bubbles.

A more rigorous derivation would be extremely complicated and would not be justified in view of uncertainties existing with respect to basic information necessary for practical use of the result.

A complicated, more rigorous derivation is not justified when the basic information for practical use of the result is uncertain.

On the basis of the foregoing discussion it is apparent that
This discussion shows

## Upside-Down Subordination

Most speakers and writers place their principal ideas in the main clauses and the subordinate ideas in dependent constructions such as clauses and phrases. Some writers slip occasionally and put main ideas into a subordinate clause or phrase or the subordinate idea into a principal clause.

> INCORRECT: The building caught fire after he saw the first flames.
> CORRECT: He saw the first flames after the building had caught fire.
> INCORRECT: A student mixed the wrong chemicals when he blew up the building.
> CORRECT: By mixing the wrong chemicals, a student blew up the building.
> INCORRECT: Although she was eager to invest in the corporation, its profits were lower than those of last year.
> CORRECT: Although the profits were lower than those of last year, she was eager to invest in the corporation.

## Parallel Construction

A major device for sentence emphasis is parallel construction. Equal thoughts demand expression in the same grammatical form. Repetition of structure within a sentence is a most effective device for making the long sentence easy to read, and repetition of structure in two or more sentences connects them. An understanding of parallelism is therefore essential for emphasis and coherence.

To illustrate the various kinds of parallelism, we have deliberately chosen mundane material of no literary pretensions. We use three units in parallel construction because three is a most satisfying rhythmical number.

### Single Words in Parallel Construction

> NOUNS: The location, size, and price of the factory are satisfactory.
> ADJECTIVES: The factory is convenient, large, and inexpensive.

Any single words—verbs, adverbs, participles, gerunds, etc.—may be used in a series in parallel construction. The important point is to maintain the parallelism. It is easy to permit a construction to slip out of parallel construction with untidy, ineffective results like

The factory is large and convenient, and it doesn't cost very much either.

## Phrases in Parallel Construction

PREPOSITIONAL PHRASES: They will be interested in the convenience of the factory, in its size, and in its price.

INFINITIVE PHRASES: To find a factory with a convenient location, to make sure that the size is right, and to arrange a low price, send Jack.

## Subordinate Clauses in Parallel Construction

SUBSTANTIVE CLAUSES: Jack made clear in his telegram that the factory was conveniently located, that it was large enough, and that the price was satisfactory.

CONCESSIVE CLAUSES: Although the factory was large enough, although it had a convenient location, and although the price was low, Jack was too shrewd to show much interest.

CAUSAL CLAUSES: Because the factory was large enough, because it had a convenient location, and because the price was low, Jack bought it.

## Parallelism in Main Clauses and in Sentences

MAIN CLAUSES: The factory is located conveniently two miles from the city on a main road, it is divided into five areas suitable for our research activities, and it is priced three hundred thousand dollars below our budgeted cost.

SENTENCES: Located conveniently on a main road two miles from the city, the factory can be reached easily by train or car. Divided into five areas suitable in size and arrangement for our research activities, it has adequate truck entrances to the areas and more parking space than we need. Priced three hundred thousand dollars below our budgeted cost, it is two million dollars less than any other suitable property in the selected area and requires fewer structural changes and repairs than any other site we have considered.

### Repeating Introductory Words
When writers repeat the introductory words of a parallel construction, they emphasize each thought; when they use the introductory words only once, they emphasize as a unit the ideas in parallel construction.

A. The Development Committee recommends that the president appoint a committee to consider new systems because the present system duplicates work, because it creates unnecessary delays in the work of several divisions, and because it fails to use effectively the machines bought last month.

B. The Development Committee recommends that the president appoint a committee to consider new systems because the present system duplicates work, creates unnecessary delays in the work of several divisions, and fails to use effectively the machines bought last month.

329

Most writers would choose Sentence *B*, but Sentence *A* is useful if the paper discusses each weakness of the old system. For stronger emphasis in preparation for the discussion, a writer may indent or indent and number:

> *C.* The Development Committee recommends that the president appoint a committee to consider new systems for these three reasons:
> 1. The present system duplicates work.
> 2. It creates unnecessary delays in the work of several divisions.
> 3. It fails to use effectively the machines bought last month.

But Sentence *C* is inappropriate if the following text does not discuss these weaknesses.

### Parallel Construction for Lists

Items in lists, such as those in Sentence *C*, should be parallel in construction, and the ideas should be equal or nearly equal divisions of a subject. The following sentence illustrates the disadvantages of constructions that are not parallel.

> The present system has the following five disadvantages:
> 1. It causes delay in the distribution of incoming mail.
> 2. Too expensive.
> 3. That it has been disapproved by two illustrators in advertising.
> 4. It fails to use effectively the machines bought last month.
> 5. No inspection of efficiency of workers.

Additional discussion of consistency in lists appears in Chapter 4.

### Balanced Constructions

Two sentences in balanced, or parallel, construction connect so closely that conjunctions and other connectives may be few. When a pattern of words, phrases, or clauses is repeated for equal ideas in two sentences, the sentences are balanced.

> EXAMPLE THAT NEEDS BALANCE: Chemical literature is doubling every eight years in volume. Every ten to twelve years there is twice as much biomedical material published.
> EXAMPLE OF BALANCE: Chemical literature is doubling in volume every eight years. Biomedical literature is doubling every ten to twelve years.
> EXAMPLE THAT NEEDS BALANCE: It is acknowledged at the outset that detailed requirements for all possible isotope laboratories will not be described by this standard. Rather the general requirements which are applicable to the categories of radioisotope laboratories as further defined in the next paragraph will be identified.
> EXAMPLE WITH UNNECESSARY WORDS AND UNNECESSARY PASSIVES REMOVED AND BALANCE SUPPLIED: This standard does not describe the requirements for all

isotope laboratories. It describes the general requirements for the radioisotope laboratories defined in the next paragraph.

Similar details about several subjects are suitable for parallel construction—job descriptions, comparisons of bookkeeping entries, comparisons of drugs or chemicals. When the ideas are presented in parallel construction, a reader can compare the details easily. But when the information is jumbled into a variety of structures, it is troublesome to compare details. If information suited to tabular arrangement must be presented in sentences, it should be expressed in parallel construction:

> Secretaries Grade A have typing speeds of 75 words a minute and shorthand speeds of 135 words a minute; they are able to use a word processor, to write good letters, and to assume responsibility; and their salaries range from $450 to $550 weekly. Secretaries Grade B have typing speeds of 60 words a minute and shorthand speeds of 120 words a minute; they are able to use a word processor, write adequate letters, and supervise file clerks and typists; and their salaries range from $325 to $400 weekly. Secretaries Grade C have typing speeds of 45 words a minute and shorthand speeds of 100 words a minute; they are able to file and proofread; and their salaries range from $200 to $275 weekly.

In the prose that treats material less pedestrian than routine comparisons of drugs, chemicals, and jobs, parallelism is usually more subtle, the members being varied sufficiently in length or beginnings to provide the advantages of balance without monotony.

## Climax

The arrangement of three or more units in climactic order is another device of emphasis. The crescendo of louder and louder beats ending in a big bang is agreeable. A haphazard arrangement is disagreeable, and a number of louder and louder beats succeeded by a soft one (anticlimax) is an unpleasant surprise unless it is amusing or witty.

> CLIMACTIC: Miss Jones has been a competent file clerk, a good typist, and an excellent secretary.
>
> HAPHAZARD: Miss Jones has been an excellent secretary, a competent file clerk, and a good typist.
>
> ANTICLIMACTIC INEFFECTIVE: Miss Jones has been a competent file clerk, a pleasant receptionist, a very good secretary, an excellent assistant manager, and a good typist.
>
> ANTICLIMACTIC FOR HUMOR: Grandchildren are anticipated with eagerness, greeted with loving pride, spoiled with boundless enthusiasm, and then avoided with desperate lies and tricks.

## SENTENCE RHYTHM

The relationships among parallelism, climax, and sentence rhythm, are close, for parallel construction provides sentence rhythm, and a series of three climactic parallel constructions is common and may be the most satisfying of English rhythms. Three parallel sentence elements provide a number of rhythms, such as the following:

1. Three parallel elements with the third expanded:

    *a.*  Man, tools, and extinct animals
    *b.*  Number, sex, and average age
    *c.*  Profits of thousands, of tens of thousands, and of tens of millions of dollars

Example *c* illustrates a popular form of this arrangement; each member increases in length thus: __ ___ ____
Example *b* illustrates the long first member, short second, and longer third: ___ __ ____
Example *a* has roughly equal first and second members (the second has a slightly longer vowel) and an expanded third member: __ __ ____

2. Three parallel elements with the third the shortest:

    This reorganization will increase the efficiency of the Shipping Department, speed by at least twelve hours all shipping east of the Mississippi, and save money.

The stress on *save money* is increased by the contrast with the length of the second member.

3. Three parallel members of equal length:

    *a.*  Red, white, and blue
    *b.*  Patients, nurses, doctors
    *c.*  Knowledge, training, and practice
    *d.*  By land, by sea, by air

If parallel elements are long and number more than three, the final member is usually the longest:

    When doctors break appointments, try to make another appointment with their secretaries. When doctors refuse to see you, inquire tactfully about their reasons. When doctors limit the length of interviews, stop your talk at exactly the time set or earlier. When doctors see you promptly, try to estimate whether they are meticulous about appointments and would like a brisk talk or whether they are relaxed and have time for leisurely conversation.

The rhythm of two balanced sentence elements is also common in English. Balance is effective when the elements contrast; antithesis is a good device for emphasis as well as for rhythm.

> The inspector said that we were late submitting samples but he was late in examining them, and he implied that we were not testing accurately but he was not recording accurately.
>
> The chairman of a meeting should be patient but not docile, strict but not discourteous, informed but not informative, fluent but not longwinded.
>
> A good business letter, like a chat, may be personal and informal, but it should not be slangy or discursive.
>
> He was a good leader in the presence of followers and a good follower in the presence of leaders.
>
> He was a good assistant and a good worker, but he was a poor executive and a poor source of ideas.

Careful attention to rhythm and to euphony is most important when one is writing speeches and letters. These should follow closely the rhythms of a writer's normal speech. To test their work, writers of speeches should read their work aloud and mark any sections that are difficult to say. Actors sometimes find such passages in plays and complain that certain groups of words do not fit in the mouth. When writers find groups of words that do not fit in the mouth or roll from the tongue, they should rephrase. The only good test is reading aloud. When a secretary reads a speech aloud from notes, an executive can repeat any phrases that the secretary stumbles over. If the executive finds them difficult to say, the solution is to replace them. Usually it is not the secretary's shorthand that is at fault but the rhythm or euphony of the passage. It is essential that the person who is to deliver a speech read it aloud beforehand. The speech patterns of a writer may be stumbling blocks for a speaker, as writers for presidents of the United States have sadly discovered. The criterion must always be the ease with which the speaker can utter the words. Reading aloud sometimes helps letter writers who have been criticized for bookish or unfriendly styles. Unless they are extraordinarily pompous speakers, saying the words helps them to rephrase passages to achieve conversational rhythms. Such rhythms are the basis of the success of good writers and speakers.

## REVISION THAT CHEERS

Writers who are not in the habit of revising may at this point be surprised by the time and effort involved in correcting and improving sentences. The only comfort for them is the fact that intelligent attention to the suggestions given will bring satisfying results and that some parts of the work will be fun. Beginners often learn by writing each sentence ten different ways; successful authors may spend hours on the revision of one sentence. If writers in the professions will take their best report and try writing the sentences as many different ways as possible, they will be pleasantly surprised not only by the improvement in their writing but by the enjoyment they find in revising. Revision is dull only when writers work without any idea of what to look for and how to improve. As soon as they understand a few basic principles, they work with a purpose, and the encouragement they gain from their progress keeps them going even when a sentence proves especially recalcitrant. In fact, at that point they will dig in, determined to lick the sentence if it takes all night.

## NOTE

1. Porter G. Perrin, *An Index to English* (New York: Scott, 1939) 545.

# CHAPTER 12

# Style and Paragraphs

*The whole is greater than its part.*
                                    *Euclid*

Good paragraphs are pleasant to read because they vary in length, development, and organization. They move their readers quickly through simple material and explain and illustrate more slowly any difficult points. Good paragraphs are carefully connected, and when there is a marked change in thought, there are enough indications to help readers follow the shift. Good paragraphs do not repeat unnecessarily or digress; instead they cover their subjects thoroughly and briefly. While their readers are still interested, the writing ends in a satisfactory final paragraph and leaves readers wishing for more such reading.

## LENGTH

### Logic and the Paragraph

A paragraph, like a stage of a journey, is a logical division of the whole; or it may be a complete work in itself. As a unit of a larger work, a paragraph marks divisions of thought, thus enabling readers to perceive easily the structure of the whole and to observe the progress of their reading. Such a logical, coherent paragraph signals a shift in thought and a step toward comprehension of the whole. A paragraph that stands alone requires an appropriate beginning and ending and the same attention to logical organization that any complete work demands (Chapter 3).

Logic is a principal influence on paragraph length. Some subjects have natural divisions that it would be foolish to ignore. A discussion of three characteristics of a polymer may occupy one paragraph or three. A generalization and an example may be developed in one paragraph or two or more: one paragraph presenting the generalization and the two examples; two paragraphs, one for the generalization and the other for the two examples; three paragraphs, one for the generalization and one for each example; any number of paragraphs for generalizing and any number of paragraphs for presenting each example. But it is illogical to place the generalization and one example in one paragraph and the other example in another paragraph. A writer must clarify for the reader the logical relationships.

Obviously, even though a subject has inherent divisions, writers still have some choice. If they exercise this choice intelligently, they will not break a weighty subject into slivers that are difficult for readers to perceive as a whole, nor will they build mountainous, discouraging paragraphs. Rather they will use paragraphing as a device to stress the inherent divisions of a subject effectively for their readers. For example, the following long paragraph of Alice S. Rossi is long to show the unity of its four ideas and is skillfully broken by indention and numbering to look less forbidding.

> If we want more women scientists, there are several big tasks ahead:
>
> 1) We must educate boys and girls for all their major adult roles—as parents, spouses, workers, and creatures of leisure. This means giving more stress in education, at home and at school, to the future occupational roles of girls. Women will not stop viewing work as a stopgap until meaningful work is taken for granted in the lives of women as it is in the lives of men.
>
> 2) We must stop restricting and lowering the occupational goals of girls on the pretext of counseling them to be "realistic." If women have difficulty in handling the triple roles of member of a profession, wife, and mother, their difficulties should be recognized as a social problem to be dealt with by social engineering rather than be left to each individual woman to solve as best she can. Conflicts and difficulties are not necessarily a social evil to be avoided; they can be a spur to creative social change.
>
> 3) We must apply our technological skill to a rationalization of home maintenance. The domestic responsibilities of employed women and their husbands would be considerably lightened if there were house-care service firms, for example, with teams of trained male and female workers making the rounds of client households, accomplishing in a few hours per home and with more thoroughness what the single domestic servant does poorly in two days of work at a barely living wage.
>
> 4) We must encourage men to be more articulate about themselves as males and about women. Three out of five married women doctors and engineers have husbands in their own or related fields. The views of young and able women concerning marriage and careers could be changed far more effectively by the men who have found marriage to professional women a satisfying experience than by exhortations of professional women, or of manpower specialists and family-living instructors whose own wives are homemakers.[1]

## Format and Paragraph Length

Without the indentions, the preceding paragraph would have seemed formidable in the narrow columns of the letters in *Science*. Writers should remember that paragraph length should be adjusted to format. Paragraphs that look pleasing in handwritten manuscript may be long in the narrow columns of a newspaper, short in typescript, and fragmentary in a printed book. A beginner can learn the adaptation of paragraph length from journalists. In the narrow columns of a newspaper, where paragraphs of many sentences would appear cumbersome, a reporter uses very short paragraphs, many of them one-sentence paragraphs. When columns are wider, as on an editorial page, newspaper paragraphs are longer.

Many writers of informational prose never visualize their paragraphs as readers see them; consequently they write forbidding paragraphs three and four typewritten pages long. When these are published in narrow-column periodicals, they seem interminable. And some authors in business and industry who are used to writing short paragraphs in letters do not realize that the brief paragraphs suitable to a letter look choppy in type. Writers should anticipate the final appearance of their paragraphs and plan them to be neither ponderous nor scrappy, even though planning may require counting the words of a particular format to estimate how paragraphs will look.

### The Reader and Paragraph Length

Paragraphs written for poor readers should be shorter than those directed to competent readers. To poor readers short paragraphs indicate that they will not be asked to grasp too much at one time, that the subject has been divided into small segments for easy understanding. Competent readers may consider such short paragraphs baby food and feel that they want to use their teeth on ideas that are not predigested. But they do not want paragraphs of three pages or more, like those in some scholarly articles. Average readers of the literature of science, business, and technology want normal portions, and they recognize that some pieces will be larger and some smaller than others because of the material and the need for diversity. Readers who reject baby food do not necessarily want to bite into a whole joint.

Most readers of literature on professional subjects expect paragraphs of one hundred to three hundred words. If readers are not well educated, writers may plan shorter paragraphs. If it is impossible to shorten long paragraphs, they may break them by numbering internal divisions, by indenting, by using several methods of development, by the use of dialogue, or by an adroit placement of a topic sentence near the middle of a paragraph. Unbroken paragraphs longer than three hundred words should be infrequent, even in papers directed to experienced readers. Long paragraphs that slowly drag their lengths along like slow freight trains delay readers discourteously.

### Variety and Emphasis

Contrast in paragraph length may be used to good effect. After a number of short paragraphs, a long one receives strong stress; after a number

337

of long paragraphs, a short one is emphasized; after a succession of paragraphs of medium length, a long or a short one is prominent. By placing an important idea in the paragraph that differs, a writer can highlight that idea and subordinate unimportant ideas.

Ignorance of this principle accounts for some failures to communicate effectively. If a writer of a memorandum of four paragraphs, each of one hundred to two hundred words, places a minor thought in a twenty-word paragraph, readers are likely to note and remember the unimportant information of the twenty-word paragraph. While revising, a writer should be certain that any paragraph markedly different in length from the preceding paragraphs contains material suitable for major stress. And a writer wondering how to emphasize a division of the subject might consider placing it in a paragraph distinctly longer or shorter than the surrounding paragraphs. Skillful use of contrasting lengths provides economical and effective stress and variety. And it is neither difficult nor onerous to write paragraphs of contrasting lengths.

Yet many writers do not vary paragraph length sufficiently. If their monotonous paragraphs are long, their style is likely to seem slow moving, heavy, difficult, even turgid. A reader may consider the writing too deep. If monotonous paragraphs are short, the style will seem choppy, rapid, light, or even childish. A reader may think the work superficial. But even long paragraphs mingled with short ones can become monotonous. A writer's paragraph lengths should not be only elephants and mice. There were also animals of intermediate size in the ark.

Achieving effective paragraph length is part of the art of writing. For each work a writer should weigh the relative importance of the demands of the subject, format, readers' preferences, need for variety, considerations of emphasis, and effect on style. Many writers judge these subconsciously. Those who have never thought about paragraph length may have to weigh and choose consciously for a time, but soon they too will plan effective lengths without much conscious effort.

## UNITY

Whether it is long or short, a paragraph develops a single idea. This idea may be in one sentence—the topic sentence—or it may not be stated anywhere in the paragraph. Often it opens or closes a paragraph, but sometimes the topic sentence is so obvious that it is not expressed.

If everything in a paragraph concerns the topic sentence, the paragraph has unity, which readers expect in a paragraph. Including bits and pieces of other ideas confuses them and weakens or even destroys the central idea; therefore writers who are inclined to digress should test the unity of their paragraphs when they revise. Anything that does not enforce the topic sentence should be removed entirely or placed elsewhere. This clarifies and sharpens the main point of a paragraph and is thus a major step toward a clear, clean-cut style.

## METHODS OF DEVELOPMENT

Paragraphs are usually more than mere statements of central ideas; they are developments of central ideas. Each paragraph offers variations on a theme. Good writers do not just state an idea and then jump to some other subject, because that would confuse readers, who expect and need the development of ideas. There are many methods of developing paragraphs, and writers often use several methods in one paragraph to display facets of a topic effectively or to vary a long paragraph. Some of the common methods of paragraph development are the following: abstract and summary, analogy, analysis and classification, cause and effect, comparison and contrast, definition, details and particulars, examples and illustrations, questions, scope and qualifications, and the straw man.

### Abstract and Summary

In scholarly journals, a paragraph abstract opens most articles, and a summary closes many.

Informative abstracts, which present as much as possible of the contents of a paper, are the most useful kind for most purposes. Organizing the abstract in the order of the paper is helpful, as the following example from the *American National Standard for Writing Abstracts* shows:

> Because damage to sweet potatoes by root-knot nematodes makes it difficult for some growers in Mississippi to produce marketable grades, the Truck Crops Branch Experiment Station in 1967 conducted off-station tests with nematocides (including fumigants) on three- or four-row replicated and randomized field plots known to be infested with the nematodes. Both known and experimental nematocides were employed. The commercial fumigants Vorlex, Dow W-85, and DD significantly increased yields and quality in the treatments of rows. Vorlex or Dow W-85 should be applied at 2.5 gal/acre and DD at 9–10 gal/acre, 8–10 inches deep in the center of the row, 14–30 days prior to planting. Broadcast fumigation was also effective, but required higher fumigant levels. Among the experimental solid nematocides, Bayer 68138 and Dasanit showed promise. More information is deemed necessary than was obtained from this one-season field test.[2]

An abstract for the same paper is organized to emphasize the findings:

> The yield and quality of sweet potatoes can be increased by soil fumigation or the addition of solid nematocides in some areas of Mississippi. The commercial fumigants Vorlex, Dow W-85, and DD significantly increased yields and quality in the treatments of rows. Vorlex or Dow W-85 should be applied at 2.5 gal/acre and DD at 9–10 gal/acre, 8–10 inches deep in the center of the row, 14–30 days prior to planting. Broadcast fumigation was also effective, but required higher fumigant levels. Among the experimental solid nematocides, Bayer 68138 and Dasanit showed promise. This study of control of root-knot nematodes was conducted by the Truck Crops

Branch Experiment Station in 1967 on three- and four-row replicated and randomized field plots known to be infested with the nematodes. More information is deemed necessary than was obtained from this one-season field test.[2]

Indicative abstracts, which are similar to tables of contents expressed in sentences, used to dominate the writing of abstracts but are now used mainly to reduce costs. For comparison an indicative abstract of the same paper on nematode control follows:

> Problems caused by root-knot nematodes in growing sweet potatoes in Mississippi are discussed. Experiments with commercial and experimental nematocides, conducted in 1967 by the Truck Corps Branch Experiment Station, are described. Methods of application, including imbedding in rows and broadcasting, are compared. Results are given for specific nematocides, including the commercial fumigants Vorlex, Dow W-85, and DD, and the experimental solid nematocides Bayer 68138 and Dasanit.[2]

The indicative abstract is far less effective than the informative, which most writers prefer.

When there are restrictions on the length of the abstract or other limitations, the hybrid informative–indicative type is sometimes employed. Examples of this type and other examples of informative abstracts are in the *American National Standard.*

Paragraphs of summary may appear in any appropriate place in a document; ending with a summary is a frequent use.

Chapter 1 of *Only One Earth: The Care and Maintenance of a Small Planet,* which has a wealth of detail, ends with sweeping summary statements:

> In short, the two worlds of man—the biosphere of his inheritance, the technosphere of his creation—are out of balance, indeed potentially in deep conflict. And man is in the middle. This is the hinge of history at which we stand, the door of the future opening on to a crisis more sudden, more global, more inescapable, and more bewildering than any ever encountered by the human species and one which will take decisive shape within the life span of children who are already born.[3]

Similarly, Elizabeth Janeway ends Chapter 17 of *Man's World, Woman's Place: A Study in Social Mythology* with a broad summary paragraph:

> To sum up the trend behind the headlines and slogans, what is happening appears to be this. Many married couples include a working wife. Her job is an understood thing between them, either because she was working when she married and kept on or—as in the more easily analyzed cases we've been considering–because she decided to go back to work after staying home for some years as a housewife and mother. Implicitly in the former situation

340

and explicitly in the latter, her husband agrees to her doing so. Her job thus becomes a part of the marriage and its context, for both husband and wife actively (if sometimes unconsciously) adjust their marriage to take account of the wife's outside activities. Her decision to work is ratified and the old pattern of wedlock changes and expands to contain it. A new equilibrium has appeared, and what we think of as being a normal and usual sort of marriage has become more varied and less restricted.[4]

A final summary is written for those who have read the preceding material; it need not represent only the contents of the book, paper, chapter, or section it follows. Simone de Beauvoir completes the concluding chapter of *The Coming of Age* with a summary paragraph that has a challenging final sentence:

> We are far from this state of affairs. Society cares about the individual only in so far as he is profitable. The young know this. Their anxiety as they enter in upon social life matches the anguish of the old as they are excluded from it. Between these two ages, the problem is hidden by routine. The young man dreads this machine that is about to seize hold of him, and sometimes he tries to defend himself by throwing half-bricks; the old man, rejected by it, exhausted and naked, has nothing left but his eyes to weep with. Between youth and age there turns the machine, the crusher of men— of men who let themselves be crushed because it never even occurs to them that they can escape it. Once we have understood what the state of the aged really is, we cannot satisfy ourselves with calling for a more generous "old-age policy," higher pensions, decent housing and organized leisure. It is the whole system that is at issue and our claim cannot be otherwise than radical—change life itself.[5]

A final paragraph expressing a summary may be difficult to write; it requires careful attention to organization, to concision, and to expression that enforces statements in the paper without repeating the wording. The organization should be such that the summary does not end with a whimper. Concise, lively expression should keep the reader from feeling any tediousness in the ending. The presentation of ideas should seem fresh and should add to the first presentation something new—force in expression, a change of pace, a fresh view, an element of poetic truth, the welcome clarification that is possible only after all else has been said, or a clever, memorable twist of thought or expression. Such an ending requires careful editing and polishing. Still more difficult is selecting the thoughts for a final summary. If the work is long, if the research is still incomplete, if the project has taken a long time, the author, the person most involved, may have trouble recapturing the main points for a final summary. If that is still so after writers have refreshed their memory with a reading of their research objectives and early notes, the writers are well advised to use some other ending.

Experienced writers have less trouble with the final summary though they often spend much time polishing it. Arden Neisser ends *The Other Side of Silence* with a forceful summary of main points:

> Educational policy in the past has done much to discourage deaf culture, and now appears actively trying to destroy it. But deaf people have deep convictions about themselves; they understand deafness completely. They also have a long history of resistance to theories and policies that threaten those convictions. The deaf have dug in their heels, cried out with their hands, and preserved the better part of their humanity: their spirit, their intelligence, and their language.[6]

Summary beginnings (Chapter 5) are common; many of them illustrate how well summary combines with other types of openings.

### Analogy

Frequently used to explain ideas, particularly to acquaint readers with unfamiliar concepts by comparison with familiar ones, analogy is helpful, for example, in explaining technical matters to a layperson. Much depends on the choice of the familiar concept, which is usually a simple one with which readers are well acquainted; it should be original, appropriate, and consistent. Sometimes a comparison with a concrete subject vividly clarifies an abstract subject, or an abstraction grows from concrete examples. An analogy may also impress upon the memory of a reader something that has just been explained. Analogy has the appeal of the picturesque and the imaginative; it may please readers by its cleverness, help writers to simplify, and provide a change from more mundane reasoning. Analogies may be poetic or almost poetic.

Loren Eiseley in an effective analogy describes a naturalist looking backward:

> The door to the past is a strange door. It swings open and things pass through it, but they pass in one direction only. No man can return across that threshold, though he can look down still and see the green light waver in the water weeds.[7]

This says much that would be difficult or impossible to state clearly without the analogy, as does the following paragraph by Marjorie Hope Nicolson:

> For three hundred years men have vainly tried to put together the pieces of a broken circle. Some have been poets, some philosophers, some artists. They have shared a common desire for a unity that once existed, and have sought a "return to medievalism," when life seemed integrated about a strong center, whether of the Church or of a monarch. Except for an occasional individual who has found peace in old religion, their efforts have proved fruitless. Poets and artists have deliberately revived old styles, but

these attempts have been equally abortive. Modern critics have kidnapped to our times poets like Donne, in whom they find a "unified sensibility" of feeling and thinking. Philosophical poets—Pope, Wordsworth, Tennyson—have tried to express a world view, as did Lucretius for the ancients, Dante for the Middle Ages, Milton for the seventeenth-century Protestant. But all the king's horses and all the king's men cannot put Humpty-Dumpty together again. Mere fitting together of pieces may remake the picture in a jigsaw puzzle; it will not remake an egg. Nor can we reconstruct the old Circle of Perfection, broken by modern science and philosophy. Donne spoke truly when he said: "Nothing more endless, nothing sooner broke."[8]

John R. Platt presents a clarifying analogy to describe strong inference:

> It is like climbing a tree. At the first fork, we choose—or, in this case, "nature" or the experimental outcome chooses—to go to the right branch or the left; at the next fork, to go left or right; and so on. There are similar branch points in a "conditional computer program," where the next move depends on the results of the last calculation. And there is a "conditional inductive tree" or "logical tree" of this kind written out in detail in many first-year chemistry books, in the table of steps for qualitative analysis of an unknown sample, where the student is led through a real problem of consecutive inference. Add reagent A; if you get a red precipitate, it is subgroup alpha and you filter and add reagent B; if not, you add the other reagent, B'; and so on.[9]

Later this writer uses analogy for one of its principal advantages—its contribution to the coherence of a whole section. In the paragraph that follows the section just quoted, he echoes the analogy with "to proceed to the next fork," and in the next paragraph he writes, "Strong inference, and the logical tree it generates. . . ." He concludes this part of his discussion with another appropriate analogy:

> The difference between the average scientist's informal methods and the methods of the strong-inference users is somewhat like the difference between a gasoline engine that fires occasionally and one that fires in steady sequence. If our motorboat engines were as erratic as our deliberate intellectual efforts, most of us would not get home for supper.[10]

Timothy Ferris in "Physics' Newest Frontier" presents a wealth of analogies in his effective treatment of a difficult subject:

> The confluence of unified theories and the newest ideas in cosmology is producing a new physics, one that sees matter, energy and the laws according to which they behave not as immutable and eternal, but as evolving processes in an evolving universe. Like archeologists unearthing the ruins of an ancient city, theorists are uncovering the outlines of a new, more profound and, in some ways, simpler physics, a physics that could reveal and explain much of the history of the cosmos.

343

Physicists have hit upon this new vision of a unity of large and small not through the mystical transports of a William Blake, who saw the world in a grain of sand, but while trying to do little more than put their scientific house in order. Their goal is not to rewrite Genesis, but only to craft a theory that would explain the interactions of the myriad particles—from the now-familiar quarks to obscure moons—that inhabit the subatomic world.

. . .

The first of the new unified theories, known as the Weinberg–Salam theory, views electromagnetism and the weak interaction as aspects of a single "electroweak" interaction. Formulated in the 1960's, it was named for the American physicist Steven Weinberg, then at M.I.T. and now at the University of Texas at Austin, and the Pakistani physicist Abdus Salam of the International Center of Theoretical Physics in Trieste, Italy, who arrived at it independently. They shared the 1979 Nobel Prize in Physics with Sheldon L. Glashow of Harvard, for which reason the electroweak theory is sometimes called the Glashow–Weinberg–Salam theory. (There is so much interaction and collaboration in theoretical physics that the list of names appended to major papers can rival those dangling from the letterheads of Wall Street law firms.)[11]

Analogies of the familiar with the unfamiliar, of the simple with the complex, even of the concrete with the abstract put the reader at ease, clarify concepts, and vivify ideas to make them memorable. More writers of informational prose should use this efficient and effective method of presenting and developing ideas—analogy.

## Analysis

Analysis, or classification, occurs frequently in writings on science and technology. In explaining a method or process, a writer may divide a subject into such topics as materials, apparatus, steps, and results. Part of a paragraph, one paragraph, or several paragraphs may analyze each subject. Robert L. Sproull ends a paragraph as follows:

> Of all the types of accelerators we discuss only two, the cyclotron and the synchrotron. References that describe the betatron, linear accelerator, synchrocyclotron, and other accelerators are listed at the end of the chapter.[12]

David V. Becker begins "Choice of Therapy for Graves' Hyperthyroidism" with a paragraph of classification:

> Surgery, antithyroid drugs, and radioiodine are all effective for the treatment of adult Graves' hyperthyroidism. However, reports in the medical literature present widely varying and often conflicting data with regard to cure, relapse, frequency of side effects, cost, and convenience. The differences are probably related to differences in patient classification and selection, quality and duration of follow-up, and immunologic and nutritional

344

status. Furthermore, the natural tendency for patients with Graves' disease to become hypothyroid eventually and the unpredictable nature of surgical complications, adverse drug reactions, and radiation sensitivity may perplex the therapist.[13]

A common paragraph development in papers on science and technology is analysis or discussion of a process or procedure. In its most unadorned form this resembles a recipe or other simple instructions. The arrangement should be chronological for such subjects as in the order of the steps in an experiment, in assembly-line activities, in a simple procedure like recharging a battery. For larger concepts the paragraphs may be steps, or each paragraph may contain a few related steps.

A general analysis of major points is also common. Barbara Ward and René Dubos open *Only One Earth* with a helpful analysis of a major point of the book:

> Man inhabits two worlds. One is the natural world of plants and animals, of soils and airs and waters which preceded him by billions of years and of which he is a part. The other is the world of social institutions and artifacts he builds for himself, using his tools and engines, his science and his dreams to fashion an environment obedient to human purpose and direction.[14]

## Cause and Effect

Cause and effect—which should come first? This is a question that troubles some writers. It is usually easier for readers if a paragraph begins with whichever is more familiar to them. A paragraph may state a cause or causes, an effect or effects, or both causes and effects.

The following illustrative paragraph lists a cause, the general effect, and then the specific effects:

> If the 50,000 control devices in the oil refineries of the U. S. should go "on strike," we would be faced with social disaster. The refineries would become lifeless industrial monuments. If we undertook to replace them with old-fashioned, manually operated refineries to supply our present motor-fuel needs, we would have to build four or five times as much plant, cracking and some other modern chemical processes would have to be eliminated, yields of motor fuel from crude petroleum would drop to a quarter of those at present, costs would skyrocket, and quality would plummet. Automobile engines would have to be radically redesigned to function with inferior fuel. And because of lower motor-fuel yields, we would need to produce crude petroleum several times as rapidly as we produce it now. Technology in refining would be set back to the early 1920s.[15]

C. P. Snow develops a clear, convincing paragraph by stating a result and then two causes:

I said earlier that this cultural divide is not just an English phenomenon: it exists all over the western world. But it probably seems at its sharpest in England, for two reasons. One is our fanatical belief in educational specialisation, which is much more deeply ingrained in us than in any country in the world, west or east. The other is our tendency to let our social forms crystallise. This tendency appears to get stronger, not weaker, the more we iron out economic inequalities: and this is specially true in education. It means that once anything like a cultrual divide gets established, all the social forces operate to make it not less rigid, but more so.[16]

Discussing preferences for well-done and for rare lamb, Craig Claiborne with Pierre Franey states the causes succinctly:

To my mind there is a simple explanation for these cooking preferences. The lamb of France is quite tender and can be eaten with pleasure if it is not cooked to the well-done state. The lamb of Haiti and Greece requires that it be tenderized through one long-cooking technique or another. The lamb of America is more like that of France and comes off exceedingly well when cooked to the rare or medium-rare stage.[17]

## Comparison and Contrast

If a subject can be clarified by stating its resemblances to or differences from other subjects, comparison or contrast is effective. The most satisfactory comparisons treat concepts that seem much alike but on close examination show differences, or they treat concepts that seem obviously different but also have important similarities. Although a lengthy comparison does not lend itself to a treatment of first one concept and then the other, followed by a comparison of the two (Chapter 3), a single paragraph may be organized in this way. It is less likely than is a long work to fall apart where a writer shifts to a second concept, and a reader is less likely to forget the details of the first concept before reaching the comparative details of the second. In a paragraph one might discuss first Pension Plan A and then Pension Plan B, or one might compare the details of A with those of B point by point. When a comparison is to be presented point by point, a writer should choose an appropriate order— chronological, climactic, spatial, etc. (Chapter 3); if the writer treats first one complete concept and then the other, the treatments should be parallel and the order should be similar as though in a table. A tight plan is important in comparison and contrast, for helter-skelter presentation confuses readers and makes it difficult for them to know where they have been and where they are going next and arduous to comprehend the similarities and differences on which the paragraph development depends.

To analyze variations on the ammonia process, I. M. LeBaron offers a general comparison:

Continuing investigations have developed a number of ammonia processes. The principal ones, identified by the men or group of men developing them, are the Badische Anilin und Soda Fabrik, Nitrogen Engineering, Claude, Casale, Fauser, and Mont Cenis processes. These processes are fundamentally the same in that nitrogen and hydrogen react at elevated temperatures and pressures in the presence of a catalyst. They vary in arrangement, construction of equipment, and the pressures and temperatures used. A number of processes have been developed for producing the synthesis-gas mixture which depend primarily on the source of the gas, whether it be coke, natural gas, coke-oven gas, or by-product hydrogen.[18]

In her review of *Setting Limits: Medical Goals in an Aging Society*, Margery Cunningham states that the author, Daniel Callahan, wants to bring his readers promptly "to an understanding of death as the closing entry in a natural life span," and then she describes the subjects of more interest to him and clarifies them by a running use of contrast:

> Of more concern to him are the goals of medicine as they relate to the elderly, and he sees those goals being mis-set and misunderstood. Over time, the concept of medicine has gone from "we care" to "we cure," the concept of health from "we hope for" to "we are entitled to."
>
> As these concepts have been re-defined, the primary goal of medicine, which according to Callahan should be health, has become the indefinite extension of life at all costs. Callahan sees those costs jeopardizing not only the young who must bear them, but also the old who, often unwilling, incur them.[19]

Robert C. Cowen writing in the narrow columns of *The Christian Science Monitor* presents contrast effectively in short informal paragraphs. His use of the negative illustrates a common device in expressing contrast:

> What a contrast between the symposium on genetic engineering held at the National Academy of Sciences last week and the first such meeting eight years ago!
>
> There was no banner this time sarcastically quoting Adolf Hitler on the desirability of genetically perfecting the human race. No placards sporting skull and crossbones warned darkly of undefined "biohazards." No one sang, "We shall not be cloned." And, instead of an overflowing audience packed with what the harassed genetic pioneers called "the vocal public," half the seats in the auditorium were empty.
>
> This time the scientific, legal, and industrial leaders of the new biotechnology were left in peace to review the progress of research on how to alter organisms genetically and warn of the technology's still considerable limitations.[20]

## Definition

A formal definition places the term to be defined in a class and then distinguishes it from the other members of the class, but in literary definition, other methods may supplement this procedure or even substitute for it. Almost any method of paragraph development may be used to amplify a definition. Literary definitions may include examples, the name of the discoverer or inventor, derivation of the word or words, synonyms and comparisons with synonyms, semantic history, description of the appearance, names of the parts, an explanation of what the term is not, an explanation of how it works, an explanation of how it was made; even the location of the object or term being defined may augment the definition. From these, writers select the topics most helpful to readers. Definitions for technical readers, for example, need not be clear to other readers but still may benefit from the topics and methods common in less specialized definitions.

In *Science* J. Schöler and J. R. Sladek, Jr., begin an article with definition:

> Oxytocin and vasopressin are magnocellular neurosecretory peptides of the mammalian hypothalamo-neurohypophyseal system. They play an important role in the onset of parturition and milk ejection and help regulate water and electrolyte balance within intra- and extracellular fluid compartments. These hormones are synthesized in and transported by neurons of the paraventricular nucleus (PVN) and supraoptic nucleus (SON) of the hypothalamus. In addition, the hypothalamic nuclei receive a dense input from brainstem noradrenergic neurons of the locus coeruleus and pontomedullary reticular formation. These noradrenergic neurons, through their hypothalamic innervation, are thought to play an important role in regulating the release of vasopressin and oxytocin.[21]

Writing in *Scientific American,* A. K. Dewdney describes and defines a Galton board so clearly that a reader need not turn two pages to view the drawing of a Galton board (What strange chance decrees so frequently the separation of text and illustration?):

> The third simulation involves a Galton board. This device, named after the pioneering Victorian statistician Sir Francis Galton, is a sloping surface studded with a triangular array of pegs [*see top illustration on page 26*]. Marbles released just above the top peg make their way, helter-skelter, down through the array. Below the pegs are channels that collect the marbles. When the last marble falls into place, the collected columns take on a characteristic shape, not like the New York skyline but something quite different. Those attempting this piece will discover the shape for themselves.[22]

J. G. Shanthikumar and R. G. Sargent in the opening paragraph of an article in *Operations Research* define modeling in problem solving with emphasis on the uses:

> Modeling is a widely used approach in problem solving. It consists of

building a model of the system or problem entity, developing a solution procedure for the model, and using the model and its solution procedure to solve the problem. Modeling is used in a variety of ways, for example (i) in analysis, where modeling is used to obtain an output for a given system and input, (ii) in optimization, where the model and its solution procedure are used to find the values of the decision variables to optimize an objective function, (iii) in synthesis, where a model is developed to convert a set of inputs into a set of desired outputs, (iv) in gaining insight into a system's behavior by developing a model of it and using its solution procedure to explore its behavior, and (v) in the comparison of alternative systems, where modeling of various alternative systems are carried out to determine the "best" one.[23]

Ernest O. Ohsol includes many of the topics of literary definition when he writes about coke-oven gas—analysis of contents, value and some history of value, variation in yield, properties, uses and some history of uses:

> Coke-oven gas can be defined as containing those by-product materials with boiling points below room temperature. Ammonia may therefore be included in this discussion. Coke-oven gas has been and still is the most valuable by-product from the production of coke from coal; however, its value has been decreasing along with that of the other by-products as the years have gone by. In 1937, for example, gas was 19 per cent of the total value of the products from the coke oven and in 1950 11 per cent. Presently in many plants where the gas can no longer be sold to city gas companies the value has decreased to 5 per cent. Even at 5 per cent it is well above that of any of the other by-product components. Of the materials made in the high-temperature coking of coal, the first five in weight percentage are either gases or coke. For example, 2,000 lb. of coal yields 1,400 lb. of coke; 132 lb. of methane; 43 lb. of carbon monoxide; 30 lb. of hydrogen; 20 lb of ethylene; 11.7 lb. of benzene, a liquid product; 6.7 lb. of hydrogen sulfide; 6.2 lb of naphthalene, a solid; and 5.2 lb. of ammonia, a gas. Also of some interest in a consideration of the gas would be the propylene yield at 3.4 lb., butylenes at 2.8 lb., and HCN at 1.7 lb. There is, of course, some variation in these numbers with different coals and different coking operations. The properties of coke-oven gas are as follows: heating value, 550 B.t.u./ cu. ft.; specific gravity, 0.38; volume composition, 57 per cent hydrogen, 28 per cent methane, 7 per cent carbon monoxide, and some nitrogen, illuminants, and carbon dioxide. Coke-oven gas was a very important product for addition to manufactured gas for city use; however, as natural gas came north, it displaced manufactured gas. For some time coke-oven gas was added to natural gas by city gas utilities, but this practice has been partially abandoned with the continued stripping of higher hydrocarbons from the natural gas in the south. The primary use of coke-oven gas today is for fuel purposes inside the plant itself. This is still not true in Europe: the lack of hydrogen and methane in Europe has made coke-oven gas a valuable raw material for the production of ammonia, methanol, and the like.[24]

Charles W. Palmer defines a written procedure by stating its functions:

The written procedure is a document designed to
  1. Direct employees to perform specific operations.
  2. Help employees understand the significance of operations.
  3. Serve as a reference of how and why things are done.
  4. Provide for supervisory review and control.
  5. Expedite job completion by allowing several employees to work on a single job.
  6. Help assure compatibility between employees and their jobs.[25]

But not all definitions can be so simple and forthright. Delicate shades of meaning, nuances, connotations from the past that veil words, atmosphere more significant than dictionary meanings—all these must be conveyed for some words. Elizabeth Janeway's discussion of myth illustrates the development of two paragraphs of definition by the presentation of such material:

> I shall have to ask my readers to consider mythology and mythmaking from a quite different point of view. They are, in the first place, to be taken very seriously indeed, because they shape the way we look at the world. The urge to make, spread and believe in myths is as powerful today as it ever was. If we are going to understand the society we live in, we shall have to understand the way mythic forces arise, grow and operate. I do not believe we shall ever get rid of them and, in fact, I do not believe that we could get on without them: they are the product of profound emotional drives, drives that are basic to life. Sometimes these drives are able to act directly and effectively on the world of events. Sometimes they succeed in gaining their ends rationally and by logic. But sometimes (and particularly when they are thwarted) they substitute for action a will to believe that what they desire exists—or should exist. That is mythic thinking. It is illogical—or, at least, pre-logical; but from this very fact it gains a certain strength: logic may disprove it, but it will not kill it.
>
> How do we think about it then? How do we manage the double vision that refuses to believe in myth but still takes it seriously? I have quoted Erik Erikson's advice on beginnings not only because his sibylline style makes him sound rather as if he had just come from a visit to the Oracle at Delphi, but because what he says makes great sense—as his words very often do. In investigating a psychological manifestation, he is telling us, don't try to proceed in a straight line, hand over logical hand. "A myth, old or modern, is not a lie," he points out in another passage from *Childhood and Society*. "It is useless to try to show that it has no basis in fact; nor to claim that its fiction is fake and nonsense. . . . To study a myth critically . . . means to analyze its images and themes." When we think about myth, that is, we are to think about its purposes—its themes—and the material it snatches at to express, or clothe, those themes, for such material must have special, emotional significance.[26]

## Details and Particulars

Details and particulars should be carefully selected, for they can add life and vigor to writing or can bore and discourage readers. Nothing is more useful in developing a generalization than specific details. Nothing is less useful than unnecessary details.

If details are essential but repetitious, climactic order may hold a reader's interest. Sometimes details are combined effectively with a generalization. Details and particulars may be arranged in many different plans—simple to complex, known to unknown, chronological, spatial, or (if they accompany a generalization) inductive or deductive (Chapter 3). Any series of paragraphs of details or any long paragraphs of details signal a reviser to explore the possibility of relegating the particulars to appendices or of discarding some items. It is often necessary to discard unessential particulars to secure proper emphasis on main points, for paragraphs of details should never be permitted to overwhelm the main thoughts. The question is how many details must the reader have to understand the main point? not how many details would the writer like to include?

Details both clarify and vivify the examples in the following passage from Tobias Dantzig's *Number: The Language of Science:*

> But, also, wherever a counting technique, worthy of the name, exists at all, *finger counting* has been found to either precede it or accompany it. And in his fingers man possesses a device which permits him to pass imperceptibly from cardinal to ordinal number. Should he want to indicate that a certain collection contains four objects, he will raise or turn down four fingers *simultaneously;* should he want to count the same collection he will raise or turn down these fingers *in succession*. In the first case he is using his fingers as a cardinal model, in the second as an ordinal system. Unmistakable traces of this origin of counting are found in practically every primitive language. In most of these tongues the number "five" is expressed by "hand," the number "ten" by "two hands," or sometimes by "man." Furthermore, in many primitive languages the number words up to four are identical with the names given to the four fingers.[27]

In *The Other Side of Silence* Arden Neisser uses three paragraphs of interesting details to present information that is less technical:

> The year 1980–81 marked the thirteenth season for the National Theatre of the Deaf (NTD). Some critics consider NTD the best company of actors in the country. They have performed in all fifty states; in legitimate houses on Broadway, on the mezzanines of shopping malls in the Middle West, in regional theaters and college auditoriums. They regularly tour Europe, have been to Asia and Australia, appear on television, and have generated two spin-offs: the Little Theatre of the Deaf, a smaller one-truck group that gives performances mostly for children; and the Theatre in Sign, an ASL company that performs without voice interpretation for deaf communities.
> The National Theatre of the Deaf is not a mime theater. The charm of mime lies in the audience's ability and willingness to decipher the actor's gestures in the absence of language. NTD is a linguistic theater. The au-

351

dience is presented with language twice, in two modes: speech and sign. In some productions, it comes very close to being a bilingual theater.

In a typical season, NTD employs thirteen or fourteen actors; two or three are hearing, the rest deaf. Because the hearing actors are also called readers, I had expected that they would stand at the side of the stage like classroom interpreters and merely relate to the audience the meaning of the signs. Nothing so crude occurs. The hearing actors are onstage, in costume, playing particular characters.[28]

## Examples and Illustrations

Examples and illustrations can advance the thought of a section vividly, dramatically, and convincingly. Paragraphs of examples and illustrations are useful for clarifying generalizations and for adding meaning and force to abstract statements. They are helpful in teaching; a student who forgets a general point may recall a specific example and thus regain the generalization.

The textbook *Foundations of Psychology* is, as is every good textbook, a mine of paragraphs developed by examples and illustrations. Discussing the need for control in experiments, the authors illustrate as follows:

> On the other hand, hunch comes into this business too. No one can keep all the conditions constant, and the experimenter has to guess which conditions are the most important. Suppose you discovered on a Tuesday that a certain percentage of automobile drivers cannot tell a red traffic light from a green, except by knowing that the red is on top. (You could do it by interchanging the red and green in one signal, provided you prevented accidents in some other way.) Well, that was Tuesday. Would you have to repeat the experiment on Wednesday and all the other days? No, you assume that the day of the week makes no difference, that eyes see the same on Tuesdays and on Wednesdays. Nor does the phase of the moon matter, nor the last name of the driver. It is by hunch that you leave these matters out of control. You hope they make no difference. Sometimes, when a long-accepted generalization turns out later to be wrong, it is because some such essential condition was not controlled when the original generalization was formed. For instance, most people would expect sex to make no difference in observing traffic lights, but it does. Very few women are color-blind.[29]

Caroline Bird in a project of the Business and Professional Women's Foundation develops a paragraph of examples in the following discussion of the work of Jayne Baker Spain:

> It is hard to evaluate from the balance sheet all that the handicapped workers accomplished for Alvey-Ferguson but there were obvious tangible gains: Their absentee rate was negligible. They came to work on time every day, rain, snow, or shine. They did not jump from job to job but stayed so long on the job that training costs were lowered. They had no accidents, as they were extremely safety conscious, so insurance rates came down. They

were so proud to be productive members of society that the quality and the quantity of their work were excellent, and, says Jayne, their cheerful outlook, despite their physical handicaps, raised the morale of all employees. They enabled Alvey-Ferguson to attract as well as hold a loyal work force.[30]

## Questions or Questions and Answers

Questions or questions and answers can pique interest, direct readers' attention to particular ideas, and present problems briefly and impersonally.

Timothy Ferris uses question and analysis of some answers effectively in "Physics' Newest Frontier":

> Where did the universe come from? That question may be unanswerable by science—it may be unanswerable, period—but scientists do try to investigate it, and their speculations deepen our appreciation of the nature of science, whatever they may tell us about Genesis. Among the most provocative and promising of these are theories proposing that the universe sprang into existence from little or nothing, as if from a vacuum.[31]

A paragraph of questions or questions and answers is not an unusual opening for an article or chapter. In "Small Systems of Neurons" in *The Brain*, Eric R. Kandel poses a number of questions to guide readers from the start:

> Many neurobiologists believe that the unique character of individual human beings, their disposition to feel, think, learn and remember, will ultimately be shown to reside in the precise patterns of synaptic interconnections between the neurons of the brain. Since it is difficult to examine patterns of interconnections in the human brain, a major concern of neurobiology has been to develop animal models that are useful for studying how interacting systems of neurons give rise to behavior. Networks of neurons that mediate complete behavioral acts allow one to explore a hierarchy of interrelated questions: To what degree do the properties of different neurons vary? What determines the patterns of interconnections between neurons? How do different patterns of interconnections generate different forms of behavior? Can the interconnected neurons that control a certain kind of behavior be modified by learning? If they can, what are the mechanisms whereby memory is stored?[32]

The development of a paragraph mainly by questions is less common but does occur. Sometimes the questions are introduced by a one-sentence statement, as in "The Spread of Economic Growth to the Third World: 1850–1980":

> The most intriguing question in growth economics is *why* a turning point occurs in a particular country at a particular time. Why, under the favorable world conditions of 1850–1914, did some countries embark on intensive

growth while others did not? Why was this true once more during the boom of 1945–1973? Why have turning points been spread out over more than a century, and in some countries have not yet occurred?[33]

Sometimes a writer implies the question in the answer that opens a paragraph:

> Yes, there does seem to be a war between the sexes. In our research on sexual agression, it has become readily apparent that a good deal of hostility exists between men and women. In a recent study of 305 college men at the University of Manitoba, for example, we found that more than one-third basically agreed with the statement, "Women are usually sweet until they've caught a man, but then they let their true self show." And the hostility is not all one-sided. Almost 40 percent of a sample of 278 college women agreed with the statement, "A lot of men talk big, but when it comes down to it, they can't perform well sexually."[34]

Rhetorical questions also serve informational prose well, especially if the writing has an element of persuasion, as Ward and Dubos demonstrate in the ending of *Only One Earth:*

> Alone in space, alone in its life-supporting systems, powered by inconceivable energies, mediating them to us through the most delicate adjustments, wayward, unlikely, unpredictable, but nourishing, enlivening, and enriching in the largest degree—is this not a precious home for all of us earthlings? Is it not worth our love? Does it not deserve all the inventiveness and courage and generosity of which we are capable to preserve it from degradation and destruction and, by doing so, to secure our own survival?[35]

Questions are common in instructional material because their effectiveness in brief, clear, interest-arousing communication is well known.

In "The Recovery Skips Middle Managers" Jeremy Main asks, "Why after four successive quarters of economic recovery hasn't the market for middle managers revived?" and then answers at length.[36] In the *Harvard Business Review* H. Edward Wrapp also uses question and answer:

> Why does the good manager shy away from precise statements of his objectives for the organization? The main reason is that he finds it impossible to set down specific objectives which will be relevant for any reasonable period into the future. Conditions in business change continually and rapidly, and corporate strategy must be revised to take the changes into account. The more explicit the statement of strategy, the more difficult it becomes to persuade the organization to turn to different goals when needs and conditions shift.[37]

## Reiteration

Reiteration requires that one repeat a topic in different words. It is essential that the idea repeated be important enough to justify the emphasis. To develop a long paragraph by reiteration requires special skill. Although in art the repetition of good patterns is enjoyable, the repetition of poor patterns and excessive repetition are painful. For example, some variety may be achieved by viewing the subject from different angles, by variations in wording, by adding a new subordinate thought with each repetition, by expressing the idea more vividly or forcefully each time it is stated and arranging the statements in climactic order. If each sentence does not add force or vividness to the idea being repeated, then reiteration seems careless rather than purposeful and becomes irritating. But excellent writers use reiteration beautifully to stress ideas and impress them firmly upon readers.

Loren Eiseley, a most skillful writer, opens a chapter of *Darwin and the Mysterious Mr. X: New Light on the Evolutionists* with brilliant reiteration:

> If the record of the rocks had never been, if the stones had remained closed, if the dead bones had never spoken, still man would have wondered. He would have wondered every time a black ape chattered from the trees as they do in the Celebes where, of old, simple forest people had called them ancestors of the tribe. He would have wondered when he saw the huge orangs pass in the forest, their bodies festooned with reddish hair like moss and on their faces the sad expression of a lost humanity. He would have seen, even in Europe, the mischievous fingers and half-human ways of performing monkeys. He would have felt, aloof in religious pride and the surety of revelation though he was, a vague feeling of unease. It is a troubling thing to be a man, with a very special and assured position in the cosmos, and still to feel those amused little eyes in the bush—eyes so maddeningly like our own.[38]

Writing more informally and personally, Lewis Thomas appropriately uses reiteration for an idea that haunts him:

> In this circumstance, anyone can become overwhelmed by his own true belief, and I confess here to mine. I am persuaded by the connection, thin as it is, to mycoplasma infection in animals, and by the gold story. I cannot count the hours that I have wasted in my own laboratory, during the last twenty years, concocting one baroque broth after another, trying to grow mycoplasmas from arthritic joint fluid, always with negative results. I have been obsessed with the possibility, unable to give it up. In any other circumstance, I suppose my behavior might be classed as paranoid. My laboratory notebooks contain an intermittent but endless harangue on this one topic; in between more or less respectable experiments on endotoxin, streptococci, papain, whatever, this business of mycoplasmas and arthritis keeps popping up, like King Charles's head. I cannot leave it alone.[39]

Although some writers employ reiteration with skill and taste, many beginners who try to imitate them produce dullness and monotony. Nothing is drearier than inept writers repeating poor sentences three times. It is, therefore, regrettable that many willing but ill-advised counselors of writers urge them to repeat ideas three times. High school instructors teach this fallacy in the hope that it will prevent students from breaking the unity of paragraphs. They trust that if students state the idea at the beginning of a paragraph and repeat it in the middle and again at the end, they will remember what they are supposed to be writing about. But it is much easier to teach young writers to unify their paragraphs without this repetition than to teach them the subtleties of effective repetition. Many sensible writers do not try to develop written paragraphs by reiteration because of the time and skill this demands.

In a study interestingly entitled *The Metaphorical Brain*, Michael A. Arbib opens his preface by repeating in anticlimactic order instances of the working of a brain. He enhances the reiteration by the repetition of sentence patterns—simple and compound

> A tourist is accosted by a thief in a dark alley. A moment's indecision and his brain has committed him to a mode of action—to fight or to flee.
>
> A frog sits on a lily pad. Two flies buzz into view and the pattern of electrical activity passing from retina to brain changes. The brain integrates this new pattern and selects one of the flies for a target. The frog turns and "zaps" and only one fly remains.
>
> A robot is in a cluttered room. Its "eye"—a TV camera—feeds data to its "brain"—a computer. The computer resolves its input into a representation of objects and their locations. It then charts a path, and the robot moves to perform an assigned task, avoiding all obstacles as it does so.[40]

## Scope and Qualification

As methods of developing paragraphs of a report, an article, a book, scope and qualification are familiar in informational prose. Any reader of scholarly journals knows that the opening paragraphs of many articles state the scope.

Sherrie Kossoudji and Eva Mueller present the scope at the beginning of an economic and demographic study:

> International bodies and numerous Third World governments have stated that improvement in the status of women must be part of the development effort. Yet in many Third World countries rural poverty and the unequal distribution of available productive resources are causing heavy rural-urban and international migration by men, which in turn is creating a large number of female-headed households. The major purpose of this article is to analyze the demographic and economic status of female-headed households in a rural area in Africa where insufficient earnings opportunities force many men to live and work away from home. Whyte's recent work has shown that various aspects of the status of women—political, social, legal, religious, and demographic—are not closely integrated. Ac-

cordingly, we shall focus on women's *economic* status without dealing in depth with other aspects of status. The main body of this article demonstrates that in rural Botswana female-headed households are poorer than others and explores the reasons for their poverty. It examines the extent to which social customs, economic institutions, and the economic behavior of the women themselves alleviate or aggravate their economic problems. In concluding we shall consider some larger implications of our findings, including the consequences of male out-migration for changes in the status of women, for familial support systems, and for reproductive behavior.[41]

In book reviews, opening with the scope of the book is particularly appropriate if the book is mainly informative. In his second paragraph Jack De Forest emphasizes an important feature of the work he is reviewing and reiterates the thoroughness of the author:

> More than 300 outstanding color photographs of free-living butterflies (Papilionoidea) and skippers (Hesperioidea)—both males and females in representative poses—add a valuable dimension to the text, which is sensibly organized and packed with useful information. George Krizek is an entomologist whose expertise in photography can be fully appreciated only by the amateur who has struggled to get these elusive creatures into focus.[42]

Paragraphs developed by stating the scope of a book, of a conference or other meeting, or of a projected study are also frequent. They are most useful if they are placed early in the writing. The "Report of the Conference on Uses and Possible Abuses of Biosynthetic Human Growth Hormone" presents the scope in the third paragraph:

> The conferees were organized into three working groups. One group focused on the biomedical aspects of growth hormone therapy, another addressed ethical and psychological issues, and the third discussed regulation of the use of growth hormone and concern about the possible side effects of therapy. This report details some of the deliberations and conclusions of the conference participants.[43]

Discussing the scope of a book early is particularly helpful if the book is a collection of articles or chapters by various authors. In the opening paragraph of the foreword to *The Brain,* David H. Hubel does so:

> This book does not explain how the brain works. That remains the most alluring and baffling of all questions on the frontier of understanding. The occasion for the publication of this book is provided by the convergence of work in many fields that has begun to show us how to frame questions about the brain that yield accumulative answers. Knowledge about the brain, gathered at an accelerating rate in recent years, shows this organ to be marvelously designed and capacitated beyond the wonders with which it was invested by innocent imagination.[44]

This paragraph of qualification opens strongly with a negative. In this beginning there is a suggestion of a straw man: the sentence is denying an idea that may be in readers' minds, the idea that the book will explain how the brain works. But that thought is not expressed. The straw man has no identity; his existence is only implied by his negation.

A rarer use of scope is to provide transition. The opening paragraph of section 2, "Synergistic Integration," of "The Integrated Simulation Environment (Simulation Software of the 1990's)," reviews the scope of section 1 and defines the scope of 2:

> Thus far, we have confined our attention to software tools for program preparation, compilation, and execution. While these tools are very important, they are only a part of the environment needed to properly support programming activities. The skeletal logical structure of such an environment is shown in Figure 4, which contains the three components of Figure 1, augmented to include software design and a central data base. In the following paragraphs, we consider the components of a software development environment and their interconnections.[45]

## The Straw Man

The straw man, or setting up an argument against the writer's point and then disproving that argument, appears in the opening paragraphs of works and sections of works. It serves a number of purposes including destroying common misconceptions, answering well-known objections to the writer's views, belittling opposing views, and showing how times have changed. It works best when the ideas, persons, or things attacked are weak in themselves, natural straw men.

In *Technology Review* Robert C. Cowen attacks a vulnerable subject, long-range weather forecasting:

> Long-range weather forecasting has been an exercise in near frustration. Until now, its practitioners could only claim a marginal degree of skill and hope to do better one day when they had a firmer scientific grip on the problem.[46]

Yves Dunant and Maurice Israel combat a misconception of the past:

> For many years there was a wide consensus among neuroscientists that acetylcholine is released from small, spherical organelles called synaptic vesicles, which are found inside the nerve terminal. It was thought that when the nerve terminal is stimulated, the vesicles fuse with the terminal membrane and release their contents into the space between the neuron and the tissue with which it communicates.
>
> Our recent investigations contradict this simple picture. They suggest that although the vesicles do indeed store acetylcholine and play a role in

its regulation within the cell, the acetylcholine released by the nerve terminal does not originate in the vesicles. Instead the released acetylcholine is derived directly from the cytoplasm, which makes up the ground material inside the neuron. The releasing mechanism appears to be operated by a compound, most likely a protein, that is embedded in the membrane of the nerve cell. The protein may act as a valve, enabling acetylcholine to pass through the membrane.[47]

Marjory Blumenthal and Jim Dray begin "The Automated Factory: Vision and Reality" with the straw man anticipated in the title of their article:

> Advocates of automation often paint a utopian vision of a factory run almost completely by computers. Machining centers hew metal, robots transport workpieces, and finished parts are placed in storage automatically, their whereabouts duly noted in the computer's memory. Proponents argue that such factories will make U.S. manufacturing competitive.
>
> This view is, in the end, simplistic and misleading. Manufacturing operations may seem simple—until you try to reduce them to computer programs. Then cleaning up metal shavings, aligning odd-shaped parts, and routing work around broken equipment do not turn out to be so easy. Automating many tasks may not be technically or economically practical.[48]

If a writer must weaken the presentation of the opposing arguments in order to make them easier to attack, the method is likely to have unfortunate results, such as loss of the readers' confidence; the implanting of a seed of doubt in readers as to the writer's honesty, fairness, judgment; or in a prolonged argument the subjection of the writer to the same treatment when those in opposition create a straw man of the writer's points. After using it, the writer can hardly object to this attack.

## Combined Methods of Development

Combined methods of development are so numerous that there is no practical way to illustrate fairly the wealth of possibilities. Because most paragraphs exhibit more than one method, a good paragraph plan is essential. Paragraphs, being small essays in themselves, may be organized by any of the methods suggested for the whole paper in Chapter 3, and the choice of a plan is governed by the same considerations that determine the choice of a plan for the whole work. But in choosing an organization for a paragraph, a writer also considers the preceding and following paragraphs in order to achieve harmony without monotony.

Graham Bell uses the details and particulars of history, generalizations, examples and illustrations, and questions in the following paragraph:

> Sex is the queen of problems in evolutionary biology. Perhaps no other

359

natural phenomenon has aroused so much interest; certainly none has sowed as much confusion. The insights of Darwin and Mendel, which have illuminated so many mysteries, have so far failed to shed more than a dim and wavering light on the central mystery of sexuality, emphasizing its obscurity by its very isolation. No doubt the roots of this difficulty lie very deep. There are problems which are not excessively difficult to solve, but which are exceedingly difficult to see: not because they are obscure or trivial, but because they are painted so large in the foreground of the canvas that the eye glides over them, taking them as the givens which can be used to solve other and more important problems whilst not themselves requiring solution. After more than a century of Darwinism, during which time most of the conspicuous details in the background have yielded their secrets, we are too close to the canvas to appreciate that large areas in the foreground are still uncharted, still less explored. It seems that some of the most fundamental questions in evolutionary biology have scarcely ever been asked, and consequently still await an answer. Every student knows that homologous chromosomes usually segregate randomly during the division of the nucleus; no professor knows why. Every layman knows that all the familiar animals and plants have two sexes, but never more; few scientists have thought to ask, and none have succeeded in understanding, why there should not often be three or many sexes, as there are in some ciliates and fungi. The largest and least ignorable and most obdurate of these questions is, why sex? Or, to put this more technically, what is the functional significance of sexuality, which leads to its maintenance under natural selection in biological populations?[49]

Daniel R. Vining, Jr., uses a question, an answer that is a straw man, and then details and particulars to knock down the straw man:

What is the most pressing demographic problem in the Third World? The answer generally given to the question is rapid population growth. Yet in many developing nations there is a demographic problem with far more serious immediate implications: the increasing concentration of the population in the major cities. South Korea provides a dramatic example. In 1955 the population of South Korea was 21.5 million, 18 percent of whom lived in or near Seoul. During the next 25 years more than half of the population growth in the country took place in the region of the capital. By 1980 the total had reached 37.4 million and 36 percent of all South Koreans lived in or near Seoul. Furthermore, in 1980 more than half of all the economic production in South Korea took place within 25 kilometers of the center of the capital.[50]

A summary statement, details, and contrast develop a paragraph in the *New England Journal of Medicine:*

At present, the medical profession emphasizes that modification of the traditionally accepted risk factors should receive a high priority in preventive cardiology. Understandably, cigarette smoking, hyperlipidemia, and hypertension are reasonable and perceptible targets for therapeutic inter-

360

vention. Yet such risk factors are found in less than half of those who succumb to ischemic heart disease. Psychological stress, on the other hand, has been viewed as a vague, subjective entity that cannot be quantified and has therefore not been accorded much attention in terms of prophylaxis.[51]

Additional examples of these and other ways of developing paragraphs appear in Chapter 5, which presents ways to begin papers. The number and variety of ways to develop paragraphs seem endless, the only limitation being the inventiveness and imagination of an author. Because many writers are surprised by a study of paragraph development, it obviously offers them chances for improvement. Writers who have been painfully grinding out unplanned paragraphs benefit greatly from even a few minutes spent organizing the development of paragraphs, and good writers begin to capture some feeling for practicing writing as an art, not just a skill.

## COHERENCE

Regardless of how it is developed, every paragraph must have coherence. Supplying coherence in paragraphs improves style so markedly that when we consult with discouraged writers we often work on coherence first. Then their readers will comment on how much easier and smoother their style is or on how polished it seems and thus will encourage the writers to further improvement. Coherence is easy for writers on science or technology to achieve because their work is likely to be well organized, and the basis of coherence is unity and a plan.

Writers use some transitions when they write their first drafts, but coherence is a minor concern at that time. During revision writers can easily provide transitions where they are missing and improve weak ones. This is a better procedure than trying to write all the transitions in a first draft, for worrying about coherence may impede the flow of ideas.

If a paragraph is well organized, providing transitions is a simple step in revision. Revisers may supply the following kinds: (1) pronouns and demonstrative adjectives, (2) conjunctions and other transitional words and phrases, (3) word echo and synonyms, (4) parallel constructions.

### Pronouns and Demonstrative Adjectives

Pronouns and demonstrative adjectives referring to antecedents in the preceding sentence connect sentences efficiently, especially if they appear near the beginning of the second sentence. They are common and useful transitions, seldom obtrusive or annoying. Usually most of the necessary pronouns, such as *you, he, she, it, they, I, we, one, each, either, this, who,* and *which,* appear in the first draft. Sometimes demonstrative pronouns may be added during revision: coherence between loosely connected sentences may be strengthened by changing *the* to *this, that, these,* or *those.*

## Conjunctions and Other Transitional Expressions

Conjunctions and other transitional expressions not only connect elements within sentences but also connect whole sentences. A precise use of conjunctions is essential in functional prose. The right conjunction makes reading easier and clearer; the wrong one may stress incorrectly, necessitate rereading, or permanently confuse. Advice on the accurate use of conjunctions appears under specific conjunctions in Appendix B. Conjunctions vary in meaning and tone; therefore writers who wish to choose well should know many of them and should avoid monotony by varying their choices. Some commonly used conjunctions and transitional words and phrases follow.

ADDITION: *and, again, also, besides, equally important, finally, first (second, third), further, furthermore, last, moreover*

COMPARISON AND CONTRAST: *after all, but, however, in comparison, in like manner, on the other hand, on the contrary, likewise, nevertheless, notwithstanding, still, whereas*

NOTE: If a sentence contains a comparative or superlative adjective or adverb that compares or contrasts a word or thought with one in the preceding sentence, the adjective or adverb is transitional. For example, in "If *x* has a greater value," *greater* connects clearly with the value stated in the preceding sentence of the writing.

RESULT: *accordingly, consequently, hence, therefore, thus, wherefore*

EXEMPLIFICATION, REPETITION, SUMMARY, INTENSIFICATION: *as has been stated, as I have said, as well as, for example, for instance, in any event, in brief, indeed, in fact, in other words, in particular, in short, in summary, obviously, of course, that is, to be sure, to sum up*

PURPOSE: *for this purpose, to this end, toward this objective, with this goal*

## Three Major Connectives

Some expressions are as directly and strongly connective as conjunctions. Numbers and letters help readers keep track of ideas, and connecting one place with another or one time with another stresses continuity.

NUMBERS AND LETTERS: *1, 2, 3; first, second, third; I, II, III; a, b, c; A, B, C.* Labeling by numbers or letters should not be overworked, lest the section or sections sound or look like an outline.

References to place or time are excellent transitions—effective, unobstrusive, organic. And like labeling, these transitions please readers, who like to know where they are.

362

PLACE: *adjacent to, beyond, here, near, on the other side, opposite to*
TIME: *afterward, at length, after an hour (day, week, year), immediately,
in the meantime, meanwhile*

Long and obvious transitional expressions burden a paper and may
irritate readers. A wordy mechanical transition ("Now let us consider the
features just mentioned") and sentences that repeat long sections of preceding sentences are usually avoided today because they give a Victorian
flavor. Organic connections, such as word echo, are preferred. Nor do
writers of today place heavy transitional words and phrases at the beginnings of sentences unless the connecting thoughts are so difficult that
transitions are needed immediately. Transitions are more often placed
after the first few words of a sentence, where they are less conspicuous
and receive less stress than at the beginning.

## Word Echo and the Use of Synonyms

Also called *the repetition of key words,* word echo is a common transitional method in informational prose. When main words are repeated,
the central idea of a paragraph is stressed, and thus coherence and emphasis reinforce each other. Synonyms and synonymous phrases may be
used for variety instead of too many repetitions of a main word. Indeed,
even antonyms provide connections, for example, *hot* with *cold, dry* with
*wet, economical* with *wasteful, busy* with *idle.*

## Parallel Constructions

Repetitions of sentence patterns connect sentences so closely that other
transitions may not be necessary. Scientists and technologists often use
parallel constructions for similar thoughts and thus not only connect the
thoughts tightly but make them easier to comprehend because the divisions of thought follow as regularly and clearly as items in a table. The
writers of "Sex-Associated Differences in Serum Proteins of Mice" use
parallel construction:

> In several animal species certain serum proteins appear to be different
> in males and females; for example, the relative concentration of albumin
> in rats, as measured by moving boundary and zone electrophoresis, was
> found to be higher in females than in males. In cattle, males possessed less
> $\alpha$-globulin glycoprotein and more $\beta$-globulin and $\gamma$-globulin glycoproteins
> than females. In toads, the separation of some of the serum components by
> starch gel electrophoresis has been reported to be different in the two sexes.
> In mice, the concentration of agglutinating antibody to chicken and sheep
> heteroantigens, and to human erythrocytes was found to be higher in fe-
> males; and, in addition a protein fraction has recently been described as

363

missing in male mouse serum (Cal A strain) analyzed by starch gel electrophoresis.[52]

A number of sentences in parallel construction are more effective in a logical or emphatic order. A pattern of gradually lengthening parallel sentences is pleasing, parallel sentences in the order of climax are interesting, and parallel sentences in logical or chronological plans are easy to understand. But a writer should not be tempted to contort ideas or structures to make sentences parallel and should not place unequal ideas in balanced construction.

## Example of Effective Coherence

Rachel Carson links closely the sentences in the following paragraph and uses a pleasing variety of transitional methods:

> From the green depths of the offshore Atlantic many paths lead back to the coast. They[a] are paths[b] followed by fish; although unseen and intangible, they are linked with the outflow of waters from the coastal[c] rivers. For thousands upon thousands of years[d] the salmon[e] have known and followed these threads of fresh water that lead them back to the rivers,[f] each returning to the tributary in which it spent the first months or years of life. So, [g] in the summer and fall of 1953,[h] the salmon[i] of the river called Miramichi on the coast of New Brunswick moved in from their feeding grounds in the far Atlantic and ascended their native river. In[j] the upper[k] reaches of the Miramichi, in the streams that gather together a network of shadowed brooks, the salmon[l] deposited their eggs that autumn[m] in beds of gravel over which the stream water flowed swift and cold. Such places, [n] the watersheds of the great coniferous forests of spruce and balsam, of hemlock and pine, provide the kind of spawning grounds that salmon must have in order to survive.[53]

(a) The pronoun *they* in the second sentence refers to *paths* in the first sentence. (b) The word *paths* occurs in the first and second sentences. (c) *Coastal* echoes *coast* in the preceding sentence. (d) The time phrase *for thousands upon thousands of years* and the present perfect tense *have known* take the reader over a long span of time from the distant past to the present. (e) The specific word *salmon* echoes the general word *fish* of the preceding sentence. (f) *These threads of fresh water that lead them back to the rivers* refers to *the outflow of waters,* and *rivers* echoes *rivers,* the last word in the preceding sentence. (g) The conjunction *so* prepares the reader for a similar act. (h) The time phrase *in the summer and fall of 1953* moves the reader from the span of time of sentence 3 to specific seasons in a specific year. (i) The word *salmon* echoes *salmon* in the preceding sentence.

Sentences 3 and 4 are subtly parallel. In 3 "the salmon have known and followed these threads of fresh water that lead them back to the

364

rivers, each returning to the tributary in which it spent the first months or years of life." In 4 "the salmon . . . moved in from their feeding grounds in the far Atlantic and ascended their native river." Such parallelism, with each sentence slightly different from the other, is modern. The older writers used more exact parallelism.

(j) The phrase *in the upper reaches of the Miramichi* places the reader. (k) The word *upper* subtly echoes the thought of *ascended* in the preceding sentence. (l) *Salmon* again; it connects with *salmon* in sentences 3 and 4 and *fish* in sentence 2; thus a main word runs through the paragraph. There is no elegant variation here (page 285). (m) The demonstrative *that* and the word *autumn* (echoing *fall* in the preceding sentence) connect the two sentences. But these connectives would come too late in sentence 5 if they were the only connectives between it and sentence 4. (n) *Such places,* the opening words of sentence 6, connect clearly with the closing words of sentence 5, *in beds of gravel over which the stream water flowed swift and cold.* Connecting the end of one sentence with the beginning of the next is a popular technique to achieve coherence.

Tracing the transitions in an editorial, a book, or a well-written article will help writers who wish to improve the coherence of their writing. The techniques that they observe in the writing of others will prove useful when they meet problems in revising their own writing to make it coherent. Excellent writing, like Rachel Carson's, illustrates for them the kinds of transitions and the number of transitions that are to be expected in modern paragraphs. More important to an advanced writer is the fact that good writing demonstrates the close and subtle linking of the sentences of a paragraph and the varied and interesting ways that words and phrases emphasize and strengthen that linking.

The same transitional methods that are used to link sentences within a paragraph are used to connect paragraphs. The best place for connections between paragraphs is early in the first sentence of a new paragraph. Transitions between paragraphs are usually stronger and more numerous than those between sentences because a new paragraph begins a new idea, and the break between ideas is greater than the break between the sentences of a paragraph, which relate to one idea. Sometimes paragraphs are part of a larger pattern within an article or chapter, and transitions are easy. One paragraph may be the explanation of a general statement, and the next two paragraphs may be two examples. The connections in such a case are as easy to accomplish as they are in a single paragraph. But occasionally there is a marked break that must be bridged, as, for instance, in moving from that one paragraph of generalization and the two examples to another idea that is not closely related. And in a few cases part of the thought development of a paper may be a stressing of the break in thought. Then writers often use a sentence of transition or even a short transitional paragraph.

Rachel Carson followed the paragraph just discussed with a brief one that repeats the idea of the preceding paragraph and adds in a sharply contrasting short sentence a major contrast in idea:

> These events repeated a pattern that was age-old, a pattern that had made the Miramichi one of the finest salmon streams in North America. But that year the pattern was to be broken.

How smoothly this paragraph is connected by *these events* to the whole preceding paragraph. How well the *but* and the short sentence of foreshadowing prepare the reader for a complete change in action. These methods of developing a paragraph and of providing transitions are neither complex nor difficult; therefore the skillful ways in which an artist adapted them to her subject and her purpose are all the more fascinating and, we dare hope, inspiring.

## EMPHASIS IN PARAGRAPHS

Excellence of communication depends on conveying clearly to readers not merely ideas and connections between ideas but also the desired emphasis. Correct paragraphing tells readers the divisions of one's subject, and this information helps to convey emphasis.

Within a paragraph, emphasis may be secured by attention to paragraph plan, the position of sentences, proportion, repetition, and contrast. Judicious employment of techniques enables writers to convey subtle emphasis as accurately and precisely as they convey thoughts by their careful choice of words. Ignorance of techniques may cause writers to bury ideas that they have worked hard to express clearly and effectively and to stress insignificant thoughts so heavily that readers remember nothing else. It may even lead readers to the very action that the writers want them to avoid. Yet many writers are ignorant of the functioning of the principles of emphasis in paragraphs because their teachers regarded emphasis as advanced work. True, it may be advanced work, but it is essential if writers are to make their points effectively.

### Contrast

Contrast in sentences—in their length, grammatical type, and rhetorical type—may be used to make a sentence containing a main thought different, conspicuous, memorable (page 326). A sentence that attracts notice by being agreeably different from surrounding sentences is the best place for a principal thought. Rachel Carson's "But that year the pattern was to be broken" illustrates the use of a short sentence after a number of long ones, a periodic sentence in a passage that has mostly loose sentences, a simple sentence after a number of complex sentences, and the passive voice after many main clauses in the active voice. All these contrasts stress sharply "But that year the pattern was to be broken," which is the major thought not merely of the paragraph but of the chapter. Thus the device of contrast enforces presentation of thought.

## Proportion

If a major portion of a paragraph deals with an idea, that idea is stressed, bulk being impressive. This presents a problem to writers who must develop at length relatively unimportant material before their readers can grasp main points. The thoughts in the lengthy development will be stressed at the expense of the main ideas unless this is prevented by skillful use of other devices to stress the most important thoughts and thus to rectify any incorrect emphasis on the details that bulk so large in the paragraph. But usually the principle of proportion works harmoniously with the development of a paragraph topic. There is a natural tendency to write more about important ideas and to write fleetingly of subordinate thoughts; therefore in most instances material flows smoothly into this arrangement, fitting it as neatly as a round peg fits a round hole. Thus it is usually not difficult for writers to be sure that the length of their treatment of ideas is proportionate to the value of the thoughts. The emphasis that space gives to an idea is a warning against disunity in a paragraph. Any unrelated idea de-emphasizes the main thought. Therefore unity is essential.

## Repetition

Repetition requires special skill if it is used to develop a whole paragraph (page 354), but some writers of average ability can repeat main thoughts effectively. More advanced writers also find reiteration for emphasis a good device on occasion. But used too frequently, it is likely to bore readers and to make them feel that the repetition of main points they have already thoroughly grasped insults their intelligence.

Difficult points sometimes benefit from reiteration if the demands of variety are met. Each restatement must add something—more vivid language, another facet of the thought, a different point of view. An idea may be expressed positively and then negatively, in a question and then in the answer, in general and then with the addition of detail, impersonally and then personally. The possible variations are countless. So are the writers who do not know this and think that they achieve emphasis by repeating the same thought in the same words. Although even a writer's best sentences are seldom good enough for repetition, variations may be interesting and effective.

## Placement of Sentences

Positioning sentences is an easy way to provide correct stress. The important positions in a paragraph are the beginning and the end. Yet many writers carelessly ignore this principle and place a minor thought —even an irrelevant point—in the first sentence and a minor thought in the final sentence—perhaps an afterthought not properly related to the main idea. The stress offered by these positions in paragraphs is too important to discard carelessly.

Clever revisers improve their first drafts spectacularly by moving important thoughts to the first and final positions. Sometimes by this change alone, a cloudy paragraph focuses sharply. This is a technique for a writer in a hurry whose work is clear and fairly correct. How can this writer gain maximum results from a few minutes of revision? By moving major ideas to positions of prime importance—paragraph beginnings and endings.

## INTRODUCTORY AND CONCLUDING PARAGRAPHS

Just as the first and final sentences of a paragraph are the most important, so the first and final paragraphs in a work are the most important. Some writers sense this, worry about these paragraphs unduly, and work over them so fruitlessly that they create fancy paragraphs completely out of keeping with the rest of their work. Whatever else they may be, the first and final paragraphs must be appropriate parts of the whole—on the same subject, in harmonious style and tone. They must enforce the main idea.

Some of the many effective ways to begin a work are discussed in Chapter 5. In general, an opening paragraph should be interesting and informative. If it is easy to read, it encourages readers to continue because they find it satisfying to grasp an idea quickly and easily at the start of a work, especially an idea that opens the subject of the paper for them. Whether writers start with prefatory comments on the scope of their work, a definition of the theme, a review of existing knowledge, a survey of the plan of development, an anecdote, an example, a point of view, or a summary of the central idea depends on the subject and the readers. But all readers like a beginning that makes the subject seem interesting and the style clear.

Ironically, the beginning least likely to do that is the one that states baldly that the work is interesting or significant or important. Readers are not gullible enough to take a writer's word for that at the start; in fact, many of them cynically doubt it. A writer who wants to convince readers that a work is interesting, significant, or important should begin with a paragraph that exemplifies those qualities.

An introduction, like a first meeting, is a promise of what is to come. It sets the tone and style. Its very nature—clear, succinct, thorough, lively—promises what the rest will be like. And writers should not disappoint readers. Writers should not promise a wider coverage, a more scholarly study, a more interesting presentation than they provide. An introduction is not just a lure to entice a reader; it is also an augury, and it should not be a false one.

A conclusion, too, should be truthful. There are few experiences more annoying, for example, than listening to speakers who keep saying that they are ending long before they do end. By the time that they finally conclude, their audiences have given up; they stopped listening the fourth time the speakers said, "In conclusion."

A good ending is a smooth stop; readers should know that they have arrived at the conclusion and should not feel inclined to turn the page to find the ending. Nor should a writer present one ending after another, like a guest who stands at the door saying goodby but not leaving. The last paragraph, then, should be strong and final; it should round off the theme effectively.

The content of good endings varies—a restatement of the central idea, an analysis of the significance of the subject, a striking final example, a recommendation, a massing of recommendations or reasons or results, some interesting interpretations or reflections, a forecast flowing naturally from the rest of the paper, an epitome of the important ideas. A last paragraph is not the place to limit the subject—that should have been done near the beginning—or to hedge the conclusions—the conclusions should have been limited when stated—or to offer concessions—these too belong earlier in the paper. And a conclusion is the worst place for an apology. Writers should avoid apologizing for their work. If it is poor, they should rewrite it or scrap it. But if they cannot avoid apologizing for hasty work, they should apologize early in the paper. And if they can, they should apologize so indirectly that readers do not recognize the apology as such but think that the writers are explaining how the paper was written or what its limitations are.

A conclusion is the place for a good presentation, not a poor one. It is the readers' last impression of a work, and it should be strong, forceful, convincing, and final. Readers should feel that everything necessary and expected has been covered and that they are ready to have the work draw to its end. Writers may end briskly and forcefully; or they may conclude with wit, a neat restatement of an opening phrase, a clinching repetition of an important idea; or in a stately way they may bring the paper slowly to its final, well-rounded sentence.

If writers will list the errors and weaknesses in their paragraphing and work on them when they revise, they will develop better paragraph style easily. Even a writer who has time to apply only one technique will notice an improvement after a few trials, and a writer who perseveres in improving paragraphs will end with a style that not only efficiently guides readers to all the facets of the paper and the significance of each but leaves the impression that the writer is well organized.

## NOTES

1. Alice S. Rossi, "Women in Science, Why So Few?" *Science* 148 (28 May, 1965): 1201.

2. *ANSI Z* 39.14-1979, *American National Standard for Writing Abstracts* (New York: American National Standards Institute) 13.

3. Barbara Ward and René Dubos, *Only One Earth: The Care and Maintenance of a Small Planet* (New York: Norton, 1972) 12.

4. Elizabeth Janeway, *Man's World, Women's Place: A Study in Social Mythology* (New York: Morrow, 1971) 247.

5. Simone de Beauvoir, *The Coming of Age,* trans. Patrick O'Brian (New York: Putnam's, 1972) 543.

6. Arden Neisser, *The Other Side of Silence: Sign Language and the Deaf Community in America* (New York: Knopf, 1983) 282.

7. Loren Eiseley, *The Immense Journey* (New York: Random, 1957) 14.

8. Marjorie Hope Nicolson, *The Breaking of the Circle* (New York: Columbia UP, 1960) 123.

9. John R. Platt, "Strong Inference," *Science* 146 (16 Oct. 1964):347.

10. Platt 348.

11. Timothy Ferris, "Physics' Newest Frontier," *The New York Times Magazine* 26 Sept. 1982:38, 44.

12. Robert L. Sproull, *Modern Physics* (New York: Wiley, 1963) 18.

13. David V. Becker, "Choice of Therapy for Graves' Hyperthyroidism," *New England Journal of Medicine* 311 (16 Aug. 1984):464.

14. Ward and Dubos 1.

15. Eugene Ayres, "An Automatic Chemical Plant," *Automatic Control* (New York: Simon, 1955) 41.

16. C. P. Snow, *The Two Cultures* (New York: Cambridge UP, 1959) 18.

17. Craig Claiborne with Pierre Franey, "Lamb for All Tastes," *The New York Times Magazine* 31 March 1985: 77.

18. I. M. LeBaron, "Fifty Years of Chemical Engineering in the Plant Food Industry," *Chemical Engineering in Industry,* ed. W. T. Dixon and A. W. Fisher, Jr. (New York: AIChE, 1958) 52.

19. Margery Cunningham, "Coming to Grips with Death," *The Miami Herald* 8 Nov. 1987: 9F.

20. Robert C. Cowen "Wanted: Dialogue on Genetics," *The Christian Science Monitor* 7 March 1985: 25.

21. J. Schöler and J. R. Sladek, Jr., "Supraoptic Nucleus of the Brattleboro Rat Has an Altered Afferent Noradrenergic Input," *Science* 214 (16 Oct. 1981):347.

22. A. K. Dewdney, "Computer Recreations," *Scientific American* 252 (April 1985):22.

23. J. G. Shanthikumar and R. G. Sargent, "A Unifying View of Hybrid Simulation/Analytic Models and Modeling," *Operations Research* 31 (Nov.–Dec. 1983):1031.

24. Ernest O. Ohsol, "Coke By-products and Gas," *Chemical Engineering in Industry* 297.

25. Charles W. Palmer, "The Written Procedure—Welcome Relief for the Engineer," *STWP Review* (April 1965):11.

26. Janeway 27.

27. Tobias Dantzig, *Number: The Language of Science* (Garden City, NY: Doubleday, 1956) 10.

28. Neisser 248

29. Edward Garrigues Boring, Herbert Sidney Langfeld, and Harry Porter Weld, *Foundations of Psychology* (New York: Wiley, 1948) 14.

30. Caroline Bird, *Enterprising Women* (New York: Norton, 1976) 226.

31. Ferris 69.

32. Eric R. Kandel, "Small Systems of Neurons," *The Brain* (New York: Scientific American, 1979) 29.

33. Lloyd G. Reynolds, "The Spread of Economic Growth to the Third World: 1850–1980," *Journal of Economic Literature* 21 (Sept. 1983):963.

34. James V. P. Check et al., "On Hostile Ground," *Psychology Today* 19 (April 1985):56.

35. Ward and Dubos 220.

36. Jeremy Main, "The Recovery Strikes Middle Managers," *Fortune* 109 (6 Feb. 1984):113.

37. H. Edward Wrapp, "Good Managers Don't Make Policy Decisions," *Harvard Business Review* 62 (July–Aug. 1984):10.

38. Loren Eiseley, *Darwin and the Mysterious Mr. X: New Light on the Evolutionists* (New York: Dutton, 1979) 187.

39. Lewis Thomas, *The Youngest Science: Notes of a Medicine Watcher* (New York: Viking, 1983):181.

40. Michael A. Arbib, *The Metaphorical Brain: An Introduction to Cybernetics as Artificial Intelligence and Brain Theory* (New York: Wiley, 1972) vii.

41. Sherrie Kossoudji and Eva Mueller, "The Economic and Demographic Development of Female-headed Households in Rural Botswana," *Economic Development and Cultural Change* 31 (July 1983):831.

42. Jack De Forest, rev. of *Butterflies East of the Great Plains* by Paul A. Opler and George O. Krizek, *Sierra* 70 (March/April 1985):84–85.

43. Louis E. Underwood, "Report of the Conference on Uses and Possible Abuses of Biosynthetic Human Growth," *New England Journal of Medicine* 311 (30 Aug. 1984):606.

44. David H. Hubel, foreword, *The Brain* vii.

45. Charles O. Henricksen, "The Integrated Simulation Environment (Simulation Software of the 1990's)," *Operations Research* 31 (1983):1060.

46. Robert C. Cowen, "Payoffs of Long-Term Weather Forecasts," *Technology Review* 88 (Jan. 1985):6.

47. Yves Dunant and Maurice Israel, "The Release of Acetylcholine," *Scientific American* 252 (April 1985):58.

48. Marjory Blumenthal and Jim Dray, "The Automated Factory: Vision and Reality," *Technology Review* 88 (Jan. 1985):29–30.

49. Graham Bell, *The Masterpiece of Nature: The Evolution and Genetics of Sexuality* (Berkeley: U of California P, 1982) 19.

50. Daniel R. Vining, Jr., "The Growth of Core Regions in the Third World," *Scientific American* 252 (April 1985):42.

51. Thomas B. Graboys, "Stress and the Aching Heart," *New England Journal of Medicine* 311 (30 Aug. 1984):594.

52. E. Spinosa et al., "Sex-Associated Differences in Serum Proteins of Mice," *Science* 144 (24 April 1964):417.

53. Rachel Carson, *Silent Spring* (Boston: Houghton, 1962) 129.

# CHAPTER 13

# Brevity: The Soul of It

*Words are like leaves; and where they most abound*
*Much fruit of sense beneath is rarely found.*

*Pope*

Writing too much is a common fault. An overload of reading is the obvious result. A chemist grumbles that if during every working hour of a year he were to read the articles published that year in his special field, he would at the end of the twelve months be behind in his reading by ten years. A college professor complains that he receives a pound a day of notices, reports, book advertising, and articles—not only more than he can read but more than his wastebasket can hold. And a vice president estimates that skimming the material that a day brings to his desk would take him forty-eight hours. Exhausted readers take courses to improve their reading speeds, delegate to others the task of sifting the towering piles of written matter, and curse the easy and inexpensive methods of reproduction that encourage the deluge.

It used to be feasible just to tell a porter to supply more towels to the company lavatories. Now memorandums on the subject are dictated, typed, reproduced, distributed, read (by some recipients), filed, stored, and sometimes even retrieved. In industry few people talk; they write instead. In fact, some companies supply pads inscribed "Don't say it; write it." And everyone does—at length. Speeches are preprinted, printed, and reprinted. Inexhaustible writers on science and technology stretch their articles to the utmost even for journals known to be as much as a year behind in publishing. Writers vie at dashing off the windiest letters, the longest memorandums, and the most diffuse reports—and at addressing them to the most people.

Some writers broadcast every paper as though they must convince the whole company that they are working. Supervisors should prevent this discourtesy to busy readers and should themselves avoid the error of telling their whole business world what only a few people need to know. Good writers show consideration for others by selecting judiciously the readers to whom they address their pages.

An even more successful way to reduce the flood of words is to delete unnecessary material. The amount of detail in any writing for industry should depend not on the writers' infatuation with their subjects, but on their readers' needs. As Schopenhauer wrote, "True brevity of expression consists in everywhere saying only what is worth saying, and in avoiding tedious detail about things which everyone can supply for himself."

Specialists who write overlong reports cater to their own interests, and in doing so waste the readers' time. Though suffering readers may doubt it, these writers usually know that much of their material is of interest to no one but themselves—and possibly their doting mothers. They also know that parts of their reports are suitable only for fellow specialists, that those educated in science and technology can read some parts, but that little is of general interest in their company. If they are pressed hard, they can identify this material accurately.

First, they should discard the material of personal interest. Even the files do not need detailed reports of their unsuccessful attempts to build equipment or to find pure chemicals if there is nothing to be learned from those attempts.

Then they should place in appendices a large part of the material that interests only specialists. Detailed graphs and tables, for example, belong there, and only representative data need appear in the body of the report. But some writers are reluctant to relegate any material to appendices. When a national engineering society began publishing only representative data, loud screams and groans haunted the editors, who had sent the details to an information center, from which they could be retrieved easily if anyone needed them. But only the writers screamed and groaned—and just about their own articles; each writer saw clearly the advantage of printing only representative data in the articles of other engineers. And readers expressed pleasure. The material filed was seldom requested; yet for many years this society had published the superfluous data on pages needed for other articles. The more material writers can place in appendices the better. They save the time of busy readers who wish to skip such material. They gain readers because their shortened reports are more attractive. And they achieve a reputation for thinking clearly and knowing their subject well when readers, freed from the clutter of unnecessary details, can understand the main points more easily.

Writers should be careful not to supply background and explanation that their readers know, especially after telling the readers that they know it. Experienced scholars are less prone to this weakness than are recent products of graduate schools. A little learning can make a very tiring writer. A reader of a medical report written by a new M.D. is lucky if it mentions a BMR without supplying an introduction to all the ductless glands. A young botanist may summarize the general information about nuclear reactors before even hinting at what happened to her irradiated stringbeans. And the new Ph.D. in chemistry will not only explain everything mentioned in his paper but will regale his readers with his favorite professor's theories—even those only remotely connected with the subject. This tendency to squeeze out all the information one knows is probably left over from filling as many pages as possible of college examination books.

In functional prose, explanations are welcome and useful where they are needed. If the explanations relate to point three of a paper, they should appear at point three. If a few sentences will make result two easier to understand, the writer should place those few sentences just where the reader needs them. A writer who places explanations where they are helpful is less likely to explain more than is necessary. When discussing point three or result two, a writer shares the reader's eagerness to get on with the subject and explains briefly. But if the introduction is a blank in the author's mind, the writer may seize this explanatory material and detail it until the introduction seems long enough. Such long-windedness is a disfavor to one's subject and a discourtesy to readers. As Jerome B. Wiesner points out, "More information does not automatically mean better information . . . . The user should be informed, not overwhelmed."[1]

Even after a writer has reduced the subject matter, the report may still be too long because of a wordy style. Therefore, after discarding unnecessary details and adapting the material to the reader, the writer should examine the work for verbosity and rigorously delete unnecessary words.

Much informational prose expresses a single thought in words enough for ten thoughts. In many reports words obscure the sense so thoroughly that readers have to unearth the meaning and rephrase it to understand it. One supervisor, for example, wrote the following instruction, "Prior to proceeding to the further and succeeding step of written communication, make every reasonable effort to plan ahead in orderly fashion and to place your plan in a written form that clearly marks major and minor divisions." Her readers eventually translated this to "Before writing, outline." Laboriously they had winnowed the meaning in spite of the writer.

Readers of professional journals, government documents, textbooks, and industrial reports develop through practice a facility for this kind of translation. But we never heard of anyone who grew to like it. Being forced to sift a paragraph of words to get one sentence of meaning is always annoying, and when a reader is busy or tired, it is exasperating. Yet writers, showing no consideration for their readers, dilute their meaning in oceans of words. The only extenuation that we have ever found is that they are totally unaware of their wordiness. Among the most verbose writings in government and industry are complaints about wordiness. No one, apparently, minds his own prolixity, but everyone objects wordily to the prolixity of others.

The first step, then, in acquiring a brief style is awareness of one's own lack of concision. We have seen instructors impress this upon students by eliminating one third of the words on a page without sacrificing any thoughts. A public demonstration leaves a lasting impression on the victim whose page was reduced, but many in the amused audience do not apply the lesson to their own work. Like the verbose complainers, they see only the wordiness of others.

A less painful way to achieve concision is to study examples of common kinds of wordiness and to eliminate them from one's writing. Functional prose is prone to seven kinds of wordiness, which we discuss under the headings *Tautology, Dilute Verbs, Hiccups, Roundabout Constructions, Hedging and Intensifying, False Elegance,* and *Pointless Words and Phrases*. These seven faults of style are not unrelated to the seven deadly sins, particularly Pride, Envy, and Sloth.

Writers who remove such wordiness from their writing gain effectiveness. Cogency is a Siamese twin of brevity. Some verbiage may be struck out as soon as it is found; removing other kinds requires rephrasing. At first writers may spend hours revising for brevity. But shortly wordiness disappears from their first drafts because after they have repeatedly taken out unnecessary words and phrases, they will not write them even in first drafts. Finally wordiness will disappear even from their speech, the improvement in speech being, as those master tautologists, advertisers, write, "an extra added bonus."

## TAUTOLOGY

Tautology—the unnecessary repetition of an idea in other words—results from failure to consider meaning. If those who write *round in shape, twelve feet by twenty feet in size, beige in color, at this point in time,* and *in the range of ten to seventy-five pages* will think about meanings, they will write instead *round, twelve by twenty feet, beige, now,* and *ten to seventy-five pages*. Examining the correction of the following common examples of tautology should increase a writer's sensitivity to the fault.

| Example | Discussion and Illustration |
|---|---|
| about | When *about* precedes what is clearly stated as an estimate or approximation, omit *about*.<br>WORDY: She estimated the profits as about one million dollars.<br>IMPROVED: She estimated the profits as one million dollars.<br>WORDY: He reads about twenty to thirty articles each week.<br>IMPROVED: He reads twenty to thirty articles each week. |
| a.c. current, d.c. current | Use *a.c.* and *d.c.* or *alternating current* and *direct current* to avoid the repetition of *current*.. |
| adequate enough | Use either of these words, but not both. |
| advance forward | Use *advance*. |

| | |
|---|---|
| advance planning, advance warning | Omit *advance* because the meaning is contained in *planning* and *warning*. |
| and etc. | *Et cetera* means *and others*. Omit *and*. |
| any and all | Use *any* or *all*.<br>WORDY: Any and all employees are invited to the picnic.<br>IMPROVED: All employees are invited to the picnic.<br>WORDY: Any and all accidents should be reported on form A.<br>IMPROVED: Any accidents should be reported on form A. |
| as a usual rule, as a general rule, generally as a rule, as a rule usually | Use *as a rule, generally,* or *usually.* |
| at an early date | Avoid this wordy business cliché. There are many acceptable words that mean *soon.* A specific date is most helpful to a reader. |
| at a price of | Omit *a price of.*<br>WORDY: He sold the house at a price of $90,000.<br>IMPROVED: He sold the house for $90,000.<br>WORDY: She valued the painting at a price of $300,000.<br>IMPROVED: She valued the painting at $300,000. |
| at hand, in hand | These are business clichés.<br>COMMERCIAL AND WORDY: I have your letter of July second at hand (*or* in hand). I am sending the samples you requested.<br>CORRECT: I am sending the samples you requested in your letter of July second. |
| at present, at the present time | Omit these business clichés when the verb indicates present time. When a time word is needed, the simple word *now* is useful.<br>COMMERCIAL AND WORDY: She is reviewing the matter at the present time.<br>IMPROVED: She is reviewing the matter. |
| attached hereto | Use *attached* alone.<br>REDUNDANT: Please file the notices attached hereto.<br>IMPROVED: Please file the attached notices. When it is feasible to omit *attached,* do so for brevity and simplicity: *Please file these notices.* |
| attach together | Use *attach.* |
| autobiography or biography of his life | Omit *of his life.* |

377

| Example | Discussion and Illustration |
|---|---|
| basic fundamentals | What other kind? |
| be in receipt of | In many sentences this business cliché is unnecessary; in others it should be replaced by a simple reference to what the writer has received. |

COMMERCIAL: I am in receipt of the samples I requested. Thank you for sending them promptly.

We are in receipt of your letter of March 16. We will ship the order therein immediately.

CORRECT: Thank you for sending promptly the samples I requested.

We are shipping your order of March 16 today.

before in the past — Use *before* or *in the past* or neither.

WORDY; This division has never before in the past used form B.

IMPROVED: (1): This division has never before used form B.

IMPROVED: (2): This division has never used form B.

(Note the difference in meaning. Improved sentence 1 implies that form B is being used or may be used. Improved sentence 2 does not imply this.)

| bisect into two parts | Omit *into two parts*. |
| brief in duration | Use *brief, quick,* or *fast* without *duration*. |
| but that | Omit *but*. |

REDUNDANT: I had no doubt but that he would fail.

IMPROVED: I had no doubt that he would fail.

by return mail — This business cliché comes from an era of infrequent mails, when *return mail* meant something. Today it is meaningless. If an indication of time is necessary, use *today, next week,* or a specific date.

by the use of — Use *by* when it conveys the meaning.

REDUNDANT: He proved his point by the use of examples from his experience.

IMPROVED: He proved his point by examples from his experience.

| circle around | Use *circle*. |
| circulate around | Use *circulate*. |

| | |
|---|---|
| collect together, combine together, connect together, consolidate together, cooperate together, couple together | Omit *together* in expressions where it is implied.<br>REDUNDANT: He combined their suggestions together in his report.<br>IMPROVED: He combined their suggestions in his report. |
| complete monopoly | Omit *complete*. |
| consensus of opinion | *Consensus* means a collective opinion; therefore this popular phrase is redundant.<br>REDUNDANT: The consensus of opinion was that we should act immediately.<br>IMPROVED: The consensus was that we should act immediately. |
| consequent results | *Results* are consequential.<br>REDUNDANT: The consequent results are discussed on page twenty.<br>IMPROVED: The results are discussed on page twenty.<br>The consequences are discussed on page twenty. |
| contents duly noted | Omit this business cliché. Your correspondent assumes that you have read the letter you are answering. |
| continue to remain | *Remain* conveys the idea of continuing.<br>REDUNDANT: Regardless of the treatment, the spot continued to remain.<br>IMPROVED: Regardless of the treatment, the spot remained.<br>REDUNDANT: The report indicated that the problem continued to remain unsolved.<br>IMPROVED: The report indicated that the problem remained unsolved. |
| currently | Omit when the present progressive tense of the verb is used.<br>REDUNDANT: She is currently checking the file.<br>IMPROVED: She is checking the file. |
| desirable benefits | Because a benefit is usually desirable, the adjective is unnecessary; however a writer may wish to describe undesirable benefits: *He refused the demeaning benefits offered by the Society for the Assistance of the Incompetent Needy.* |
| disappear from sight | Omit *from sight* unless there is a need to distinguish sight from other senses. |

| Example | Discussion and Illustration |
|---|---|
| early beginnings | Because a beginning is the earliest step, the start, or the first stage, the word *early* is unnecessary.<br>REDUNDANT: He discussed the early beginnings of the company.<br>IMPROVED: He discussed the beginnings of the company.<br>He discussed the founding of the company |
| enclosed herein, enclosed herewith | *Enclosed* means that something has been placed in the same envelope or package.<br>REDUNDANT: Enclosed herewith is the report of the meeting. You will be interested in page three.<br>IMPROVED: You will be interested in page three of the enclosed report of the meeting.<br>(Note that *enclosed* is nearly always unnecessary in business letters because enclosures are listed or noted at the end. *Enclosed please find* is a hackneyed beginning.) |
| end result, final result | Use *result* unless there is a need to distinguish. |
| endorse on the back | *Endorse* means to write on the back. |
| equally as good as | *As good as* and *equally good* are the same.<br>REDUNDANT: Our machine is equally as good as this one.<br>IMPROVED: Our machine is as good as this one.<br>The machines are equally good. |
| excess verbiage | Omit *excess*. |
| fast in action | The attribute *fast* implies action unless the context indicates that *secure* is meant.<br>REDUNDANT: We need not worry about his completing the job on time because he is fast in action.<br>IMPROVED: We need not worry about his completing the job on time because he works fast.<br>We need not worry about his completing the job on time because he is fast. |
| few in number | Use *few*.<br>REDUNDANT: The samples were too few in number for the experiment.<br>IMPROVED: The samples were too few for the experiment. |

| | |
|---|---|
| final completion, final ending, final upshot | The many steps involved in completing legal and financial procedures do make one long occasionally for the *final completion* or *final ending*. But because a *completion* or *ending* is *final*, the adjective is unnecessary. |
| first beginnings | See *early beginnings*.<br>REDUNDANT: She discussed the first beginnings of her plan.<br>IMPROVED: She discussed the conception of her plan.<br>She discussed the first step in her plan. |
| following after | Use *following* or *after*.<br>REDUNDANT: Following after the dinner came the reception for honored guests.<br>IMPROVED: After the dinner came the reception for honored guests. |
| funeral obsequies | Omit *funeral*. |
| hopeful optimism | Use *hope* or *optimism*. |
| if and when | Use *if* or *when*. |
| important essentials | Essentials are important, vital, indispensable.<br>REDUNDANT: The important essentials of this plan are . . .<br>IMPROVED: The essentials of this plan are . . .<br>The important aspects of this plan are . . . |
| in my opinion I think | Use *in my opinion* or *I think*. |
| in the course of | Use *during* or *in* when it conveys your meaning. |
| in the form of | This phrase is usually unnecessary.<br>WORDY: They get their profits in the form of tax rebates.<br>IMPROVED: They get their profits as tax rebates.<br>ACCEPTABLE: He referred to energy in the form of heat. |
| in the range of ten to twenty million | As *ten to twenty million* is a range, there is no need to say so.<br>REDUNDANT: He has in the range of ten to twenty million dollars in cash.<br>IMPROVED: He has ten to twenty million dollars in cash. |

| Example | Discussion and Illustration |
|---|---|
| in the same way as described | Use *in the same way* or *as described,* but not both.<br>REDUNDANT: The second group was injected in the same way as described.<br>IMPROVED: The second group was injected in the same way.<br>The second group was injected as described. |
| in the shape of | This is often unnecessary.<br>REDUNDANT: He gets his recreation in the shape of fishing.<br>IMPROVED: His recreation is fishing.<br>ACCEPTABLE: He ordered a cake in the shape of a horseshoe. |
| in this day and age | Use *today.* |
| increasingly more complex | Omit *more.* |
| join together | Use *join.* |
| joint cooperation | Cooperation is a joint effort.<br>REDUNDANT: The work represents their joint cooperation.<br>IMPROVED: The work represents their joint efforts.<br>They cooperated on this work. |
| joint partnership | Use *partnership.* |
| just exactly | Use one or the other.<br>REDUNDANT: Their solution looks just exactly like ours.<br>IMPROVED: Their solution looks exactly like ours.<br>Their solution looks just like ours. |
| large in size | See *larger-sized.*<br>REDUNDANT: The samples were large in size.<br>IMPROVED: The samples were large. |
| larger-sized, smaller-sized | *Larger* and *smaller* usually clearly indicate size.<br>WORDY: a larger-sized beaker, a smaller-sized package<br>IMPROVED: a larger beaker, a smaller package |
| main essentials | See *important essentials.* |
| melt down, melt up | The verb is *melt.* |
| merge together, mingle together, mix together | Use *merge* or *mingle* or *mix.* |
| minimize as far as possible | Use *minimize.* |
| modern methods of today | Use *modern methods* or *methods of today.* |
| more preferable | Use *preferable.* |

| | |
|---|---|
| mutual cooperation | *Cooperation* is mutual. |
| | REDUNDANT: She asked the divisions to give their mutual cooperation to the project. |
| | IMPROVED: She asked the divisions to cooperate on the project. |
| necessary requisite | There is no other kind of requisite. |
| | REDUNDANT: The bachelor's degree is a necessary requisite for matriculation. |
| | IMPROVED: The bachelor's degree is a requisite for matriculation. |
| | The bachelor's degree is required for matriculation. |
| | REDUNDANT: Larger equipment is a necessary requisite for the experiment. |
| | IMPROVED: Larger equipment is necessary for the experiment. |
| | The experiment requires larger equipment. |
| neurogenic in origin | Use *neurogenic.* |
| 9 A.M. in the morning | Use *9 A.M.* or *nine o'clock in the morning.* |
| obsequies for the dead | For anyone else? |
| one and the same | Use *the same.* |
| personal friend | Use *friend* unless it is necessary to distinguish. |
| plan ahead, plan in advance, plan for the future | Use *plan.* Futurity is inherent in *plan.* |
| prolong the duration | Use *prolong.* |
| repeat the same (story, idea, etc.) | Omit *the same.* |
| resultant effect | Omit *resultant.* |
| seems apparent | Use *seems* or *is apparent.* |
| separate and distinct | One is enough or too much. *The seven steps* is usually better than *the seven separate and distinct steps.* |
| single unit | Use *unit.* |
| state the point that, stress the point that | Omit *the point.* |
| still continue | Use *continue.* |
| summer months | Use *summer* or *June, July, and August.* |
| surrounding circumstances | *Circumstances* are surrounding conditions. |
| | REDUNDANT: The circumstances surrounding his dismissal were the basis of the union complaint. |
| | IMPROVED: The circumstances of his dismissal . . . |
| ten miles distant from | Use *ten miles from.* |
| three hours of time | Use *three hours.* |

| Example | Discussion and Illustration |
|---|---|
| throughout the entire | Use *throughout* or *in the entire.* |
| to the northward | Use *to the north* or *northward.* |
| total effect of all this | Use *total effect* or *effect of all this.* |
| true facts | Use *facts* or *truths.* |
| ultimate end | Use *end.* |
| veritable | Omit it when it is unnecessary. It usually is. |
| ways and means | Writers usually mean one or the other, not this tired combination, unless it is the name of a committee. |

Some writers do not even bother to repeat a meaning in other words; they use the same words. This strange repetition occurs when advertisers, having destroyed the meaning of a word, repeat the word in an attempt to restore its meaning. If they had not used *portable,* for example, to describe what only a derrick can lift, they would not need the phrase *portable portable TV sets.* Writers of functional prose should not mangle language in this ridiculous way, which makes them sound like tellers of children's stories: "And so then the teensy, weensy, tiny Little Bear said . . . ."

Ineffective repetition of words also occurs through failure to revise carefully. In the following sentence the second *that* should have been omitted:

> The results indicated that, at least for this problem, that a different module was needed.

If writers who have a propensity for careless repetition do not train themselves to eliminate it in revision, they will write sentences like this one from a chapter news letter of a technical writing society:

> Before we meet to check a rough draft or to brief you on developments, try to have your questions ready beforehand.

The repetition and the tautology illustrated in this section are pitfalls for teachers and public speakers, who necessarily repeat when lecturing. When they write, they should revise carefully to avoid such tautology as appeared in an entertainer's letter to a network executive: "If I have acted hastily on an impulse . . . ."

## DILUTE VERBS

Also common in functional prose is the weakened, or dilute, verb. Some writers avoid a specific verb like *consider*; they choose instead a general verb of little meaning like *take* or *give* and add the noun *consideration* with the necessary prepositions, as in *take into consideration* and *give consideration to, devote consideration to,* and *expend consideration on.* Thus they not only use three words to do the work of one, but they also take the meaning from the strongest word in a sentence, the verb, and place the meaning in a noun that has a subordinate position. They flout the nature of the English sentence, and thus their wordy writing has poor emphasis. Such writers never *study* a subject; they *make a study of* a subject. Or when they want to impress readers, they *undertake a study of* a subject. They do not *analyze*; they *make an analysis* (or *analyzation*) *of*; they never *prove*; they *give proof of*; they do not *measure*; they *take the measurements of*; they never *approximate*; they *make approximations of*; their data do not *agree* but *are found to be in agreement.* By using a dilute verb in the passive voice they effect awkwardness and poor emphasis; they do not *examine*; *an examination is made* or *performed* or *carried out*; and they never *purify* but in their pompous, long-winded way write that *purification was achieved.*

Dilute verbs and use of the passive voice instead of the active voice (Chapter 9) give to some scientific and technological writing a characteristic verbose, awkward, and vague style. Using specific active verbs, one might write, "The members of the committee agreed to examine secretarial salaries and report to the president by March first." But this is too simple, direct, and clear for a recording secretary who translates it into dilute passive verbs: "The members of the committee were found to be in agreement that an examination of secretarial salaries should be made by the committee and a report should be submitted to the president by or on March first." Unpalatable as a jigger of Scotch in a pitcher of water, this is neither good liquor nor good water.

Such dilute verbs as the following weaken style or, in the words of this debilitated style, *give a weakness to it.*

| Dilute Verb | Discussion and Illustration |
|---|---|
| achieve purification | Use *purify*. |
| | WORDY: An expert will achieve purification of this water. |
| | IMPROVED: An expert will purify this water. |
| are found to be in agreement | Use *agree*. |
| | WORDY: The values are found to be in agreement. |
| | IMPROVED: The values agree. |
| analyses were made | Use *analyze*. |
| | WORDY: Analyses were made of each sample. |
| | IMPROVED: Each sample was analyzed. |
| is applicable | Use *applies*. |
| carry out the work of developing | Use *develop*. |

| Dilute Verb | Discussion and Illustration |
|---|---|
| carry out, has been carried out | Avoid this phrase when nothing is carried. |

carry out, has been carried out — Avoid this phrase when nothing is carried.

WORDY: The installation of the television station has been carried out.

IMPROVED: The television station was installed.

carry out experiments — Use *experiment*.

carry out mixing — Use *mix*.

carry out purification — Use *purify*.

is characterized by, has the character of — Use *be* or *have* or *resemble* or *look like* when suitable.

WORDY: Her work is characterized by errors.

IMPROVED: Her work has many errors. Her work is inaccurate.

WORDY: This element has the character of several less common elements.

IMPROVED: This element resembles several less common ones.

bring to a conclusion — Use *conclude, complete, end, finish*.

connection is made — Use *connects*.

WORDY: The connection is made by pipes . . . .

IMPROVED: Pipes connect . . . .

is a corrective of — Use *corrects*.

arrive at a decision — Use *decide*, at least sometimes. Occasionally one must decide without the delay suggested by *arrive at*.

determine detection — Use *detect* or *determine*.

WORDY: The detection of $x$ is determined by the method.

IMPROVED: We detect $x$ by the method. The method determines $x$.

failed to find — Use the negative, at least occasionally.

WORDY: They have failed to find an answer.

IMPROVED: They have not found an answer.

is found to be — Use *be*.

WORDY: The recommendation is found to be preferable.

IMPROVED: This recommendation is preferable.

give an indication of — Use *indicate*.

give proof of — Use *prove*.

give a weakness to — Use *weaken*.

WORDY: This finding gives a weakness to his conclusions.

IMPROVED: This finding weakens his conclusions.

is indicative of — Use *indicate*.

WORDY: This is indicative of carelessness.

IMPROVED: This indicates carelessness.

| | |
|---|---|
| institute an improvement in | Use *improve*.<br>WORDY: His method would institute an improvement in the process.<br>IMPROVED: His method would improve the process. |
| are known to be, is known to be | Use *are* or *is* except in rare instances.<br>WORDY: The reaction time is known to be unreliable.<br>IMPROVED: The reaction time is unreliable.<br>EXCEPTION: Although the readings of this operator are known to be reliable, he insists upon checking them. |
| make adjustments to | Use *adjust*. |
| make an approximation of | Use *approximate*. |
| make an examination of | Use *examine*. |
| make an exception of | Use *except*. |
| make mention of | Use *mention*. |
| make out a list | Use *list*. |
| make a study of | Use *study*. |
| obtain an increase or decrease in temperature | Use *raise* or *lower* temperature. |
| are of the opinion that | *Think that* and *believe that* are briefer.<br>WORDY: We are of the opinion that the process is economical.<br>IMPROVED: We think that the process is economical. |
| perform an examination of | Use *examine*. |
| present a conclusion | Use *conclude* if that is what you mean. |
| present a report | Use *report*. |
| present a summary | Use *summarize*. |
| proceed to separate | Use *separate*. |
| put to use in (building, measuring, purifying, etc.) | Use *build; measure; purify; used for building, measuring, purifying*.<br>WORDY: The old wood was put to use in building the garage.<br>IMPROVED: The old wood was used to build the garage.<br>He built the garage of the old wood. |
| was seen, was noted | In many sentences these are unnecessary and weak.<br>WORDY: A large increase in volume was seen.<br>IMPROVED: Volume increased by 2,500 sales.<br>WORDY AND WEAK: The temperature was noted to be 40°C.<br>This suggested that . . . .<br>IMPROVED: The temperature of 40°C suggested that . . . . |

| Dilute Verb | Discussion and Illustration |
|---|---|
| succeed in doing, in making, in measuring, in estimating, etc. | Use *do, make, measure, estimate,* etc. |
| is suggestive of | Use *suggests.* |
| take cognizance of | Use *note, notice, heed.* |
| take into consideration | Use *consider.* |
| undertake a study of | Use *study.* |

Many entries in this list needlessly use *by, in, of, out.* The next section illustrates the resultant choppy effect.

## HICCUPS

Closely related to the dilution of verbs is the insertion of unnecessary prepositions and adverbs. This clutters writing with useless words and creates a jumpy effect as though the writer had the hiccups. "Up to the time he came out from the office to follow up the problems outside of the plant, we never met up. Later on we cooperated together checking up on speeding up the sealing off of pipes, and the number of problems was reduced down. And every job ended up with a good write-up when we were paid off." Such speech or writing sounds uneducated in some sections of the United States, and in other sections it suggests a motion picture gangster. A study of the following examples should help writers to avoid these hiccups.

| Phrase with Unnecessary Preposition or Adverb | Discussion and Illustration |
|---|---|
| as to, as regards, in regard to | Use *about* or some other simple preposition. POOR: Will they question him in regard to the patent? IMPROVED: Will they question him about the patent? POOR: She will speak as regards taxes. IMPROVED: She will speak on taxes. |
| as to whether | Omit *as to* whenever possible. POOR: They will decide as to whether the experiment will continue. IMPROVED: They will decide whether the experiment will continue. |
| at about eight o'clock | Use *about eight o'clock.* POOR: They will meet in the auditorium at about eight o'clock. IMPROVED: They will meet in the auditorium about eight o'clock. |
| at above, at below | Use *above, below.* POOR: This reaction occurs at above 70°C. IMPROVED: This occurs above 70°C. |

| | |
|---|---|
| back of | Use *behind* when that is the meaning. |
| | POOR: The garage is back of the research building. |
| | IMPROVED: The garage is behind the research building. |
| call for | Use *demand, require.* |
| | POOR: The project calls for $5,000. |
| | IMPROVED: The project requires $5,000. |
| check into, check upon, check on | Use *check.* |
| climb up, close down, close up | Use *climb* and *close.* |
| connect up | Use *connect.* |
| | POOR: He will connect Pipe A up with Pipe B. |
| | IMPROVED: He will connect Pipe A with Pipe B. |
| | POOR: She connects this response up with a neurosis. |
| | IMPROVED: She connects this response with a neurosis. |
| count up | Use *count.* |
| | POOR: He counted up the money in the safe. |
| | IMPROVED: He counted the money in the safe. |
| debate about | Use *debate.* |
| decide on | Use *decide, select.* |
| descend down | Use *descend.* |
| divide up | Use *divide.* |
| empty out | Use *empty.* A tank is just as empty when it is emptied as when it is emptied out. |
| end up | Use *end.* |
| enter in, enter into | Use *enter.* |
| | POOR: The wires enter into the thermocouple at B. |
| | IMPROVED: The wires enter the thermocouple at B. |
| face up to | Use *face.* |
| | POOR: He faced up to his shortcomings. |
| | IMPROVED: He faced his shortcomings. |
| figure out | Use *understand.* |
| | POOR: He could not figure out the meaning. |
| | IMPROVED: He could not understand the meaning. |
| file away | Use *file.* The redundancy may reflect the common experience that any wanted paper seems to be filed far away, probably in Outer Mongolia. |
| | POOR: Please file away this report. |
| | IMPROVED: Please file this report. |

389

| Phrase with Unnecessary Preposition or Adverb | Discussion and Illustration |
|---|---|
| first initiated | Use *initiated*. |
| follow after, follow up | Use *follow*. There is no escaping the doctor's *follow-up treatment* and the *follow-up form* of the Personnel Department, but otherwise avoid the unnecessary *up*. |
| free up | Use *free*. |
| generally agreed | Use *agreed*. |
| go into | Prefer *investigate* or *examine* if that is what you mean. |
| go on with | Use *continue*. |
| head up | Use *head*. |
| hoist up | Use *hoist*, also *raise* or *lift* without *up*. |
| in between | Use *between*. |
| inside of | Use *inside* or *within*. |
| | POOR: She placed the wires inside of the housing. |
| | IMPROVED: She placed the wires inside the housing. |
| | POOR: He will leave inside of three hours. |
| | IMPROVED: He will leave within three hours. |
| later on | Use *later*. |
| level off | Use *level*. |
| link up | Use *link*. |
| lose out on | Use *lose*. |
| meet up with | Use *meet*. |
| miss out on | Use *miss*. |
| of between | Use *to*. |
| | POOR: widths of between two and three inches, speeds of between sixty and eighty miles. |
| | IMPROVED: widths of two to three inches, speeds of sixty to eighty miles. |
| of from | Omit *from*. |
| | POOR: heights of from fifty to sixty feet, distances of from ten to twenty miles |
| | IMPROVED: heights of fifty to sixty feet, distances of ten to twenty miles |
| off of | Omit *of*. Sometimes replace *off of* by *from*. |
| | POOR: She took the apparatus off of the table. |
| | IMPROVED: She took the apparatus off the table *or* from the table. |
| | POOR: He removed the name off of his accounts. |
| | IMPROVED: He removed the name from his accounts. |
| open up | Use *open*. |

| | |
|---|---|
| out of | Use *of*. |
| | POOR: Only two out of the five drugs were approved. |
| | IMPROVED: Only two of the five drugs were approved. |
| outside of | Use *outside* or *besides* or *except*. |
| | POOR: He placed the extinguisher outside of the laboratory. |
| | IMPROVED: He placed the extinguisher outside the laboratory. |
| | POOR: No one outside of my supervisor knows that I am leaving. |
| | IMPROVED: No one besides my supervisor knows that I am leaving. |
| | No one except my supervisor knows that I am leaving. |
| over with | Use *over* or *ended*. |
| | POOR: The recruiting is over with. |
| | IMPROVED: The recruiting is over. |
| | The recruiting is ended. |
| pay up | Use *pay*. |
| penetrate into | Use *penetrate*. |
| plan on | Use *plan*. |
| protrude out | Use *protrude*. |
| raise up | Use *raise*. |
| recall back | Use *recall*. |
| recur again | Use *recur*. |
| reduce down, refer back, remand back, repeat again, return again, return back, revert back | Use *reduce, refer, remand, repeat, return, revert*. |
| rest up | Use *rest*. |
| resume again | Use *resume*. |
| retreat back | Use *retreat*. |
| seal off | Use *seal*. |
| speed up | Use *speed, accelerate*, or *hasten*. |
| spell out | Use *explain, detail*. |
| still remain | Use *remain*. |
| succeed in doing | Use *do*. |
| take off | Use *remove*. |
| | POOR: He took off the extra charge. |
| | IMPROVED: He removed the extra charge. |
| termed as | Use *termed*. |
| try out | Use *test* except for the theater and sports. |
| up to this time | Use *before*. |
| weigh out | Use *weigh*. |
| win out | Use *win*. |
| write up (*verb*) | Use *write*. |

## ROUNDABOUT CONSTRUCTIONS

Circuitous phrasing and weak indirect constructions waste a reader's time. Even worse, they enfeeble thoughts, frequently the very thoughts that the writer wishes to stress. "There is a treatment available. It consists of . . ." is a typical example, for many roundabout constructions begin with *there is* or *there are* and waste a sentence to state that something exists. Most of these constructions are easily improved; eight words become four words: "The treatment consists of . . . ." The following examples also illustrate the gain in directness and strength achieved by eliminating weak indirect constructions.

1. There are three active ingredients in these tablets. They are . . . .
   The three active ingredients in these tablets are . . . .
2. There are six causes that produce poor morale in our division. They are . . . .
   The six causes of poor morale in our division are . . . .
3. There were at least a hundred chemicals that were added to the list.
   At least a hundred chemicals were added to the list.
   The F.D.A. added at least a hundred chemicals to the list.
4. It is with the third process that this paper is concerned.
   This paper concerns the third process.
   The third process, the subject of this paper, reveals that . . . .
5. This is an important section of your proposal and it should be prepared carefully.
   This important section of your proposal should be prepared carefully.
   Prepare carefully this important section of your proposal.
6. It might be expected that there would be some exceptions to this treatment.
   Some exceptions to this treatment are expected.
   Some exceptions to this treatment are likely.
7. It appears that the new formulation is better than the old.
   The new formulation seems better than the old.
8. In the case of polymer studies it was shown that . . . .
   Polymer studies showed . . . .
9. In the case of experiments with dogs it was proved that . . . .
   Experiments with dogs proved that . . . .
10. The fact that the chemicals were impure caused the delay.
    The impurity of the chemicals caused the delay.
11. If the chemical is not one listed on the chart . . . .
    If the chemical is not listed on the chart . . . .
12. There is no doubt that he will be promoted.
    Undoubtedly he will be promoted.
13. It is obvious that the law will pass.
    Obviously the law will pass.
14. In a previous article by Cope and Dodge, it was shown by them that . . . .
    Cope and Dodge showed that . . . .

The writers of indirect constructions 1 to 5 wasted opening sentences or clauses on stating the existence or importance of their subjects and added their principal thoughts in other sentences or clauses. Presenting the main thought succinctly in the main clause gives a more effective, neater statement.

Examples 6 and 7, which have oblique wording, waste the main construction on unimportant thoughts and bury the main thoughts in subordinate constructions.

Examples 8 and 9 begin with the unnecessary and somewhat misleading phrase *in the case of,* a common cause of wordiness, as are its counterparts *the fact that* (sentence 10), *in this respect, with regard to, from the point of view of, in regard to, with reference to, in reference to, insofar as . . . is concerned, because of the fact that,* and *with relation to.*

The writer of sentence 11 subordinated the main verb, *listed,* by introducing an unnecessary *one* and by employing *listed* as a participle. These examples illustrate upside-down subordination (Chapter 11) as well as wordiness, as do sentences 12 and 13, in which a main clause performs the work of an adverb.

Sentence 14 contains the cumbersome main clause *it was shown that,* which has some equally burdensome relatives: *it was stated that, it was proved that, it was argued that, it was thought that* (or *felt that*), *the statement was made that.* Such wordy and awkward constructions are so common in journals of science and engineering that a reader receives the impression that no one ever states anything although the book or article has an author: "In Bacon's essay the statement is made that . . ."; "In Van Antwerpen's letter the point is made that . . . ." Apparently gremlins place these statements, arguments, and points in the writings of innocent authors, who incidentally are never mentioned except in the possessive case, as though they owned the articles or books but had not written them. One suspects that the next step will be: "In Wiley's book on designing servomechanisms and regulating systems it is stated that . . ."; and the names of the authors will not be mentioned. Some writers in industry and government have evaded responsibility so often in their writing that they evade for other writers too.

## HEDGING AND INTENSIFYING

"It seems that it might possibly be very wise to follow this procedure if no better one is proposed," states (if we may use so strong a word about so noncommittal a sentence) a writer, who not only adds hedge to hedge (*seems, might, possibly, if*) but then ironically intensifies the idea by *very.* The best advice for the writer of such a passage is—"Cut those hedges; prune those intensifiers."

An antipathy to forthright statements may reflect a scientist's honest caution. But placing hedge after hedge in a single sentence suggests a timid, self-protective, indecisive writer. This Milquetoast returns from a meeting that voted unanimously for a holiday on October twelfth and tells his secretary, "We may perhaps have a holiday on October twelfth if nothing happens to interfere." He writes to his supervisor, "We expect to complete the runs on the new machine tomorrow if there are no absences or unanticipated interruptions." We knew one writer who qualified even the inevitable. Asked to name his beneficiaries, he wrote, "If by some chance I should die, distribute my company benefits as follows . . . ."

Like other kinds of wordiness, hedging is contagious. An aggressive, self-confident executive who was used to making decisions for her own company took an important post in a large corporation. She conferred with a consultant weekly about her writing.

At the end of the second meeting she said, "Well, I'll see you at the same time next week if nothing interferes."

"Do you expect something to interfere?"

She banged her fist on the desk. "No, I don't expect anything to interfere. I'm beginning to hedge and qualify like everyone else around here. I sound like a green kid. It's all that stuff I read and listen to. I'll see you next week—without fail."

Some of the most prolific hedgers in industry are not timid. One of them was the most aggressive person we have ever taught. But as soon as he wrote, he sounded as awkward and unsure as a shy boy at his first party. Later he told us that he had been taught to hedge by a timid supervisor. Writing without excessive hedging would have cost him his job. By the time he left that supervisor, the hedging that he had practiced for years had become a deep-seated habit, and he was about to lose an important position because of lack of forcefulness in his writing.

Overuse of intensives is also contagious, and it too creates a poor impression. Qualifying absolutes—*most unique, absolutely perfect, more paramount, very essential*—gives an impression of illogic or carelessness (Appendix B). Conversational intensives, which are inappropriate in formal writing, make writers sound too casual and relaxed. For example, "The color did not change" is preferable to "The color did not change at all." And "It was impossible to obtain results" is clear and strong, but "It was quite impossible to obtain results" suggests to some readers that the writer is attempting to conceal laziness.

Sometimes intensives worn out by use retaliate by weakening instead of strengthening the meaning. Hollywood has so overworked *colossal, supercolossal,* and *epic* that when a theater advertises the coming of a supercolossal epic, experienced moviegoers avoid it. In the same way *very* and *extremely* may weaken the words they modify. When a high school English teacher tells her students that an assigned book is *interesting,* they accept her statement with only slight reservations about the peculiar reading tastes of English teachers. But when she says that a book is *very interesting* or *extremely interesting,* their eyes glaze with the resistance of a customer being told by a dealer in secondhand cars, "This 1980 Dodge is just like new." Writers of informational prose should use words like *very* and *extremely* cautiously. The sentence "His work is of a very uneven character" does not mean more than "His work is uneven." "The work was extremely difficult and . . ." may antagonize a supervisor, for she suspects that if the work has not been completed, the writer is concealing incompetence or laziness and that if it has been completed, the writer is bragging.

## FALSE ELEGANCE

Many writers attempt to impress with their learning, efficiency, or intellect by substituting words for thoughts. Their simplest ideas emerge in jargon, business clichés, gobbledygook, or bookish terms. To them *pretentious* and *impressive* are synonymous. The poorly educated who admire false elegance in diction are pathetic. One woman asked, "But who'll know I'm educated if I use easy words?" Other offenders are well educated— but too recently. After interviewing a new Ph.D., an executive said, "Well I suppose we'll hire him, and he'll be unbearable for a couple of years until he's displayed every term he learned in graduate school. And then after two years he'll begin to talk and write like a normal human being and we'll begin to think that maybe we can live with him after all. You know what will happen then? He'll get another job." Suffering readers of long dissertations are often inclined to think that Ph.D. stands for Phony Diction but Ph.D. readers of business writing reply that M.B.A. stands for Master of Business Argot.

Like a fully inflated balloon, writing puffed up with words tempts critics to prick it into ludicrous collapse. The pin is easy to use. A *New Yorker* writer in his article on teenagers quoted a distinguished sociologist on the characteristics of teenage culture as follows:

> 1. Compulsive independence of and antagonism to adult expectations and authority. This involves recalcitrance to adult standards of responsibility.
> 2. Compulsive conformity within the peer groups of age mates. It is intolerable to be "different."
> 3. Romanticism: an unrealistic idealization of emotionally significant objects.

Then the *New Yorker* writer applied the pin: "Practically anybody would agree with this summary, and practically anybody could have made it, though in less stately language: Teenagers are disobedient, group-minded, and unrealistic."[2]

Pretentious professors must be assigned some of the blame for the reverence of young graduates for fancy language. Although college handbooks of writing stress simplicity and brevity, some professors—and this includes English professors—set poor examples in their lectures and textbooks and atrocious examples in their articles in learned journals. They foster esoteric cults that make it difficult and sometimes impossible for biologists to talk with sociologists, for chemists to talk with engineers, for specialists of one school of linguistics to talk even with those of another school. Their students complain, "Well, he may be a great scholar, but he can't teach." Colleagues shrug their shoulders and comment sardonically that wordy, pretentious writing weighs heavy on the scales that measure writings for promotion. Thus this professional use of pompous language is carried on by coteries of likewinded students.

Some supervisors in industry have earned an even larger share of the responsibility for the cult of false elegance. They write, "We will endeavor to ascertain" instead of "We will try to find out" even when all they are investigating is whether their staffs want meat or fish at a dinner for retiring members. They write, "Select the main course of your choice" when they mean, "Choose your main course." When they mean *this shows,* they write, "As stated above, observations seem to indicate." They prefer *in the nature of, in the neighborhood of, in the order of,* and *in the order of magnitude of* when they mean *about.* Every profession has writers of poor style training the young: lawyers use *in view of the fact that, in view of the aforementioned,* and *in view of the foregoing circumstances* instead of *therefore*; and it is better for the sensitive to avoid thinking of the examples presented to those who are learning to write federal tax forms, especially when the writers call it simplifying federal tax forms.

Anyone who wishes to recover from the influence of pompous professors, pretentious supervisors, and falsely elegant reading should study the following list of long-winded phrases and simple replacements for them. Fancy language like *transpire* for *happen, methodology* for *methods,* and *hypothesize* for *assume* is poor taste. The words in the false-elegance column below are often used in an attempt to disguise the absence of thought. I have never met a writer who could make them sound natural in writing for government or industry. They are fancy dress, and government and industry are not costume parties.

Writers who consistently use the words in the left column should strenuously revise their work. Writers who think that they seldom offend by such use should check their writing carefully to be sure that they have not been blind to these faults. We ask the writers that we coach to try to eliminate fancy language entirely because even then some of it will escape their editing. Occasional use of a fancy or longwinded phrase may not be the most objectionable characteristic of poor style. But it is one of the most dangerous, for the habit of using such phrases is easily acquired and strengthened, especially if one reads and hears them.

| False Elegance and Wordiness | Suggested Improvement |
| --- | --- |
| abide at | live at |
| above, abovementioned | the, this, that, those, these |
| accounted for by the fact that | due to, caused by |
| add the point that | add that |

| | |
|---|---|
| advent | coming, arrival |
| afford an opportunity | allow, permit |
| after this is accomplished | then |
| aggregate | total |
| a great deal of | much |
| along the line of | like |
| a majority of | most (when *most* is meant) |
| analyzation | analysis |
| an example of this is the fact that | for example, as an example, thus |
| another aspect of the situation to be considered | as for |
| answer is in the affirmative (negative) | answer is yes (no) |
| a number of | several, many, some |
| approximately | about |
| as of now | now |
| as per | Avoid this phrase. |

POOR: I am sending you the information as per your letter.
IMPROVED: I am sending you the information requested in your letter.
I am sending you the information you requested.

| | |
|---|---|
| as regards | about |
| as related to | for, about |
| assist | help |
| assistance | help |
| assuming that | if |
| at this time | now |
| based on the fact that | due to, because |
| beg leave to say, beg to differ, beg to say | Omit. |
| by means of | by ⎫ |
| by the use of | by ⎭ when *by* is clear |
| by occupation, by profession | Unnecessary in phrases like *a lawyer by profession, a bookkeeper by occupation* |
| come to an end | end |
| commence | begin. Save *commence* for official ceremonies and procedures. |
| communicate | Prefer a specific verb like *write, telephone, telegraph, cable.* |
| compensation | pay |
| concerning | about |
| conclude | end |
| construct | build |
| contemplate | intend (when *intend* is the meaning) |
| demonstrate | show, prove |
| due to the fact that | because, due to |

| False Elegance and Wordiness | Suggested Improvement |
|---|---|
| duly noted | noted |
| during the time that | while |
| dwell | live |
| effect, effectuate, institute a change | change (when *change* is meant), achieve, cause |
| encounter | meet |
| endeavor | try |
| eventuate | happen |
| except in a small number of cases | usually |
| exhibit a tendency to | tend to |
| final completion | completion |
| firstly | first |
| floral offering | flowers |
| for the purpose of | for, to |
| for the reason that | because, since |
| for the simple reason that | because, since |
| forward | send (when *send* is meant) |
| for your information | This is usually superfluous. |
| give up the ghost | die |
| having reference to this | for, about |
| if at all possible | if possible |
| in case, in case of | if |
| in close promixity | near |
| in conjunction with | Use *with* alone whenever *in conjunction* is unnecessary. |
| in connection with | Use *about* or *of* whenever possible. |
| in favor of | for, to |
| initial | first |
| initiate | begin |
| in lieu of | instead of, in place of |
| in order to | to |
| interpose an objection | object |
| in the course of | during |
| in the first place | first |
| in the vicinity of | near |
| inquire | ask |
| in rare cases | rarely |
| in reference to, with reference to | about |
| in regard to | about |
| in relation with | with |
| in short supply | scarce |
| intermingle | mingle |
| in terms of | in |
| in the amount of | of, for |
| in the case of | for, by, in, if |
| in the event of, in the event that | if |
| in the instance of | for |
| in light of the fact that | because |
| in the majority of instances | usually |

| | |
|---|---|
| in the matter of | about |
| in the neighborhood of | about (except in descriptions of a locality) |
| in the not-too-distant future | soon |
| in the proximity of | near, nearly, or about |
| in view of the above, in view of the foregoing circumstances, in view of the fact that | therefore |
| involve a great deal of expenditure | be expensive |
| involve the necessity of | require |
| is defined as | is (at least occasionally) |
| it is incumbent on me | I must |
| it is often the case that | often |
| it stands to reason | Omit. |
| it was noted that if | if |
| it would not be unreasonable to assume | I (we) assume; I (we) think |
| kindly (in a request) | please |
| leaving out of consideration | disregarding |
| make the acquaintance of | meet |
| meet his maker | die |
| modification | change |
| necessitate | require, need |
| not of a high order of accuracy | inaccurate |
| notwithstanding the fact that | although |
| nuptials | wedding, marriage |
| objective | aim, goal |
| obligation | debt (when *debt* is meant) |
| of a dangerous character | dangerous |
| of considerable magnitude | big, large, great |
| of very minor importance | unimportant |
| on account of the fact that | because |
| on a few occasions | occasionally |
| on behalf of | for |
| on the grounds that | because |
| one and the same | one *or* the same (whenever possible) |
| pass away | die |
| pre-owned car | used car |
| presently | now, soon |
| prior to, prior to the time that | before |
| proceed | go |
| proceed to investigate, study, analyze, etc. | Omit the unnecessary *proceed to.* |
| reimburse | repay |
| relative to this | about this |
| remains | body (when that is the meaning), corpse |
| remuneration | pay |
| reside | live |
| ruination | ruin |

| False Elegance and Wordiness | Suggested Improvement |
|---|---|
| similar in character to | like |
| spells | is (when that is the meaning) |
| subsequent | next |
| subsequent to | after |
| sufficient | enough |
| take appropriate measures | act |
| take cognizance of | note, notice |
| taking this factor into consideration, it is apparent that | therefore, therefore it seems |
| terminate | end |
| termination | end |
| the foregoing | the, this, that, these, those |
| the fullest possible extent | Omit or use *most, completely*, or *fully*. |
| the only difference being that | except that |
| the question as to whether or not | whether |
| there are not many who | few |
| there is very little doubt that | doubtless, no doubt |
| to be cognizant of | to know |
| to summarize the above | in summary |
| transmit | send |
| transpire | happen, occur |
| usage | use (except for other meanings such as *custom, practice in language*) |
| within the realm of possibility | possible, possibly |
| with reference to | Omit (or use *about*). |
| with the exception of | except |
| with this in mind, it is clear that | therefore |

## POINTLESS WORDS AND PHRASES

Many words and phrases are just fillers: they fill time and space until an idea appears. And their sedative effects on writer and reader are great. The writer dictates them without thinking, and the reader skims them with an equal lack of attention. Some common examples are

> It may be said that
> It might be stated that
> In this connection the statement may be made that
> It is interesting that
> It is interesting to note that
> You will find it interesting to know
> You will be interested to learn

Not all of these are tranquilizers for all readers. Tell some readers that they will find information interesting, and their first response is, "What's interesting about that?" Executives reading for the thousandth time "As you may recall," "You may no doubt recall," "You will no doubt remember," and "As you know," may explode. "If he thinks I know it, why is he taking two paragraphs to tell me about it?" demanded one busy executive. We have heard many faculty treasurers read reports like, "As you may remember, our balance a year ago was $1,129.64." The faculty was delighted when one treasurer reported, "As you may remember, though I doubt that you do, our balance a year ago was $473.26." Meaningless fillers become so irritating that harassed readers wish writers would think about what they are writing and interpret it literally, particularly that common phrase *needless to say*, which usually introduces a long, unnecessary discussion.

More treacherous are time-wasters like *angle, aspect, case, character, element, field, situation, type,* and *nature* used pointlessly. Many sentences containing *in the case where* or *in the case when* are better sentences if *in the case* is omitted. "These examples are taken from the field of chemistry" does not need *the field of*. "He is the type of employee who is frequently late" means nothing more to the Personnel Department than "He is frequently late." The phrase *with respect to* is a favorite filler in writings on science and technology. "The pills with respect to their effects cause headaches" is clearer as "The pills cause headaches." Removing such fillers improves clarity and speeds comprehension.

Teachers and other public speakers are prone to use pointless words and phrases. Afraid of silence, they fill in with such expressions while they are phrasing or planning their next sentences. Such fillers may not be completely undesirable in public speaking, for they do give the audience a chance to catch its breath. (However, a transitional phrase would be more useful.) In writing, fillers serve no useful purpose; and if they are frequent, they may ruin an otherwise good style.

In some examples in the preceding list wordiness results from the use of a phrase instead of a single adverb or adjective: *He presented his opinion in a brief statement* instead of *He presented his opinion briefly* or *Wines having a Spanish origin are competing with wines made in France* instead of *Spanish wines are competing with French wines* or *He annoyed her by acting in a foolish manner* instead of *. . . by acting foolishly.*

Other writing is prolix because the diction is not precise: for example, *men and women who are working here for the first time* instead of *new employees* or *He said that they should consult someone who really knew about security problems* instead of *. . . an expert on security* or *Those who are serving on the committee are of the opinion that ignoring even small rules and regulations designed to prevent injuries can result in accidents, many of which are serious* instead of *The members of the committee believe that infractions of minor safety rules can cause serious accidents.*

Some verbose tendencies of United States writers may be due to teachers' demands for papers of five thousand words from students who have enough thoughts on the subject for only fifty to five hundred words. This leads to the belief that longer papers are given higher grades, an error encouraged by some advisers to students writing graduate theses. The misconception that wordiness succeeds becomes so ingrained that professors and managers who know the value of concision have great difficulty convincing writers to be brief. One vice president of a major corporation acquired a reputation as an eccentric because he insisted on concise style. And one of our students in industry, despairing of ever getting his long-desired permission to purchase an expensive TV camera for photographing operations, was too depressed to write another twenty-page request. He submitted one sentence stating why the company needed the camera "described in the enclosed brochure." When he received the necessary approval and check the following afternoon, he could hardly believe it.

"The general manager must have been in a good mood yesterday," he said.

"Maybe he was glad to get a brief memo for a change," one of us suggested.

He looked at the check. "All this money for one sentence? I can't believe it."

"Believe it and remember it the next time you write."

## NOTES

1. Jerome B. Wiesner, *Where Science and Politics Meet* (New York: McGraw, 1965) 158.
2. Dwight Macdonald, "A Caste, a Culture, a Market," *The New Yorker* 34 (29 Nov. 1958): 76.

# Part Four
# ADVICE ON COMMON FORMS

# CHAPTER 14

# Advice on Some Short Forms

*"The horror of that moment," the King went on,*
*"I shall never, never, never forget!"*
*"You will, though," the Queen said, "if you*
*don't make a memorandum of it."*

*Lewis Carroll*

## MEMORANDUMS

Memorandums convey information and opinions within an organization. Even when sections of a company or government agency are in different parts of the country, memorandums are useful. Letters are generally used for communicating in a more personal way with individuals or for communicating with persons in other organizations or countries or both. To those who, like field representatives, work independently at a distance from company centers, letters seem more personal than memorandums and therefore link such workers to their companies. Memorandums are suited to the communication of information, suggestions, recommendations, etc., because they do not contain the friendly material of letters and are therefore briefer and because they may be formal or informal.

## Standard Forms

Standard forms for memorandums save time and effort; a reader or file clerk finds certain information—name of sender, names of those addressed, subject of the memorandum, etc.—in the same place in each memorandum regardless of its source. The statement of the subject of a memorandum should be meaningful and specific; not, for example, *Change in Bookkeeping Form* but *Change in Items 1 to 5 of Bookkeeping Form 104.*

Some writers waste time because they do not develop memorandum forms with blanks for routine information but expend effort writing sentences and even trying to vary the structure and the diction in routine daily reports. To prevent such unnecessary writing, attendance reports and other routine matters, as well as some progress notes, like those on construction, should be entered on forms. Sometimes writers working at a site enter them in handwriting. Supervisors can increase the efficiency of their staffs by ensuring that material suitable for memorandum forms with spaces for stated information has not become the subject of memorandums or letters that consume too much of a writer's time. For example, forms stored in the disks of word processors can be edited to effect changes; thus rewriting entire works can be avoided.

## Selection of Readers

In addressing their memorandums writers should select their readers carefully. Reading memorandums in industry and government is costly, and writers who carelessly direct memorandums to readers who do not need them waste expensive time. Often the expense is not justified by the value of the memorandum even to the people it should reach. Some lightening of the excessive burden of reading in government and industry today is in the hands of everyone who selects readers for memorandums. Unless readers are fools, they have a poor impression of writers who waste time with pages the readers should never have had to read and the writers should never have addressed to them.

## Length

The length of memorandums is controversial. Everyone agrees that they should be as short as possible, but not everyone agrees that they should be as long as necessary. Some supervisors and some companies insist that memorandums should never be longer than a page. Such a ruling may mean that many memos are incomplete and others are divided into two or more.

Writers can easily shorten memorandums by removing superfluous material and superfluous words when they revise. And they should organize memorandums carefully to avoid repetition. But how far beyond that they should go in struggling to shorten their work depends on common sense. If they are addressing one person whose time is more valuable than their own or if they are addressing many people, writers should revise scrupulously for brevity. But supervisors should not expect writers to produce brief memorandums as fast as they produce long ones. It takes time to be brief—especially at first.

If shortened memorandums are not clear, the achievement of brevity has little point. If the brevity makes them difficult to read, the writers have confused saving paper with saving their readers' time. If memorandums are so incomplete that readers must ask questions to get the information the memos should have supplied, the omission of information has wasted time. A good memorandum is as brief as is consistent with completeness, clarity, and readability.

## Style

The principles of clarity and correctness that apply to memorandums are the same as those for other writing, and, as in other writing, the language and subject matter of memorandums should be adapted to the readers. Both information and writing should be as specific and concrete as the subjects permit. A writer's choice of organization is particularly important in memorandums; therefore a writer should consider the advantages and disadvantages of all possible organizations before selecting one. The first sentence of a memorandum should be the most effective beginning for the readers. Useful as summary beginnings are, they have been employed so often in some companies that they have become trite there. Companies that store on disks memorandums for all occasions are particularly guilty of this offense. If writers do not wish to reveal their main points before they have prepared the readers, they should choose openings that attract readers.

The style of a memorandum may be formal or informal, depending on the subject and the reader. A memorandum of technical information may be as formal as a technical report. A memorandum of changes in procedure may be formal if it is addressed to a pompous executive and informal if it is addressed to a breezy field representative. Any tendency to make all memorandums routinely formal and cold should be fought. A memorandum to a friendly person with whom one works closely may be as warm and informal, though not so wordy, as a friendly telephone conversation. Although memorandums do not usually contain the social digressions of a friendly business letter, they may in their own way be friendly too. Supervisors who write informal friendly memorandums to their staff will get better response and establish better relationships than will those whose writing is stilted and cold. Busy supervisors should not overlook this opportunity to make employees feel like more than robots. Good management courses train supervisors so successfully to display warmth and friendliness that they never pass an employee without a greeting. But much of the value of this is lost if they send to those employees memorandums that sound as though they were written by a machine to some cogs.

## Carelessness and Inadequacy

Carelessness is a common fault in hurried memorandums. Some writers read an inquiry incorrectly and supply in their memos the wrong information or only part of the information. Others write memorandums less correctly than reports, and their errors annoy and distract readers. And some who write hurriedly and neglect revision do not say what they mean. Part of this carelessness is due to a failure to consider memorandums important. Readers who write and receive superfluous memorandums are hardly likely to attach much significance to anything written in that form; therefore, company improvement requires the elimination of superfluous memorandums and stress on effective ones.

Incompleteness, ambiguity, and even utter lack of meaning characterize many memorandums sent to immediate supervisors familiar with a subject.

"Only my boss reads this, and he knows all about it anyway," is the writer's explanation.

"Then why do you send a memo?"

"Well, for the files, I suppose, or in case someone sends for information when I'm not here."

Pity the reader of those filed confusions! A substitute for the supervisor or the writer, a replacement for one of them, an executive seeking information while they are away, an expert reorganizing systems, an examiner of files trying to decide what to discard—any one of these finds only meaningless or confusing words. Thus the fallacious conviction that "only my boss reads this" is expensive to industry and government. Occasionally justice prevails. Such an incomprehensible memorandum is requested unexpectedly by an executive whom the writer wishes to impress favorably, and then the writer suffers more than the reader. The writer suffers too if months after filing a memo he needs the information in it and cannot understand his own expression. Inability to understand later the memos written in careless haste is frequent in business, science, and technology. If writers cannot understand their own memos six months after writing them, who can?

If writers do not have sufficient pride to write understandable prose, their supervisors should be responsible for seeing that every paper filed can be understood by a competent reader. And that does not mean the supervisors or the writers who from their experience can supply missing material and translate meaningless sentences; it means readers who might have to consult the paper in an emergency. Otherwise writing, filing, and storing such papers is wasteful. It would be less expensive and less confusing for the supervisor to state, "I don't believe in files. We don't keep them. If you want to know what we're doing, ask. If we're not here, guess."

## LETTERS

Why do many people who write good letters of introduction and good social letters write poor business letters? is often asked. There are three principal causes of poor business letters. The instruction in writing given to undergraduates today stresses informal writing like that in freshman themes and formal writing like that in technical reports. Training in letter writing is usually offered only to business majors. When it is given to students specializing in science, the liberal arts, or engineering, it concentrates on how to write a letter applying for a job, that being the letter of most immediate concern to students. Some courses do touch briefly on the nuts-and-bolts letters of business—the letter written to order merchandise and the letter correcting an error in an order. But students who are not business majors are seldom trained in the writing of more difficult business letters. Most handbooks addressed to scientists and engineers concentrate on the format of letters and offer little help with writing business letters. It is clear that many graduate students acquire while writing their theses a style that is particularly unsuitable for letters and not much good for anything else. Therefore when graduates who have had instruction only in essays, theses, and technical reports must write business letters, they approach the task with little experience or instruction.

What do government and industry do about this? They usually assign to these novices the least experienced, least competent secretary or typist available. Together these neophytes write letters for their agency or their company, and they may even write them without any supervision. At the same time the government and industries of the United States spend billions on what their public relations departments term "improving the image," the image created at least in part by poor letters.

Many a supervisor makes correspondence unnecessarily difficult for beginners by asking new employees to write letters in the supervisor's style for the supervisor's signature. Making ghost writers of young scientists and engineers is one way to prevent good performance. To compose a letter in the style of another person requires special skills usually acquired only through long practice. A few experienced writers can assume now the personality of the president of a company, now of the vice president, and now of the general manager well enough to write speeches, articles, and letters for them all. And some secretaries learn to write letters in the styles of their employers. But to expect young, relatively untrained graduates in science and technology to become ghost writers for supervisors is wasteful of the time of the company, demeaning to a professional, and frequently annoying to both writer and supervisor. Under this arrangement writers of business letters spend time trying to learn not good style but the style of their supervisors; thus beginners do not develop letter styles suited to them unless they happen to be clones of their supervisors.

Professionals should sign their own letters. A supervisor's initials or signature of approval may, and for many letters should, be required. But writers of business letters should be permitted to use and should be encouraged to develop styles of their own that reflect their personalities. Such styles will do more credit to the government or company than any copy of the style of another.

Writers who wish to learn to write good business letters will learn faster from instruction combined with their own observation than from imitation. The principles of good business correspondence are essentially those of other good writing. The few exceptions to the general rules follow, as does a discussion of some business letters that pose special problems— letters of contract; letters of technical information; and letters of adjustment, apology, gratitude, congratulations, recommendation, and condolence.

## The Style of Business and Professional Letters

### *Style in General*

Writers of good letters give even more consideration to their readers than do writers of reports and memorandums. They all adapt subject matter and language to their readers' understanding, but letter writers often drastically modify their styles for their readers. To stiff and dignified officials they write formal letters; to friendly salespeople, informal, even breezy, letters; to colleagues with good senses of humor, amusing letters. But their correspondence reflects their moods of formality or informality or humor because the styles of all their letters are their styles, not copies of their correspondents' or their supervisors' or of the styles of letters in books of model letters.

The paragraphs and sentences of good business letters are usually shorter and simpler than those suggested for other writing (Chapters 11 and 12); subjects may be broken into small units; and language tends to be informal. Although a letter of information is likely to be wriitten in the language of a technical report, other business letters may use contractions and informal language if these are appropriate to the readers.

Letter writers should avoid business clichés, most of which are wordy and pointless. And they should not use old-fashioned participial closings or other phrases that are out of date. One undesirable result of the imitation of the letters of others is the copying of antiquated language by generation after generation of letter writers. In some companies the wording of letters has not changed in this century.

### *Courtesy and Friendliness*

Courtesy is more important in a business letter than is brevity; therefore letters should never be so brief as to seem curt. A sentence or two of friendly chat are appropriate in correspondence with someone who appreciates a friendly business letter. Social subjects are appropriate in a letter to an acquaintance of long standing or to a friend. Writers should no more restrict themselves to business in a business letter than they would on the telephone or in the office. The friendly material may be briefer in a letter, but those who would chat on the telephone to the reader of the letter may chat in a letter. If for some subjects, like correcting an error in a routine bill, writers would not digress from the subject on the telephone, they do not digress in a letter. But routine letters should still be friendly and courteous.

As professionals become more successful in their careers, they write fewer of these routine letters, and nearly all their letters have personal touches that only they can provide. Their letters would seem unfriendly if there were no inquiry about the health of a child who was sick when last the writer saw his correspondent, no recalling of the entertainment offered by the correspondent at their last meeting, no remark about the books or sports they have been discussing for years. These informal comments are not unimportant or inessential; they are builders of friendship and good will, which are important to the writers and their employers. Administrators, physicians, and lawyers seem to understand the value of encouraging this good will better than most scientists and engineers do. The scientists and engineers who do understand it and have acquired the knack for it are able to establish relationships that contribute to their success. Their associates like them, remember them, enjoy working with them, and want to help them.

Yet many writers who are friendly in person are cold and forbidding in their letters. Their business letters read like routine orders for merchandise. Ordering a book, requesting information from an acquaintance, or answering the inquiry of a friend, they write the same kind of letter. It never occurs to them that an acquaintance may feel rebuffed or that a friend may regret having asked a favor. Nor do they understand why others are chilly or brusque in return.

Such unhappy divergence between the personalities of writers and the styles of their writing need not exist. If the personality of their letters is too informal, writers usually have little difficulty making it more formal. But if the tone of their letters is too austere and formal, they may have some difficulty changing it. This is due to a failure to see just what needs to be changed. Usually the personality of such letters has been formalized by the absence of pronouns in the first and second persons and by the presence of too many passive verbs, as well as by the use of old-fashioned business phraseology. As soon as writers learn to use the first person, many verbs tend to become active instead of passive. Although writers who are stiff and inflexible may fight to retain their business clichés, replacing trite business phrases with simple conversational English is necessary to complete the metamorphosis from stuffy to informal, from pompous to friendly.

These changes do not occur overnight, nor are they easy for all writers. We have even had to supply dictating machines in the shape of telephones to encourage some obstinately pompous writers to dictate as though they were friendly with their readers. After a day of such dictating and of reading the results critically, even the most intractably stiff writers loosen up on paper. Only one letter writer complained. He said, "I used to dictate business letters while I was thinking of something else. Now I have to think about what I am saying." Other transformed writers considered the change well worth the trouble and effort it had required. One spoke the feelings of all of them when he said, "Now I can write like a

411

human being. I used to write like a stuffed shirt; I used to pour out the same old tired phrases like a machine."

## The Pronouns *I, We, You*

A word of caution here about the use of pronouns in the first person is necessary. Most writers worry more than necessary about using *I* too often. They need not strain to avoid beginning letters with *I* if it is appropriate to start that way. And if *I* appears in different positions in the sentences of a letter, it is seldom obtrusive or even noticeable. The people who worry about the number of *I*'s in their writing are seldom those who use too many. Writers who are anxious about their use of *I* should resist the easy but dangerous escape to *we*. For when they use *I*, they commit only themselves; a writer may commit himself as the shipping manager of a company, but he does not always commit the entire company. Writers who use *we* commit their companies or agencies. Another difficulty with *we* is its slipperiness in writers' minds and the resulting confusion to readers. In one letter of such a writer *we* is the writer alone, his company, his division, he and his wife, and he and those working with him on an experiment. But in one letter *we* is not capable of serving clearly so many meanings ("The Evasive I," Chapter 9). The confusion that results from *we* signifying first one group and then another may be major.

*You* is not a misleading pronoun in itself, but some misleading advice has been given about *you*, and some good advice has been misunderstood. To dot a letter with frequent *you*'s does not necessarily make it personal or attractive. Too many expressions like "you will agree," "of course, you see," and "you will no doubt find" draw from some readers the well-deserved response, "The hell I will." Many readers do not like to be pushed so hard, and *you* in such expressions is an obvious shove. Writers who have successfully taken their reader's point of view can write effective personal letters without a single *you*. And dozens of *you*'s do not help letters from writers who fail to see their reader's point of view; the failure in fact may be more obtrusive and hence more annoying. "You are hereby notified that your employment in the G. G. Psychology Company will terminate on April fifth," wrote a beginner for his supervisor; and when he was told to soften the blow, he replied, "But I started with *you*." He would not write, "We regret that it is necessary to tell you" or "I regret to tell you" because he considered it self-centered to begin a letter with *I* or *we*. Courteous consideration of the reader is more important than pronouns, and no pronoun can conceal a writer's lack of interest in a reader.

### *Readers in Other Countries*

Writers should be particularly careful to extend courtesies to Asiatics, Europeans, and Latin Americans. They are punctilious about social niceties, and some of them are easily offended by what seems to them the negligence of writers in the United States. The brisk style of a letter that would be considered polite enough in the United States may seem curt in another country. Writers must judge from their correspondence with other countries how formal and even repetitive their courteous letters ought to be. Writers should not jump to the familiarity of first names or an informal style, especially when writing to an older person. It is wiser to wait for the correspondent to set such a tone. And writers should be careful to avoid commands even when they are writing directions. In many countries commands offend professionals. Commands must be disguised as exceedingly polite requests. What a writer considers formality befitting only a few occasions may be the rule in other countries. Writers must be guided by the letters they receive from other countries and should depart from the principles of brevity when such a departure seems necessary in order to write effectively to their foreign correspondents.

## Various Business and Professional Letters

### Letters of Contract

Any letter that commits a writer or the company or both should be written and revised with care. Such a letter is a legal contract and therefore offers little opportunity for second thoughts after it has been signed and mailed. A writer should be sure that it includes exactly what he or she wants, that it is complete and accurate, and that it is unambiguous. Too many letters ordering merchandise and equipment lack a necessary specification, such as size, model number, or method of shipment.

It is wise to allow twenty-four hours between writing and signing letters of contract, for in the clearer light of another day writers may note that they have not said what they intended to say or that their letters are incomplete. If such a letter must be mailed immediately, the writer should ask a competent fellow worker to check it. If a hastily written contract is not read by a lawyer, it should be read by at least two experienced, competent people.

### Letters of Technical Information

A letter of technical information should contain technical language if it is sent to a technically trained reader; otherwise it should contain such terminology only when it is impossible for a writer to avoid it. Like a technical report, this letter may have headings helpful to readers. If answers to correspondents' questions are placed under headings, they are more useful to readers; and writers will find letters containing unrelated topics easier to write if they use headings instead of attempting to connect unrelated parts. Lengthy information should be presented in a brief report with a covering letter or in a memorandum. The covering letter introduces the report and may note the occasion for writing or sending it. The report is written like any other. Thus a writer may use one appropriate style in the letter and another in the report. But the writer of a long letter of information should not create a hodgepodge of two or more styles and tones. If there is much information, most readers find a report with a covering letter more useful than a report in letter form.

For internal correspondence some companies use a bastard form that handicaps a writer: they want short reports in letter form. Writers, correctly sensing that letters should be more than pages of information, struggle unsuccessfully to attach opening and closing paragraphs to these "letters." Their paragraphs are usually awkward or hackneyed, the opening paragraph acknowledging the request for information and the closing paragraph suggesting that the readers telephone or write if they need further information, actions which they will take anyway. Yet without these opening and closing paragraphs, the long section of information constitutes a cold letter.

It would be far better to send such information in a memorandum and thus to avoid these unnecessary difficulties. A memorandum neatly, appropriately, and conveniently conveys information from one person to another in the same company. By supplying in the proper spaces the date of the inquiry, the subject of the inquiry, and other information essential for filing, a writer escapes the awkward task of expressing these details in an opening paragraph. Readers have them in convenient form when they need them; they do not have to skim through the opening sentences of a letter to locate them. And the memorandum is more convenient for filing than a letter. Moreover, it has unity; there is no occasion for a writer to compose personal sentences to precede and follow the information, for the memorandum does not sound brusque without them. In fact, everyone is happier except, perhaps, the originator of the bastard form.

When letters of information are written by one person and signed by another, they may frustrate both. One may not like the opening and closing sentences of the other, and the time the two spend arguing, haggling, writing, and rewriting is a waste of highly trained people. We have known a supervisor to argue for nearly two weeks with an assistant, waste the time and patience of others by asking for their opinions, and end by disliking his assistant, who by this time hates him and the letters. When good workable forms like the memorandum of information are available, an administration that requires troublesome letters is seeking and finding difficulties. A company that has many strange requirements of this sort about writing makes poor writers of adequate ones and encourages good ones to hate writing. Only poor writers remain unaffected. New employees suffer even more than others, for they must learn how to meet the strange requirements—if anyone can learn. One engineer described the problem thus: "Everyone in our division writes poorly and hates to write because our division demands poor writing."

### Letters of Complaint

The secret of success in letters of complaint is to avoid complaining. Adjustments should be requested politely (and firmly if necessary), claims should be stated clearly and concisely, and arguments should be presented in a way that will move a reader to action. Whining, insults, and pug-

414

nacity may occasionally achieve results in conversation or argument, but they are usually blunders on paper. Without the person's voice, facial expressions, and bodily gestures, they are puerile; and they create a poor impression for the writer and the company or agency represented. Moreover, angry writers are inclined to present more details than are necessary for a routine adjustment and to cite irrelevant inconveniences at the same time that they omit some of the information essential to a claim. Adjusters cannot do what writers wish if they cannot understand the letters or find the necessary information.

There would be less vehemence and more clarity in letters of complaint if writers realized what happens to letters of indignation. The best thing that may happen is that the clerk who receives such a letter will read it; think, "This nut's mad at us"; and make the adjustment in the routine way. But clerks impressed by the length and tone of such letters, may send them to the desks of busy executives, where the complaints will rest until someone finds time to answer them and to return the answers to the clerks. Sometimes a clerk becomes confused by a long and complicated complaint, misunderstands what is wanted, and gives the writer another occasion for anger. Moreover, letters remain in modern filing systems a long time and may embarrass writers in the future when they are supervisors training writers.

Unless writers think that detailing their inconvenience will speed adjustments, they should omit the details. Of course, if inconvenience is part of a claim, the inconvenience must be described—preferably in an effective, businesslike way, not heatedly. Clarity, courtesy, and good humor produce desired results in professional correspondence; anger and tears are inappropriate and ineffective.

### Letters of Adjustment and Apology

Writing letters of adjustment and apology worries some writers. They can avoid the most common errors of these letters by remembering that adjustments, like contracts, must be unmistakable and that only one apology is necessary for one mistake. The apology should be so complete and sincere the first time that it need not be repeated. When apologizing for a company, writers should not be excessively humble or condescending. When apologizing for themselves, they should apologize in tones appropriate to the situation and their personalities.

Writers of letters of apology should be careful to refrain from promising more than they or their employers can or will perform; for instance, they may express their hope that an error will not occur again, they may mention the steps taken to prevent its happening again, but they should not promise that it will not happen again. If such a promise is broken, correspondents may consider the writers or the company insincere, or perhaps indifferent. Even though a promise is kept, correspondents may be dissatisfied, for the excessive apologizing may have led them to think the error or injury more serious than it was.

415

Usually of greater interest to readers than an apology is an adjustment. An adjustment favorable to readers should be placed at the beginning of a letter. Pleasing information at the beginning puts readers in a good mood. Pleasing information delayed may lead readers to think that the letter is refusing their requests, and their annoyance may not be allayed by the adjustment announced at the end. Letters begin most effectively with something that readers will be glad to know.

### Letters of Gratitude

Like letters of apology, letters of gratitude should not be repetitive. Thanks should be expressed fully and sincerely, and then writers should end their letters or turn to other subjects. There are many ways to develop the expression of thanks. In a letter of personal gratitude writers may hope that they will be able to return the favor; they may recall with pleasure facets of the hospitality shown to them; they may mention the good results of the help that they received; they may praise the promptness, the graciousness, the generosity of the giver; they may extend invitations. To be meaningful, such invitations must be specific as to time and place: "Mary and I hope that you will be our guest at dinner on Thursday, May first, the second night of the ACS meeting in Chicago"; "I will call you when you are in San Francisco next month to arrange lunch or dinner with Sally and George and me"; "I want you to meet the other physicists who benefited from your help. Will you have dinner with us on Monday or Thursday of the week you are in Oak Ridge?" Letters expressing gratitude for a favor to a company are usually shorter and less personal, especially if two companies exchange favors occasionally.

But they should be written. The best advice about thank-you letters is *Don't wonder whether to write them; write them.* No writer ever lost friends for themselves or their employers or were refused favors because they wrote gracious letters of gratitude. Thank-you notes are easy to postpone and easy to forget. But busy successful professionals write them promptly. The letters reflect their courteous consideration of others. Such writers never underestimate the value of the courtesies which are the foundation of good professional relations, and beginners can profit from their example.

A young corresponding secretary of a chapter of an engineering society, for example, may find himself pleasantly remembered by one of the most important engineers in the society because of a prompt and gracious letter of thanks for a speech to the chapter, or the secretary may establish for himself the reputation of a boor by neglecting to write such a letter. A geologist who was hired although some better geologists had applied was surprised to learn that her note of thanks for the interview and entertainment had weighed in her favor. Professionals who always write letters of thanks promptly will have many occasions to be happy that they do.

### Letters of Congratulation

Letters of congratulation should not be neglected. These pleasant courtesies are easy to offer colleagues at home or abroad, and the happiness of the recipients far outweighs the small trouble of writing. Some professionals are meticulous about sending congratulatory notes to those who have competed with them for appointments or promotions and been victorious. Congratulations serve other purposes than demonstrating good sportsmanship. They are particularly pleasing to subordinates, and a good administrator is never too busy to offer them. Distant colleagues are pleased and flattered when their successes, especially their small successes, are noticed with rejoicing. And colleagues close to a writer welcome a letter of congratulation, for it is an indication of personal interest in a world where many people feel that they are only punch cards to their associates.

Letters of congratulation may be sent to those about to retire, to those promoted, to those who receive degrees, to those whose articles or books are published, to anyone, in short, who achieves a success. When a personal relationship exists, congratulations may be offered on honors earned by a child or grandchild, on the birth of a child, on a man's wedding. (A bride is not congratulated; she is offered warm good wishes.) Greeting cards do not take the place of letters, for letters show that the writer had enough interest to take the trouble to write. On occasions of rejoicing, gracious letters that share the joy increase the joy.

It is not always tactful to be cheery in letters congratulating colleagues on their retirement. Not everyone is happy about retiring, and writers should express sentiments appropriate to the feelings of their correspondents. Letters of congratulation on retirement may wish colleagues good health and may hope that they will find satisfaction and pleasure in activities they are known to be contemplating—travel, hobbies, sports, professional activities, or moving to a new home. A specific statement of ways in which the writers or others will miss the persons retiring is not amiss, nor are good wishes to the wives or husbands of the persons who are retiring.

Writers who are happily anticipating their own retirement may mention that. Younger colleagues may appropriately express gratitude for any help or inspiration that they received. And jingles written for the occasion, company jokes, flattering remarks, another round of jokes about sports or hobbies—all have their place in this letter if they are appropriate to the person addressed. If people retire because of illness, it is tactful to omit comments on activities that they no longer may be able to pursue. Letters of congratulation should be designed to make retiring colleagues happy—happy in their prospects, happy in their good friends, happy in their achievements.

### Letters of Recommendation

Professionals who never have had to write a letter of recommendation are rare, but professionals who write useful letters of this type are not legion. Letters of recommendation should not be panegyrics that readers suspect must be untrue. Such letters should state fairly what the readers should know about the abilities of the person, about his or her character, and about any deficiencies. A letter mentioning a deficiency may be more helpful than one that omits it. For example, if a competent animal psychologist resents supervision and performs best when she has most independence, a company can try to assign her to an easygoing administrator if it is informed; but if this useful information is concealed, the company may assign her to a martinet with the result that the psychologist will soon be requesting another letter of recommendation. A person who has supervised an employee for two years or more should know the employee's strong and weak points well anough to present them fairly and helpfully.

Like many other letters, recommendations may reveal as much about the writers as about the subjects. Professors who do not know one graduate student from another show their lack of interest when they write letters of recommendation for students. Supervisors who over- or underestimate employees, who can present no useful comments on employees, or who do not know one employee from another also reveal this in letters of recommendation. And those letters, displaying their deficiencies as administrators, may unintentionally disclose the reason that employees are leaving. The letters of good supervisors assist inquirers in understanding and evaluating applicants, and in the end this is better for the applicant than fulsome praise. As for unloading inefficient and undesirable employees on other companies by means of letters of false praise, this blatantly advertises a company's or a supervisor's dishonesty. In today's litigious world some companies dictate policies for letters answering inquiries about employees or former employees. Such limitations make letters of recommendation company letters rather than supervisors' letters.

### Letters of Condolence

Expressing sympathy tactfully is difficult for so many writers, even for some experienced writers, that we treat letters of condolence in more detail than we have treated other business letters. As in other writing, problems here tend to solve themselves if the writers consider their reader or readers. Letters of condolence are meant to comfort the bereaved persons to whom they are addressed; they are not meant to comfort the writers by affording them an opportunity to unload their feelings. Writers should focus their attention on the recipient and keep it there until the letters are mailed. They should express their sympathy sincerely and simply. To write beyond that, they must consider what will comfort the person whom they are addressing, and that is not necessarily what would comfort the writers in similar circumstances. Any one of the following subjects or none of them may be appropriate in particular cases, or they may suggest something suitable:

418

1. The writers' appreciation of their association with the deceased and some particulars of the contribution of the deceased to that association
2. A statement that the deceased will be missed by the company, by a division of the company, by a particular group of associates within the company, or by the writers
3. A complimentary but not effusive statement of the contribution of the deceased to a field of learning, to a company or division of a company, to a professional society, to his associates, to the writer
4. A complimentary but not effusive statement of the best qualities of the deceased
5. Gratitude that the deceased did not suffer long illness and pain
6. A hope that the family or the persons to whom the letter is addressed will find comfort in the companionship and love they gave the deceased, in the excellent nursing and tender comfort they offered during an illness, in the children who are carrying on the work or qualities of the deceased
7. A recollection of the pride in his or her family or the love for his or her children expressed by the subject of the letter to colleagues
8. The excellent reputation of the deceased, the respect accorded him or her, the many friends who mourn
9. A recollection of the deceased in the situation in which the writer will remember him or her—helping a new employee, modestly receiving acclaim, concentrating intellectual powers on a problem, joking during coffee breaks, enjoying a game of golf, etc.

Whatever material they choose for their letters, writers should remember that a personal letter of condolence should not be long. More dignified and formal language may be used than for other business letters, but it should be simple. As this is the most personal of business letters, it should not be a copy of someone else's letter or of a model in a book. In this letter sincerity is more important than elegance. A condolence letter represents the writer as no other letter can and should never be written by someone else. If the writer is away and cannot be reached, a secretary may write to explain this. And a secretary may also write a condolence letter of his or her own. The family and friends will be more comforted by a separate letter than by a few sentences tagged on to a letter written for an absent employer.

A personal letter of condolence should be written by hand on stationery of good quality of a size that suits the handwriting. It should be written promptly, usually the day that one hears of the death. A letter written for a company or a letter written by a secretary for an absent employer may be typewritten on company stationery unless that stationery is not dignified or is otherwise inappropriate.

419

Letters that executives write to express the sympathy of a company or division may have to be formal; but if the executives knew the deceased, the letters should be personal. If the executives supervised the work of the deceased, their letters will be important to the family. The writers should realize that the families will cherish such letters and may show them to relatives and friends. By enabling the families to take pride in the reputation and work of the deceased, in the high regard and affection of his or her colleagues, in the esteem and respect shown, the letters will be a comfort. The writers of such company letters may include their own sympathy for the bereaved. If the letters are longer than the usual letters of condolence, it does not matter. That executives took the trouble to write at length may be a comfort in itself.

Besides offering sympathy, business letters of condolence may offer specific help. Writing for the company, executives may state that representatives of the company are ready to confer with the families or their lawyers. Letters about financial arrangements—pension or other benefits, company insurance, salary due, stock transfers, etc.—should be sent promptly if a bereaved family has any financial problems. Union officials may also write letters about financial arrangements, and they should take care that these letters are not so coldly businesslike as to be offensive.

If it seems appropriate, a colleague who is also a friend may offer assistance with specific tasks—funeral arrangements, notifying friends and business associates, helping to receive mourners, etc. If a colleague knows the family, it is probably more helpful to offer such assistance by telephone to allow for discussion of how friends can be most helpful. General offers of assistance are empty formalities. But specific offers are a comfort and a help: "I'm free all day Thursday to help you"; "Mary would like to do anything she can on Wednesday"; "We will gladly take care of the children until after the services"; "Don't put your dog in a kennel unless you particularly want to; we'll be glad to have him at our house"; "If your boy is interested in a summer job, please ask him to let me know. I am sure that I can find something for him"; "My son will gladly stay with your grandfather if he is not going to the service." Colleagues must use good judgment about the circumstances of the bereaved persons in deciding what help to offer when they write or telephone.

With obvious modifications the suggestions given here apply also to letters of condolence written to a colleague, an employer, or an employee on the death of a member of his or her immediate family. A writer knows the recipient of this letter and therefore is on firm ground in selecting the type of comfort to offer.

A warning against some common mistakes may be helpful. A letter of condolence should never be maudlin or effusive. Writers should not assume that death is welcome because of long suffering and should never in any way imply that the bereaved persons must be relieved or happy. They probably are not, and in any case they do not want to read such comments. Writers may express their own relief that the deceased is no longer in pain, but they should not imply that the family of the deceased shares their sentiments.

Writers should consider carefully before expressing religious sentiments. If the bereaved and the writer share the faith of the deceased, the writer may offer appropriate religious comfort should it seem natural and fitting. But when faiths vary, any religious statements should be carefully examined for acceptability to the bereaved. Christian comfort about an afterlife should not be addressed to a family of a religion that does not believe in the afterlife. Nor should a non-Christian express sympathy to a Christian "losing the deceased forever." And a writer should be sure that the bereaved are not atheists before offering to pray for them. The function of a letter of condolence, we repeat, is not to give the writer comfort but to make grief more bearable for the bereaved.

Some writers think of a time of death as a time to seek converts. Attempts at religious conversion have no place in a business letter of condolence. As a matter of fact, they should be made, if they are to be made, in person, not by letter. Then at least those who are trying to convert can see whether they are increasing the emotional problems of the bereaved.

In the writing of a letter of condolence, as of any business letter, common sense is invaluable. If writers with common sense attempt to allay the grief of bereaved persons, they will not go far wrong in their letters. If, in addition, they write sincerely and simply, their letters will be comforting ones.

## INSTRUCTIONS AND INSTRUCTION MANUALS

In instructions clarity, completeness, and helpful organization are of main importance. To achieve these a writer must consider the reader constantly. Directions that are clear to an experienced chemist may not be clear to a laboratory assistant, directions that the head of the financial division understands may confuse a new clerk, directions helpful to an editor or printer may puzzle a lawyer, and directions detailed enough for a new clerk will only annoy when addressed to a supervisor. The first consideration, therefore, is always the reader.

### Knowing the Reader

Instructions that are to be used many times, like those for laboratory assistants, clerks, and factory workers, must be clear to all who use them, not to just the present workers. Words alone may not be enough; illustrations may be necessary. If a laboratory worker does not know one flask from another, the required flask should be sketched and a clear caption should appear in the instructions. Moreover, sketches of all flasks, clearly labeled, should be posted near the storage space for flasks. Complicated bookkeeping instructions accompanied by a sample card or page are clearer than words alone can make them. And writers will find printers' proofs easier to correct if they have a sample marked proof as well as a list of printers' symbols. Directions to factory workers that tell them not to remove the fabric until it is dark blue should include a sample of that color. The first step in preparing directions is, then, to know the readers well enough to estimate what words and illustrations and samples they will need in order to follow the instructions correctly.

## Knowing the Subject

The second step is to be sure that the writers know the subject completely. Complete understanding of an operation usually requires that writers perform it themselves several times. Many mistakes in directions come from a failure to grasp exactly what the readers are supposed to do. Writers who wet their hands in the laboratory sinks, soil them with machine oil, or use them to copy documents will understand the operation as somebody who has never tried it cannot possibly understand it. And they will find that if they are friendly, operators and clerks and other workers will supply them with valuable suggestions that they could not get any other way. The best writers of directions that we know say that they never write instructions; they just record them as workers tell them how to do a job.

## Organizing and Outlining

The third step is to outline instructions in the order of performance and to check the outline to be sure that it includes all the steps and lists them in the correct order. An astonishing number of printed directions have caused mistakes for years because they list steps out of order, for example,

> 4. Heat the mixture at 120°F. for 20 minutes. Before heating, test for acidity.
> 5. To test for acidity. . . .

These directions were rewritten as

> 4. Test for acidity by. . . .
> 5. If the mixture is acid, heat it at 120°F. for 20 minutes. If it is not acid. . . .

"Can't understand it. We haven't had a spoiled batch all week with the new directions," said the supervisor. "Quite a saving in chemicals and time."

Writers of poorly organized directions are inclined to blame their mistakes on their readers. "The workers don't read the instructions through as they're supposed to," complain the writers. Writers should never rely on workers' reading all the directions before starting to follow them or on workers' remembering all the points if they do read the complete directions. For this reason any equipment, supplies, apparatus, machinery, etc., needed for the operation should be listed first. Otherwise those following the instructions may discover at an awkward moment that they need equipment that someone else is using or material from a distant storeroom.

Any statements about safety should precede the other instructions. If they are not necessary throughout the procedure, safety instructions should be repeated just before the step where precautions are necessary. Safety directions need strong emphasis; some of the visual devices that may be used to stress them are white space surrounding the safety direction, color, capital letters, underlining, asterisks, large type, pictures, and cartoons. Writers should make it impossible for workers to overlook cautions.

If workers can follow instructions better by understanding the reason, then explanations may precede a set of directions. Explanations should precede particular steps only if they will not interfere when workers are performing the task. If sweeping changes are made, explanations are essential, for workers tend to approve and defend methods that they understand. They accept change more readily and comfortably if they understand the reasons for it. If they resent changes, they may unconsciously make mistakes. Beginners who have recently managed to master a filing system may be upset out of proportion to the difficulty if they do not understand and accept the reasons for a change. Meetings that offer opportunities for questions are helpful when many workers are much affected by a change, but a meeting is not a substitute for written explanations that a worker can take away to think about. Writers should never overlook the happy possibility that a worker who reads a rough draft or outline of explanations of changes may suggest a necessary modification of a new system or a way of changing more efficiently.

## Checking and Writing

Once an outline of necessary explanations and of the instructions has been prepared, a wise writer checks it. The best way is for writers to follow the outline themselves, to have an experienced worker follow it, and to have a new worker follow it. Any insertions or other changes may be made easily when the directions are still in outline form. The next step—writing from such a carefully tested outline—is easy. The best form for directions is the simple command, also called the cookbook imperative: *Enter this sum in columns 4 and 5.*

But directions for a colleague in another country may be an exception. Usually these require more tactful phrasing: *Our bookkeeper suggests that you enter this sum in columns 4 and 5; You can assist us greatly by entering this sum in columns 4 and 5; Please help us to keep our records consistent by entering this sum in columns 4 and 5.* A writer new to foreign correspondence may find these circumlocutions strange, but writers experienced in writing to colleagues in countries where more formal courtesy is usual have found it much easier to employ such punctilious phrases than to write the apologies necessary when they do not. Beginners writing to natives of other countries should consult those who have written to their correspondents or should examine the files to learn how formal instructions should be.

The language of instructions should be simple. Any word that a worker might not understand must be fully defined the first time it is used or must be replaced by one the worker understands. Each step of the instructions should be a new sentence and should be numbered. Illustrations should appear on the page or opposite the page to which they apply, and they should be legible enough so that any lettering or numbering may be read at a glance.

Before instructions are published, a trial of a week or more is advisable. Average workers, experienced and inexperienced, should be selected to test the instructions so that the supervisor and the writer may catch any misunderstandings, omissions, or confusions, which the writer will then correct. Instructions prepared this way are not published until they are foolproof. Although they take time to prepare, they are still a great economy because of the errors, materials, and equipment they save.

Whenever safety is involved, great care is essential to prevent accidents. An operator's safety may be endangered by a procedure out of order. It is small use to place at the end of job instructions a warning to wear protective clothing. Some worker is almost certain to perform each step in order without reading all the steps and thus to risk injury.

A writer of instructions sent the following directions to be posted above machines in a college language laboratory: "To start this machine, press button B. Before pressing button B, wind all the tape onto the left spool." In the first five minutes of use, thirty of the thirty-three machines were damaged. The other three escaped because one student was absent; one student, who was slow, was warned by the mistakes of the others; and one student read and followed the instruction to read the directions from beginning to end before performing any operation.

Exact chronological order prevents such accidents. It is the best order for instructions because it leaves nothing to the discretion of the writer or reader. Many accidents might have been prevented by careful checks to be certain that directions were in exact chronology and by supervision of the operator to be sure that the operations were performed in the order listed. These two simple steps promote safety on the job by helping operators to follow directions properly and training them to observe safety instructions at the right time.

424

### Instructions for a Single Performance

Instructions to be used once by one person or by a few persons, like executives' instructions to their secretaries or assistants, are a somewhat different matter. Although the executive and the secretary or assistant may be so well acquainted that they speak and write a personal shorthand, the directions should be written so as to be clear to anyone substituting for them.

A procedure that works well for executives and those they direct follows:

1. The executive discusses the instructions with the secretary or assistant.
2. The secretary or the assistant writes the instructions.
3. The executive corrects or changes the instructions

An annoying problem is the supervisors who give oral instructions and then deny them or say that they were misunderstood. Another is the assistants or secretaries who seem to understand instructions but then ignore them or change them to suit their own ideas. The procedure described usually solves the problem of the executive who is careless or forgetful and of the assistant or secretary who is a poor listener. Executives who are merely careless or impulsive have an opportunity at step 3 to change their minds before the instructions are followed, and assistants or secretaries who have missed a point should find that out at step 3. But executives who try to weasel out of responsibility for instructions they have given and assistants who willfully pursue their own way cannot be helped by instruction in writing. Theirs are not problems in writing or reading, and it is wise to recognize that they are not.

Sometimes writers fail to indicate priorities for a list of tasks. This information is urgent when the writers will not be available for consultation about the instructions or when the persons performing the tasks are new to the work.

There is never any need to apologize for the explicitness of directions, for the explanations or definitions that accompany them, or for a statement of the order of performance. The directions that require apology are those that are not explicit, clear, and organized.

### MINUTES

Anyone who belongs to an organization, even someone who has never accepted a nomination to the post of recording secretary, may someday be asked to take minutes. Unfortunately many presiding officers choose the newest employees or committee members for their victims even though taking minutes is most difficult for those unfamiliar with the members or with the procedures usually followed. For this reason beginners in business and professional careers should have some familiarity with the contents and form of minutes. Then they can accept the assignment, "Will you take the minutes?" without confusion or panic.

Minutes vary from group to group, but basic facts in the following order appear in most minutes:

1. *Call to order*
   This includes the date, place, and time of day that the meeting was held as well as the name and title of the presiding officer and, if desired, the name of the secretary or other person who is taking the minutes.
2. *Kind of meeting* (regular, special, committee, board, executive)
3. *Attendance*
   For a small group the attendance may be a record of those present, those absent, and special guests. Larger groups may record the presence or absence of a quorum or merely list the absentees. For some shareholders' meetings it is necessary to record the number of shares of stock held by each of those in attendance.
4. *Record of the proceedings*
   a. *Minutes of last meeting*
      A record of the action taken on the minutes of the last meeting must be recorded. If the minutes are accepted as amended or corrected or both, that must be stated. The secretary is responsible for recording later, usually in red ink on the official copies, any accepted amendments and corrections to the minutes so that the official minutes will be accurate.
   b. *Reports presented or read*
      The title of a report should be recorded along with in each instance the name of the person who submitted the report and the committee action taken on each report, such as motions to append or to delete material or to accept. The record should name the person who proposed the motion and in some groups the name of the person who seconded the motion. The vote is usually recorded in one of the following ways: *Motion Carried 15 to 7, Motion Lost 5 to 17, Carried, Lost, Carried as Amended, Lost as Amended.* Motions to amend should be recorded in the same way as other motions. If an amendment is carried, the motion as amended should be stated in the minutes.
   c. *Unfinished business*
   d. *New business*
   e. *Announcement of the date of the next meeting*
   f. *Time of adjournment*
   g. *Signature of the secretary and the date of signing.*
      It is not necessary to place the old-fashioned "Respectfully submitted" above the signature. No formal closing is required.

In no case is it necessary to report verbatim everything that is said, but the wording of motions, amendments to motions, the proposers' names, and the votes must be recorded accurately. If a secretary is not sure of the wording of a motion or of an amendment or other subsidiary motion, he or she should request that it be repeated. Discussion is not usually recorded, and a secretary should not let his or her opinions intrude on the minutes, which are not a place for editorial comment. A reader should not be able to tell from the minutes whether the secretary was in favor of or opposed to a motion.

The style of minutes should be clear, brief, and precise. Report style is appropriate for most minutes. Side headings are used by some writers to make it easier to find material quickly.

If some points not detailed here, such as summaries of discussion and indexing minutes, are required of secretaries, the mechanism for them is usually already in place, and a new secretary has only to follow the work of his or her predecessor. Indeed, the first step of a new recording secretary should be to examine preceding minutes to find out whether a book of minutes is kept by the committee. If no book is kept, the secretary would be well advised to start one and to enter corrections and amendments of the minutes promptly for self-protection as well as for efficient minute taking.

The question of a nonsexist title for the presiding officer may arise. Some secretaries use *chairperson*, but that awkward noun is not necessary, as the use of the term *chair* for a presiding officer has an honorable history. At least *chair* is better than *Madame Chairman*.

It is probably wise for a new secretary to read or to review some of the problems arising from the use of sexist nouns and pronouns ("The Sexist Pronouns," Chapter 6; "Sexist Nouns," Appendix B). We recommend that new secretaries consult the minutes of the preceding secretary to learn whether and how such titles as the following were used: Miss, Ms, Mrs., Mr., Dr., Section Head, Executive Vice President. It is probably better to follow a predecessor's choices if one is only a temporary secretary. Secretaries serving full terms can wait until a second or third set of minutes to introduce their choices or in a highly traditional atmosphere to ask about doing so.

## NEWS RELEASES

Every news, or press, release submitted by a company or bureau can affect the reputation of that organization and of the writer. Poorly written releases are damaging, and well-written releases enhance a reputation for excellence and reliability or help to build one. Hence even though some professionals write few news releases, they should write them well. Large companies are likely to have public relations divisions or departments that release news daily, but even so professionals working in departments that are remote from the public relations division may be asked to write news releases on subjects like advances in science or technology in their divisions or on conventions and other meetings arranged under the chairmanship of their managers. They may also wish to send out news of a distinguished scientist or engineer who is scheduled to speak at an open meeting of the local chapter of their scientific or engineering society. The publication of such news is free publicity welcomed by employers and helpful to professional societies.

Writers should decide early which publications or which communications media should receive the news. This decision is necessary early because the choice of publication or other medium determines the kind of reader, and writers of press releases must adapt their writing to their readers even more effectively than writers of reports or letters or memorandums. A mistake in estimating how to catch the interest of readers, what language they understand, or how to hold their interest may mean that even if the story appears, audiences will not choose to read or hear or view it.

Writers may send releases to such media as the following:

Local, large city, and national newspapers

Local and national radio and television broadcasting stations

Trade journals, scholarly and professional journals, general magazines on science or technology or management

Finally, most companies have company publications ranging from inexpensively reproduced bulletins to expensive newspapers and magazines.

The news contained in company and government releases is varied. Some of the releases do concern scientific and engineering findings and developments; but releases also present many other subjects, such as property and plant improvements, new plants and offices, mergers and acquisitions, government contracts, new stock issues, reports on sales and earnings and dividends, organizational changes, elections, promotions, transfers, new appointments, ceremonies such as ground breaking and cornerstone laying, prominent visitors, honors, awards, prizes, charity and community work of employees, and their major speeches.

Regardless of the subject every news release should carry the following essential preliminary information:

> Company name and address
> Date of mailing
> Release date (This may be stated, for example, as *FOR IMMEDIATE RELEASE* or *For Release on September 15.*)
> For further information please call *name, title,* and *company telephone number and extension*

The news story is preceded by an indented dateline stating city and date. The state is included for cities with common names and for cities that are not well known. Examples of a dateline:

PHILADELPHIA, September 15 —
CRAWL KEY, Florida, September 15 —

The release begins after the dash.

Many news releases begin with a *lead*, which some textbooks insist must answer the five W's: *Who? What? When? Where? Why?* But the truth is that leads usually answer only the relevant or sometimes only the major W's and often answer *How?* (which seems to have been lost in those instructions because it does not begin with a *W*). If the story has a dateline, most writers do not answer *When?* or *Where?* again. Every word is precious in a news story, especially at the beginning, the most important position. Those who have written informative abstracts that stress findings will see that writing the lead for a newspaper is very similar except that writers must usually adapt those leads to general readers, the exception being readers of journals of science and technology, who are likely to be familiar with many technical terms.

Once the lead has been written and revised, writers will find it easy to organize the release in the order of decreasing importance as illustrated in Chapter 3. Everything important should precede less important reporting in the release, and the last sentences (usually the ones to be chopped off if space is needed) should contain the least important material.

The style of the release should be suited to the newspaper, magazine, radio, or television audience. These general readers, listeners, and viewers do not want jargon or gobbledygook. Any necessary technical terms should be defined and false elegance should be avoided. Clarity ("Second Revision," Chapter 2) and brevity (Chapter 13) should be principal goals.

Puffery of the company or of management is undesirable, and any extravagant praise will be removed by conscientious editors before publication if they have time. This may mean that if they do not have time to edit, a release will be dropped and another, ready for publication, used instead. Writers should also avoid unnecessary use of the passive voice. The active voice usually leads to shorter, less complicated sentences, which are desirable in news releases. Paragraphs written for newspapers and for narrow columns in magazines should also be short. Because of the use of short sentences and paragraphs, writers should pay particular attention to coherence in order to avoid choppy style. Photographs and drawings may accompany news releases, particularly if they add interest or make explanation briefer or more immediately clear (charts, tables, maps; drawings or photographs of equipment; photographs of a new plant). Captions for illustrations should be clear and complete. In revising, writers should be sure that names are spelled correctly, that titles are accurate, and that addresses are correct. Letters of complaint about errors make editors distrust the source of a release and hesitate to use material from that source in the future.

Scientists and engineers who take the time to follow this advice about news releases may find themselves with a new position when they retire—freelance writer on science and technology. In the meantime they can have enough control over news that they release to be sure that it is accurate and is written as well as the average article in most local newspapers—and probably better.

## EMPLOYMENT RÉSUMÉS AND LETTERS OF APPLICATION

In a departure from the contents of most professional writing, the subject matter of the letter of application and of the résumé is the writer. Both the letter and the résumé must be carefully written to persuade the reader or readers, and therefore both require an estimate of the readers' likes and dislikes and probable reactions. If the letter of application is not written for a specific position, the writer should examine the prospective readers carefully and consider how to impress on such readers their need for the writer's services.

Writers reading a blind advertisement should consider whether their experience or training might answer the stated requirements or might be close to answering them. Sometimes those who write job advertisements do not say clearly what they want, or they may fantasize about a perfect applicant, who will never appear. A pharmaceutical company, for example, advertised for a "writer with several years of experience in a pharmaceutical company." It wanted a writer experienced in writing medical proposals to the government. The company did not find a satisfactory applicant until the advertisement was rewritten. An applicant who could read between the lines and present experience in medical writing to show its value to a pharmaceutical company would certainly have been interviewed and perhaps given the position.

If a writer is asking for consideration as a candidate for a stated opening and therefore knows what the position demands, the presentation of qualifications should stress the required qualifications. The more writers of letters of application can find out about the openings in question, the better they can present the part of their experience that most closely answers the requirements and shows the readers why the applicants should be hired. The letters may refer to relevant items in the résumé and give more specific information about abilities and accomplishments briefly and effectively: "As plant manager, I increased production twenty percent by reorganizing the floor plan to save the time and effort the workers spent on obtaining raw materials." "Realizing that our packaging of an excellent product was unattractive, I found a designer to reflect the soundness of the product in its presentation. Sales increased by fifteen percent." Writing merely, "I have been a plant manager for three years" or "I have been an assistant in the marketing division" states the level of a writer's experience but does little to distinguish one eligible candidate from another.

Many employers look for people with initiative, people who see problems before they are pointed out and who seek solutions. Even those with limited experience should try to indicate initiative as long as the examples are not absurdly trivial.

Letters of application accompanying résumés may overcome deficiencies revealed by the résumés. If a résumé suggests, even if only by omission, a weakness—for example, insufficient education—the writer should try to present pertinent experience forcefully enough to induce the reader to arrange an interview. It is always easier to discuss disadvantages in an interview, where personality and other qualifications may compensate for shortcomings. When an applicant lacks sufficient education, care in writing the letter and résumé can be crucial. More than one of our students has won a position and been told, "You write very well and don't make mistakes although you are only an undergraduate. Here is a whole stack of poorly and carelessly written applications from college graduates. I'm giving you first chance at the job."

### Subject Matter of the Résumé

In addition to being correct, résumés must be complete. They usually contain the following information:

1. *Personal Data*: Name, mailing address, and telephone number should be presented briefly and clearly. If a home telephone number and a business number are given, the hours, the area code, and the extension number for each should be stated. If a temporary and a permanent address are included, the duration of the temporary address and the date when the applicant will be at the permanent address are necessary.

2. *Position Sought or Type of Position Sought*: For an advertised position the job title used in the advertisement and the source of the advertisement should be given. A large company may advertise a number of positions or may place several advertisements for the same position in various publications; therefore the reader should be told exactly which advertisement for which job the writer is answering.

If the applicant has been referred to the company by a person or an office, such as an employment office, stating the source and the job title used by that source avoids confusion.

If an application is a trial balloon, a letter accompanying the résumé may indicate the positions the applicant seeks and the reasons that led him or her to apply to the company addressed. Fulsome praise of oneself and of the company should be avoided; however, proof of the applicant's abilities or admiration of the company precisely and briefly expressed is sometimes effective.

431

If a writer wishes to use copies of the résumé for various applications, it is acceptable to omit section 2 in the résumé and to state the position sought and the source of information about that position in a letter accompanying the résumé. Then the résumé can be photocopied for inclusion with applications for several positions in various companies.

3. *Employment Record*: Employment is usually listed chronologically, the current or most recent position first. However, some applicants display more imagination and salesmanship by placing their best former positions first or by beginning with the positions most closely related to the one they seek to fill. Each entry should include the dates of employment, the title of the position, and the employer's name and address. All entries should list the data in the same order, as though they were entries in a table. Usually only employment in the field of the position applied for is included; but if an applicant's record is not overlong, job experience that developed useful qualities or provided training related to the field of interest may be mentioned. Such experience also is sometimes included to account for the time when applicants were not employed in their fields of major interest.

For all entries in a short list or for major entries in a long list, applicants may describe the main work performed, the skills acquired, the responsibilities assumed. Applicants with little employment experience may include part-time or summer work or both.

4. *Education*: Degrees and schooling are usually listed with the most recent first and the rest following in order. If the writer has an earned college degree, elementary and high school education may be omitted unless the writer wishes to name honors received or outstanding achievement in the lower schools or to use them to stress continuity of interests. Information about degrees should include the name and address of the college or university, the applicant's major and minor, any special work performed (such as assisting a professor with research for a grant, delivering recruitment speeches in high schools, serving as class or student council president).

5. *Special Abilities*: A wide variety of education, training, and skills may be listed under this heading. Training in foreign languages, in computer skills, in leadership may be stated if appropriate, but abilities appropriate for mention by an applicant for an entry-level position should be omitted if the writer is more advanced. Experienced lawyers and professors, for example, usually do not include leadership of a Girl Scout or Boy Scout troop although they may appropriately mention serving as county chairman of a committee to raise funds for the Girl Scouts or Boy Scouts. Volunteer service deserves a place under *Special Abilities,* but it should clearly indicate skills acquired and should be no more than a few years in the past.

6. *References*: The names and addresses of those willing to provide references are not usually given in a résumé. Instead a writer may state

that references will be supplied on request or that references have been filed at Blank University and are available on request.

## Writing Style

Once a writer has decided on the contents of the résumé and accompanying letter and has determined the approach, some advice on style is usually helpful. It is important to develop paragraphs so that every sentence does not begin with *I*, a style that may seem egotistical when it is only inept: "I was a management trainee for six months, and then I went into the merchandising division, where I am now employed as assistant manager. I do thus, and I do so." A writer should choose instead a style like the following: "Hired as a management trainee at Black Company, I later joined the merchandising division and after several promotions became assistant manager nearly three years ago. My present work entails. . . . I am proudest of [an accomplishment or several successes described factually]."

The letters of application may be most important in emphasizing the points that applicants wish to stress particularly. Whatever is most important to an applicant's case may be presented here away from the listing of the résumé and enlarged upon if further comment by the applicant will be useful. This emphatic presentation should be chosen for only one or two facts in the résumé. It is pointless to summarize the résumé in the letter that presents it.

The style of the application letter should be brief and businesslike. Slang, vogue words and expressions, and wordiness should be avoided. Pressing opinions on a reader ("As you no doubt know," "Of course, you will agree with me that") is also unwise. Although the style of the letter of application should be formal, it should not be pompous; and the tone should be courteous and friendly. In the rare instances when the writer has met the reader, the style may be informal if that is more appropriate to the reader, but it should not be self-consciously familiar. An executive who has to read several dozen letters may be easily annoyed by the irrelevant and the verbose, sometimes annoyed enough to throw the offending letter aside without reading more than a few sentences. On the other hand, a letter should not be so succinct that, lacking character and personality, it reads like a laundry list. A writer should remember that many employers determine on the basis of these letters which résumés to read first and consequently which interviews to schedule first.

## Common Questions

Questions about including hobbies or interests in a letter of application or résumé are frequent. "I like to read and to play tennis" is not a useful statement. "I am an avid reader of eighteenth-century history, particularly medical history," is more helpful. "I like to play tennis and have

433

been women's singles champion of the Mitchitaw Tennis Club for three years; I enjoy playing both social and competitive tennis" indicates ability to compete and to be helpful socially at athletic events. In some companies, for example, playing a good game of golf is useful for sales managers.

Questions about applicants praising their own success lead us to warn that blatant recitals of achievement and too glib a promise of prospective accomplishment are usually received skeptically. If a letter states that the writer "single-handedly saved a company from ruin," most readers will wonder why the company is willing to let the writer leave. If a letter states that the writer is "far and away the best producer in the company," readers are likely to question the caliber of the other employees. If a writer promises to increase sales by fifty percent, the readers may know that with their company's backlog of orders even Superman would do well to show a modest improvement.

Whether to discuss salary in the résumé or letter is debatable. When writers discuss salaries in letters of application, they must be sure to avoid any hint or suggestion that they are more interested in the money than the jobs. Usually the information presented in the résumé and accompanying letter give a fairly good idea of the appropriate salary range. It is unwise to state that the applicant is willing to start at a low salary with the understanding that it will improve. Unless the writer also states a reason for wanting this particular position, the mention of a low salary or offering to take a lower salary than the writer has earned suggests that the applicant is not so good as the record shows or is desperate.

It is generally acceptable, however, to ask that the writer's present employer not be informed of the application. Prudent people do not relinquish one position before being accepted for another. Most employers understand that the knowledge that an employee is actively seeking another position does not contribute to good morale in a company.

## The Quality of the Résumé and Letter of Application

The résumé and letter should be written and checked carefully. Errors in grammar, spelling, or information may remove an applicant from consideration as a candidate. Applicants should check the writing after they have cooled and can see what is on the paper, not what they meant to put there ("Steps Three and Four," Chapter 2).

# CHAPTER 15

# Reports and Style Guides

*He that hath knowledge spareth his words.*
                                    *Proverbs* 17:27

## REPORTS

Success in writing reports depends largely on identifying and understanding readers in order to satisfy their needs efficiently. We divide report readers in business, industry, and government into three categories for convenience in discussion:

Group 1: managers who are technically educated and trained and who keep informed of progress and development in their fields; managers whose training and education in science and technology is out of date; and managers who have not been educated or trained in science or technology.

Group 2: scientists and engineers working in their specializations, a few in the same specialization as the writer of the report or one close to it but most in specializations different from the writer's.

Group 3: readers in such divisions as advertising, field work, finance, marketing, public relations, or purchasing.

The needs of these readers have not changed greatly over the years, but the organization of reports has.

## The Needs of Report Readers

What do the readers in these groups want? All of them tell us that they need to be informed immediately of the department or division issuing the report, the writers' names, and the reason for directing the report to the particular reader or readers if such referral is not routine. The readers also want to know what they are requested or expected to do about the report—approve or disapprove, recommend such matters as further expenditures or consideration of other methods, vote on continuing or stopping the project, read for information and ask questions if they have any. This information is often conveyed in a cover letter.

Readers in Group 1 used to like to turn next to results and conclusions, scope, new findings, or purpose. Searching for one of these subjects, they skimmed the pages of the report. When they finally found, for instance, results and conclusions (the preferred first reading of most), many of them could not understand the material because it had been prepared for those who had read the rest of the report. To end such unhappy and unprofitable reading, members of Group 1 and Group 2 conferred and consulted with many technical writing advisers and teachers to find a way to supply readers' needs quickly at the beginnings of reports. In the consultations we had we were surprised to find ourselves discussing this matter with managers from Europe, India, South America, Saudi Arabia, as well as the United States. The problem was widespread.

The solution we liked and still like best is the informative abstract, approved by a large representative committee of the American National Standards Institute (ANSI).[1] A standard published by ANSI explains and illustrates this abstract clearly and helpfully.[2] (See also "Abstract and Summary," Chapter 12.)

The informative abstract was developed mainly for papers to be published, but it serves very well also for reports used internally in business, government, and industry. One of its great advantages is its brevity, one page being enough for long reports; a half or quarter of a page, for reports of average and shorter length. The informative abstract has another advantage: the opportunity to use it in a paper written in the conventional order, which pleases traditionalists, or in an order that emphasizes findings, which pleases most of the other readers of Groups 1 and 3. The conventional order of this abstract is purpose, methodology, results and conclusions, and recommendations. The order that stresses findings is major results, conclusions and recommendations, survey of supporting material or any other findings, methodology.

Placing an informative abstract first satisfies most management readers, especially those not up to date or educated in science and engineering. This is fortunate, for the number of such readers is growing, as indicated by thriving MBA programs in colleges and universities; expanding financial divisions throughout industry; and more readers of technical reports in divisions such as sales and marketing, public relations, and advertising. The informative abstract, or, as some call it, the management summary, provides them with much information in little space, particularly if writers revise carefully to remove unnecessary words and details (however dear to the writers those details may be) until the style is as succinct as the writers can make it.

Some members of Group 2 shared the reading preferences of the managers, but many insisted that all they were interested in was the detailed proof. Data, they said, are their main concern in their own work, and they turn to data first in the work of others. When they examined their daily reading, however, this statement proved true only for reports in the readers' own specializations, especially reports necessary to their continuation of others' work or of a large project. Such reports did not appear often in the readers' records; sometimes none would be read for six or eight months. In other reports Group 2 readers wanted to read first the answers that the readers in Group 1 requested be placed at the beginning of reports.

Group 3 readers shared the preference of Group 1 and admitted that they seldom examined the body of a report without the help of a scientist or engineer. Consequently in government and industry many abstracts of research papers soon answered such questions as the following if the questions were pertinent: What's new? How useful is it? Can it be applied? Now? Later? What action is planned or recommended? The abstracts of papers on problems answered immediately the questions What is or was the problem? How was it solved? What are the main advantages of this method? What are the major advantages of other, equally or nearly equally successful methods? Some readers said that abstracts should include a brief statement of how the company became involved in the research; and they thought that if the research goal changed, a brief description of the change or of the goal that finally governed the research should be included. The informative abstract can be written to supply whatever answers are most helpful to such readers.

The ANSI examples and the arrangement of the material in the examples are interesting and suggestive. Many readers want to decide from the information in an abstract whether a report contains material of interest to them and how urgent it is that they read the paper. The informative abstract enables users to read further according to their interests, to read only the abstract, or to stop reading after the first sentences of the abstract.

Abstracts are written in complete sentences. Short abstracts are usually only one paragraph, but full-page abstracts may be two or three paragraphs. Tables, equations, and diagrams have no place in an abstract. Any acronyms, abbreviations, symbols, or highly specialized words that readers are not likely to know should not appear in the abstract or should be explained on first appearance.

## Data and Other Detailed Information

The insistence of scientists and engineers on their need to read data alone and the complaints of other readers that data and other details interrupt their reading of the text, especially when pages of tables, charts, or graphs are not near the material they concern, led to another change in reports. Many of the best writers and journals now place data in appendices and only refer to them in the text. If some details are needed in the text, these writers supply representative charts, graphs, or tables for discussion and place the full material in appendices. This arrangement shortens reports containing large numbers of details and relieves writers of the struggle to fit large tables, charts, and graphs near the material related to them.

## Order of Reports

The following information, which is suitable for a wide variety of subjects, covers the items that may appear in reports:

Preliminary Sections
    Cover
    Flyleaf
    Title page
    Copyright page
    Letter of authorization
    Cover letter
    Preface or Foreword
    Acknowledgments
    Table of contents
    List of illustrations
    Informative abstract
Body of the Report
Supplementary Material
    Appendices
    Bibliographies or References
    Index

The pages of preliminary material are assigned lowercase Roman numerals, beginning with *ii* on the first page after the title page of a report that is typed on one side of the paper. The report and the supplementary material are assigned Arabic numerals beginning with an optional *1* on the first page or a necessary *2* on the second. If a report has chapters, the number assigned the first page of each chapter may appear or not as the writer wishes.

Each part of a report is discussed in the following sections in the order of the preceding list. Some of the preliminary and supplementary sections are optional. Some are so similar that only one of the pair need be used. The necessary information is given under the appropriate headings.

## Preliminary Sections

### *Cover*

The cover of a report contains the title and the author's name. Optional additions are a security classification, the date of the submission of the completed work, and, if the report is one in a series, the title of the series.

Covers may be paper of higher quality suited to harder use than typing paper.

### Flyleaf

Favored by advertising writers, a flyleaf is a blank page, sometimes of paper of higher quality than is used for the rest of the report, sometimes containing a decorative background design. It is, of course, optional. We have not seen a flyleaf in any unpublished papers, particularly not in scholarly reports. It is too fancy for typewritten papers.

### Title Page

The title, the subtitle if there is one, and the name of the author appear on the title page. Other items that might be included are the approval signature and date, the revision date, the series number, the security classification, and the number of the contract.

### Copyright Page

The copyright is noted in books on the back of the title page as follows: date of the copyright, name of the holder of the copyright, and a statement like "All rights reserved. This book or any part thereof must not be reproduced in any form without the written permission of the copyright owner." This is usually followed by the Library of Congress Catalog Card Number, the International Standard Book Number (ISBN), and a statement like "Printed in the United States of America."

### Letter of Authorization

A letter of authorization is necessary if the terms under which the study was made should be recorded. It is written by the person who requests or orders the study to the person who is to be responsible for the work and the report. Some letters of authorization contain detailed suggestions or instructions for the work; most do not.

### Cover Letter

A cover letter may be separate from the report. If it takes the place of the preface, it becomes part of the report. Cover letters may be addressed to the person or group that authorized the report (this letter becomes part of the report), to a special reader in order to call attention to subjects of interest to that reader (this letter is not part of the report), to a reader who has requested a copy of the report (this letter is not part of the report). Cover letters may be needed when action is requested of a reader or readers, when reports are sent to the divisions of a company (particularly those at a distance), when writers wish to remind readers of their previous connection with the project or with the writers.

A cover letter may include the main points of the report, the scope, the material of interest to the reader or readers addressed, acknowledgments, the need for and use of the report, major conclusions and recommendations, personal comments of the writer. It should not have the same wording as the abstract.

### Preface or Foreword

Prefaces and forewords contain the same or similar information; therefore only one need be used, although occasionally works have both. A foreword is written by someone other than the author who knows and is willing to admire publicly the author and his or her work. Usually both preface and foreword are not offered because they tend to include mutual admiration and thus waste readers' time. A foreword may include the background of the development of the report, the report writers' qualifications, the introduction of the subject to the readers, references to related literature, evaluation of the work. A preface is written in a personal style by the authors and may include any of the material appropriate for the foreword, the authors' reasons for writing, their ideas about using the report, and their acknowledgments of help given by people and organizations.

### Acknowledgments

Acknowledgments, however, are best presented in a separate section. Removing them from the other material of a preface gives acknowledgments more importance and is therefore more generous. This section should be revised with extra care to be certain that names are spelled correctly and titles are given accurately. The acknowledgments may be formal or informal in tone, depending on the writers' style and on the relationship of the writers and the person being thanked. Acknowledgments are, of course, optional except when writers must recognize the financial contributions of persons or organizations. It is wise to have the organization or person or both approve the expression of gratitude. We can think of occasions when for one reason or another acknowledgments have come as unpleasant shocks to those named.

### Table of Contents

A table of contents lists the topics of the report with their page numbers, thus enabling readers to locate material easily. Most short reports do not have tables of contents. An easy test of whether a table would be helpful is watching someone unfamiliar with the research and the report try to locate several topics. A table of contents should list preliminary and supplementary sections as well as the sections of the main report. Divisions of equal importance should be expressed in parallel construction. The divisions of the table of contents are often worded the same as the divisions in the text. Writers who outline fully are likely to find the wording of the divisions in the outline useful for the table of contents and for headings and subheadings.

### List of Illustrations

Illustrations are listed if a report contains many and readers may wish to locate them. They are usually listed by figure numbers and captions in the order of placement in the report, and page numbers are cited as in a table of contents. The legends, or captions, of illustrations should be brief but complete so that readers may understand the illustration without referring to the text. If parts of illustrations are referred to in the text, the parts should be numbered or lettered in the order of the references. An illustration should not be cluttered with information. If illustrations are crowded, writers should consider using two drawings or eliminating some material.

In selecting illustrations writers should consider which kind will convey the message best. Photographs are popular, but writers must be aware that reproduction may be a problem because many photographs look different in a larger or smaller size. Writers should be sure to consider whether the main points will be visible in the reproduction. Line drawings are effective for the presentation of microscopic subjects. For some purposes a diagram can be more objective than a photograph or a line drawing. Maps and charts and other suitable material should be drawn to scale. Writers should be careful, for example, to ensure that drawings compared visually are on the same scale and that figures in pictorial bar charts are the same size. Most scientists and engineers use visual aids such as graphs, histograms, pie charts, and bar charts to convey data precisely, not approximately. The use of drawings to identify the parts of animals, birds, fish, plants is a familiar and therefore an encouraging method for readers who are timid because they are not educated in science or technology. Many bird and flower lovers, for example, are well aware of how much good drawings aid description or substitute for it in their guidebooks.

As early as possible writers should examine their report outlines for material that can be presented more effectively by illustrations than by words. Drawings and photographs are useful in explaining to readers at first sight information difficult to present simply and briefly in words; moreover, seeing a picture gives readers confidence. Having seen the position of the switch, they feel that they can locate the switch when they operate in the laboratory. But having read a number of sentences about the position of the switch and never having seen it, they may feel uncertain or confused. Pictures in technical explanations save writers the trouble of writing long passages and readers the irritation of trying to understand them. Successful illustrations for reports may require extra attention and work of writers, but they will be compensated by finding the briefer text made possible by illustrations easier and quicker to write. To be successful with illustrations, writers should plan to use them where the pictures aid understanding of the text, not merely decorate it, and should present their plans early to the art department or illustrator. Rushed technical illustration is seldom as accurate or effective as it should be. Writers who allow adequate time for the work of illustrators will find themselves welcomed when they present their next requests for drawings and may quickly develop excellent working relationships with illustrators. This is not true of writers who are always pressing for last-minute work and hoping vainly to meet yesterday's deadline tomorrow.

### Informative Abstract

Discussed earlier, the informative abstract suits the needs of managers and of scientists and technologists. It is exemplified in Chapter 12. Calling this abstract a management summary has proved disadvantageous in two ways. First, some managers remove the summary for their files before passing the report to others, thus leaving recipients beneath them in rank without this useful tool and sometimes without the knowledge that it exists. Second, when scientists and technologists receive reports with management summaries attached, some of them skip the summaries because they think that the page is directed to other readers. In some companies advice to read the management summary is passed along as a valuable tip by those who have discovered its usefulness. Labeled *abstracts* or *informative abstracts*, these brief summaries, clearly intended for all readers, may precede or begin reports. If these summaries precede, the reports may begin in any way that the writers think will meet the needs of their readers (Chapter 5).

## Body of the Report

### Report Beginnings and Organization

Popular report openings are statements of problem, of necessary background, of interest-catching information, or questions and answers. The varied beginnings in Chapter 5 and a number of examples of each kind have proved particularly useful in many companies where, unfortunately, reports and report beginnings had become formularized and therefore so repetitious and dull as to be neglected by most readers. A report organization repeated frequently for the same readers also means boredom. Plans suited to the material and the readers (Chapter 3) are numerous enough to provide the variety necessary for a business, government bureau, or division of industry. Some readers become impatient, some sleepy if they know just where every report is going section by section. When they are not always familiar with the plan of reports, they are more alert and attentive, anticipating the unexpected. Pattern reports—those with the material always arranged under the same headings, which are always in the same order—may be easy writing, but, as Sheridan noted, "Easy writing's curst hard reading."

### Advice on Writing Methods

Once writers have organized and outlined their reports, they are ready to write by the method of Chapter 2, "The Flow Method." Not all the work need be completed before writers begin putting thoughts and facts into words. What they find essential is a place to work without interruption and sufficient self-control to avoid interrupting themselves once they start writing. As soon as they begin putting words on paper or tape, they should let the sentences flow until such time as they need to consult their outlines. When they have finished all that they think they can write in that session, they should put the report aside for later editing. If they want to work on the report after the flow has stopped, they may prepare the appendices, which, not usually being writing, do not lend themselves to the flow method.

When the writers have cooled and are ready to revise and edit, they should follow the program recommended in Chapter 2, which has worked for others. It helps writers to review their style objectively. For those who need the encouragement of rapid improvement, we suggest concentration on brevity and coherence (Chapters 12 and 13).

### Report Endings

"How should I conclude my reports?" writers frequently ask. Telling readers the answers to the question What next? is a good ending for many reports. The answer may be (1) that this completion of the planned work ends the project, (2) that the writer or the division or department recommends continuation of the project or of clearly delineated sections of the project, (3) that a meeting of specified officers be called to consider continuation of the project by the appropriate readers. The answer to What's next? or Now what? can be used to conclude many reports neatly, and many readers in all three groups liked and requested it. Another effective ending is the most important conclusion or conclusions if the report can be organized well with results and conclusions last and if the importance of the final conclusion will be recognized by most readers without lengthy explanation of or argument for it. This closing is preferred by many in Group 2.

### Negative Findings

In the disappointment of fading hopes of major research results, some writers and some managements do not prepare or do not file reports. Yet negative findings may be valuable; in fact, some journals publish them in every issue. How are they useful? Most obviously in eliminating certain roads to a research objective. If, for instance, scholars in a field agree that there are three possible roads to information essential for further research in a subject and the report demonstrates incontrovertibly that one of those roads does not lead to the information, researchers then, having only two roads to explore, are nearer the goal. If, in another example, a company is planning a massive advertising campaign on a promising development and the writers' research shows errors in the choice of major techniques that have cost other companies outrageously, the writers may feel that reporting this makes them the most unpopular employees in the company; but in fact they may be saving the company from the loss of millions of dollars of the advertising budget as well as a major loss of prestige. If the research in a report is sound, the results, even when negative, are worth recording.

**Supplementary Material**

### Appendices

Many writers prepare appendices before the rest of their reports. We recommend the use of appendices in reports that contain many data or other suitable details—questionnaires, checklists, computer printouts, glossaries, graphs, tables, charts. Tables in a text or in an appendix should have informative headings, as should graphs.

Wise writers move as much as possible of the data and other collections of details to the appendices and leave in the text only (1) what is essential for the discussion and (2) a mention of the appendix where further details may be found. Usually only representative data are necessary for the discussion, not pages of tables or graphs.

The figure titles, column headings, and notes in the appendices should be clear, accurate, and helpful. The material should appear in the order in which it is referred to in the text. The accuracy of this material should be doublechecked carefully, at least one check being made several days after the preparation of the data. If the tables, charts, and graphs are not computer printouts, some companies encourage writing on them numbers, letters, symbols, etc., in a black ink that reproduces well. This removes one step where errors are likely to occur, the typing of the engineers' or scientists' numbers, symbols, etc., and enables scientists and engineers to have fuller control of the accuracy of their material.

Each appendix should be given a meaningful title. In the table of contents appendices are listed by title and a letter: "Appendix A: Dolphin Research Facilities 1985, 86, 87"; "Appendix B: Glossary of Dolphin Research Terms Introduced Since 1940."

### Bibliographies or References

Usually placed in appendices, bibliographies may be separate sections if they are unusually lengthy. Short bibliographies are seldom divided into sections, but others contain such categories as books, encyclopedias, reports, government documents, periodicals or serials, articles, law cases, letters and interviews, speeches. Bibliographies not divided according to type of publication may be divided by subject matter.

Publishers, universities, companies, and government bureaus may have their own rules for entries in bibliographies and references, and writers should follow those rules. If no such standards exist, writers usually follow one of several authorities: the instructions and style guides of the prestigious journals or the associations in the author's specialization, the *Chicago Manual of Style*, and the *MLA Handbook for Writers of Research Papers*. If a writer intends to submit a report to a journal, it saves time to follow from the start that journal's instructions for notes or references and for bibliographic entries. One of the great wastes in the publishing of science and technology is the time and energy lost every time a paper is refused by one journal and has to be retyped with different notes and bibliographic forms for another. Some progress in eliminating this waste is being made through agreement among the learned societies in science and technology to accept one well-publicized form for bibliographies and another for notes or references. The Modern Language Association forms show the value of this practice. We cannot number the times we have watched a member of the MLA read a rejection slip from a journal, type a new letter, and slip the manuscript and letter into a fresh envelope to be mailed that day to another journal. The writer's peer in science and technology in the same situation may have to spend hours retyping bibliographies and notes or references. (See "Documentation," Appendix C.)

### Indexes

Indexes may be provided for long, detailed reports. Arranged in alphabetical order, not like a table of contents, which is arranged in page order, an index contains more details than a table of contents. Both a table of contents and an index refer readers only to the contents of a report.

## Optional Sections

The optional sections of reports—cover letters, prefaces and forewords, acknowledgments, tables of contents, indexes, and lists of illustrations—should be included when necessary. Writers should, for example, acknowledge help that was not paid labor and should supply tables of contents for all except the shortest and simplest reports.

## Headings and Subheadings

To help readers recognize major and minor sections of reports as well as to be useful in a search for a particular topic, headings and subheadings must be clear, complete, and specific and those more or less important must be differentiated. Outlines (Chapter 4) usually prove helpful to writers of headings. If outlines have been carefully prepared, they give the heading writer the divisions of the paper, the order or degree of importance of each division, and the wording or hints for the wording of the topics and subtopics.

The heading "Fire at Port Nevis" used in a company report is not complete or specific. "Electrical Fire at Port Nevis Plant" is specific but not complete. "Electrical Fire Destroys Port Nevis Plant" is specific and complete, illustrating the value of sentence headings for major topics.

In assigning headings and subheadings to material, report writers should make those of equal importance parallel in construction and should phrase headings of higher order in more important grammatical constructions (Chapter 4). Writers should also note that logically anything divided does not have only one part; therefore a heading should not have only one subheading.

One advantage of headings is that they break a page. This makes them of most use when paragraphs are longer than average. Writers of short paragraphs may find that headings make their pages look choppy.

Variations in the placement of headings (center or side or indented) help the appearance of a page and aid a reader's speed of comprehension. Any typewriter offers sufficient variety for most reports. In choosing from the following selections, writers, as they move from first-order heading to second-order heading and from second-order heading to third-order heading, etc., should remember to vary the appearance of the headings as much as possible.

Writers may choose one of the following headings for their main topics:

FIRST-ORDER HEADINGS
FIRST-ORDER HEADINGS
First-Order Headings

One of the following, beginning in the margin if binding permits but otherwise at the margin, is suitable for second-order headings:

SECOND-ORDER HEADINGS
Second-Order Headings
Second-Order Headings

One of the following placed where a paragraph would begin is suitable for a third-order heading:

THIRD-ORDER HEADINGS. The text states . . .
Third-Order Headings. The text states . . .

One of the following may serve as a fourth-order heading and the beginning of the sentence.

FOURTH-ORDER HEADINGS are not used in many reports.
Fourth-Order Headings are not used in many reports.

It is better not to have headings of similar kinds; for example, not all capitals, not all underlined, not all capital and small letters. Differences in type and in position help readers to keep the relationships clear in their minds.

If reports are to be printed, writers are usually shown samples of the typefaces and sizes to be used, including boldface and italic. A company or bureau art department might be willing to suggest choices, but writers should be sure to indicate what they wish the headings to accomplish—emphasis on topics, help to the reader in finding sections and in seeing interrelationships, providing white space, breaking the pages, etc. Publishers are also usually willing to help with selection, and the editor assigned to the report may be especially helpful with the appearance of pages. If reports are to be prepared on word processors, the authors should be shown the typographic options and, if necessary, offered help with the selection.

## Editing and Revising Reports

Advice about revising writing appears in Chapter 2. While conducting research, writers should remember that reports are not revised or edited well at top speed or with interruptions. Teachers and consultants see many reports that are well written and well edited at the beginning but that degenerate after the first third into pages poorly written and sloppily edited. The writers, tired of the tedium, pressed for time, and interested now in other research, stopped careful revising and editing after completing a third of the work. Editing and revising accomplish most when they are performed carefully throughout a document.

Writers do their best editing and revising when they are fresh and alert. It is wise, therefore, to work on reports as soon as possible. From the very first stages of research, writers should start thinking about the organization of the report and should work on any part of the report that can be prepared early. Preliminary material, such as the title page, copyright page, and letter of authorization, can be made ready at the start. As research moves ahead, material for appendices can also be put in final form. Notes can be started for cover letters, preface, and acknowledgments; and additions can be made as the research advances. Bibliographies and lists of illustrations also lend themselves to preparation as the research progresses. Soon writers can prepare some sections of the beginning, discussion, or ending of the report. This work should then be revised and edited carefully and be provided with headings and subheadings. Completed or nearly completed sections may be listed in a table of contents. By this time more appendices and bibliographies will probably be ready for preparation and checking.

Writing so much of the report before the research is completed means that authors need not revise or correct when they are rushed or tired, that they will have time to cool off after writing before turning to revision and editing, and that they are thus more likely to be able to present their better writing and their more accurate work. The methods used for this editing and revising are discussed in Chapter 2. What has just been said here applies particularly to long reports and complicated works. Nearly anyone can revise a one-page memorandum or letter a few hours after it has been written, but the prospect of revising a long, complex work in a short time under deadline pressure daunts even the strongest. What a relief it is to know that half or two thirds of a long work has already been revised and edited, a relief well worth a few battles with procrastination at the beginning of the project.

447

## STYLE GUIDES

In the interest of uniformity many companies, particularly those that require (perhaps exist on) long reports, issue style guides. These enable report writers to concentrate on the contents and planning of a report rather than on the physical aspects. Sometimes guides are prepared by professional writers, but often an employee is detailed to compile instructions for coworkers.

Unless the person responsible for the guide is an expert on writing, it is best to leave the choice of organization of the report to the individual writers and their supervisors, who know to whom their reports are going and what they want the reports to accomplish.

### Format and Style

The style guide should explain the format of a report: double- or single-spaced typing, width of margins, page numbering, number of copies, style of the cover page, use of and location of an abstract, types of abstract, and any other physical details deemed helpful.

The style used for the company name (perhaps it always contains an ampersand), for any company product names, for the names of company divisions and departments, for abbreviations peculiar to the company (accompanied in the report by the spelled-out term the first time the abbreviation is used) should be listed with examples. There are always new writers in a company and sometimes new readers.

If a company has any rules about tables, bar charts, pie charts, graphs, photographs, and such, these should be explained. If there are no rules, a good example of each type should be included as a guide. Many company reports contain representative tables and charts in the text and computer printouts or complete tables and charts in an appendix. Any such practice should be indicated.

If there is a preferred form for notes and bibliographies, it should be included in the style guide with examples. If there is none, a suggested form published elsewhere should be recommended, as uniformity in such matters saves time.

For technical and scientific reports, the guide should list common units of measurement and chemical formulas with their preferred usage—abbreviations, chemical symbols, spelled-out words—all with examples. It should include the style of prefixes and numbers with chemical compounds and should specify the use of the SI or British system of measurement.

A uniform practice of writing numbers is desirable: for example, always use Arabic figures, use figures only with measurements, spell out the number if it is the first word in a sentence, spell all numbers under eleven, spell the number *one* and all multiples of ten and use figures for the rest.

Whatever system is agreed upon by the company writers or decreed by someone in authority should be explained.

## A Standard for Scientific and Technical Reports

Those concerned with the production of reports will find much useful information in a recently revised standard of the American National Standards Institute, *Scientific and Technical Reports—Organization, Preparation, and Production.*[3] In considering organization and preparation, however, readers should bear in mind that the standard establishes guidelines for scientific and technical reports only as defined in the standard: "A scientific and technical report is a publication designed to convey the results of basic or applied research. The report shall include the ancillary information necessary for the interpretation, application, and replication of the results or techniques of an investigation."[4] The standard recommends and discusses format and organization, thus limiting writers to only minor variations. The sections on layout, illustrations, tables and charts, and methods of reproduction are full; and the selected annotated bibliographies, particularly those of style manuals and guides and specialized dictionaries, encyclopedias, and handbooks, are helpful.

## Grammar

Some companies include brief discussions, with corrected examples, of the most common errors in writing: misplaced modifiers, incorrectly punctuated nonrestrictive elements, overuse of the passive voice, verbosity, lack of parallel construction, lack of agreement of subject and predicate, lack of subordination, lack of transitions. Style guides should not, however, be textbooks of grammar.

The section on grammar should seldom be written by a scientist or an engineer or a secretary or a member of the advertising department or someone who happens to have free time. If the company is not willing to employ a qualified person to write this section, it is better omitted.

## Letters

If the company has prescribed forms for letters—for example, the subject on a separate line at the beginning of a letter; an information code for the files; an established list, by titles, of people who receive copies of letters; a method of indicating the second and subsequent pages; the form of salutation and complimentary closing; the signature (or whose signature); and so forth—these, together with one or two sample letters, might be included in a style manual.

## Guides for Typists

Most companies have one style guide, if any, but some issue a separate guide for typists which concentrates on the format of reports and letters and on other information for which a typist might be responsible, such as the form of company names, capitalization, spelling of numbers, and so forth. In such cases this information should be repeated in the guide for writers, in order to avoid unnecessary discussion and argument.

## Ease of Use

Style manuals range in length from a dozen or so pages to hundreds; most company guides are fairly brief, because employees are discouraged when confronted with a thick book of rules on a subject in which most of them are only marginally interested. Even a short guide, however, must have a good index. The person preparing it should be imaginative and should consult colleagues in order to help the user to find material quickly. It is time consuming as well as annoying to have to guess under what word instructions are indexed. Indexing under several appropriate words saves the user's time and temper. For a short guide, cross-referencing should be omitted in favor of placing the relevant page numbers after each word indexed.

Just how comprehensive a guide should be depends upon how insistent a company is on style and format. A guide is, as the name implies, informative and useful; it should not be a straitjacket, nor should it embody the idiosyncrasies of one person. Writers should not find it an additional burden to their report writing. They should turn to it gladly because it saves them time and effort.

## NOTES

1. ANSI Z39.14-1979, *American National Standard for Writing Abstracts* (New York: American National Standards Institute) 7.
2. ANSI Z39.14-1979, 13.
3. ANSI Z39.18-1987, *American National Standard for Information Sciences— Scientific and Technical Reports—Organization, Preparation, and Production* (New York: American National Standards Institute).
4. ANSI Z39.18-1987, 11.

# EPILOGUE

# The Editor and Supervisor and the Future Editor and Supervisor

*He taughte, and first he folwed it himselve.*

*Chaucer*

## SUPERVISING WRITING

Supervising writing is one of the most difficult and least successful kinds of supervision for a number of reasons. Some who are called upon to supervise have no training or experience in supervising or editing, some have training or experience in only one, and some are themselves poor writers. Moreover, feelings about writing are more sensitive and tender than feelings about performance in science or technology. An author considers writing, even writing memorandums and reports, more personal work and may take criticism as a reflection on the inner being. Engineers will defend at length a dangling modifier or a pronoun without an antecedent but will correct an error in engineering the moment it is pointed out to them. Conferences with such writers may take on all the more unfortunate aspects of attempts to point out personality defects. As though this were not trouble enough, editors and supervisors of writing must cope with the tendency of some writers to confuse other disappointments and annoyances with feelings about the correction of their writing and to argue interminably about insignificant details.

### The Influence of Other Problems

Many teachers of writing in industry and government are convinced that some writers cannot improve their writing unless their attitudes toward their jobs change. In fact, one editor refused to supervise the writing of engineers because the engineers needed better working conditions in the plant more than they needed editing. He thought that editors could do little good because the dissatisfaction with management would be transferred to them.

451

Such an extreme case is, of course, unusual, but it indicates clearly the influence of unrelated troubles and miseries on writing and thus may help editors and supervisors to understand the causes of the otherwise unaccountable truculence of some writers. When advancements are refused, when salary increments are withheld, when working conditions are deteriorating, editing and supervision of writing become more difficult. Even unpopular changes in the cafeteria have affected writers' attitudes. Family arguments, financial worries, and marital failures are also storm signals to supervisors of writing and to editors. Writers may direct at them the impatience and anger that they have not expressed to the person who thwarted or defeated them.

Therefore inexperienced supervisors of writing or editors who are beginners need not feel unsuccessful or incompetent should they meet contentious writers who are determined that their writing remain as it is. If the editors have not been dictatorial but have tried to be understanding and helpful, they should not blame themselves for such a response. Only a psychologist, psychiatrist, or analyst can help some writers; no one else should waste patience and energy trying to help them. Unfortunately, such writers always seem to be assigned to inexperienced editors, and inexperienced editors do not realize that they should save their patience and energy for more cooperative writers, who are bound to have difficult days and troublesome writing assignments.

If a usually cooperative writer is in a poor mood for editing and correcting, a supervisor may find it wise to devote their conference to work that is less affected by emotional attitudes, and an editor may judiciously postpone a conference. When postponement is not possible, a supervisor or editor will have to rely on psychology and tact.

### Psychology and Tact

It is good psychology at a first meeting, for example, to avoid confronting a writer with a page that seems to have more corrections than original writing. Sometimes this impression results less from the number of corrections than from the untidy, sprawling handwriting of an editor. Changes and corrections made in a small neat handwriting are less obtrusive and less likely to exasperate writers before they have a chance to read them. But a paper covered with changes puts a writer in an uncooperative mood even on a good day.

A complimentary remark never starts a conference badly. There is usually something to admire in a paper—the subject matter, if not some point of style. If the paper is hopeless, the writer may deserve praise for something else, and a competent supervisor tries to have some kind words of admiration ready at the beginning of a conference. It is hard for a writer to dislike a person who thus clearly demonstrates appreciation. But false, exaggerated compliments are much worse than unsuccessful

452

conferences. They boomerang, for they may be recalled to an editor by a writer who cannot understand how someone who once admired the work is now not even satisifed with it. And the writer who knows a supervisor is lying becomes distrustful and suspicious. We do not suggest that supervisors sugarcoat the pill of truth, but that they begin a difficult conference with a pleasant comment on the writer or the work. Supervisors who cannot find anything complimentary to say about writers should not be supervising them. Such supervisors are too unperceptive.

To end a conference well, a good supervisor has in hand a constructive view of the writing. If the present is painful, the supervisor turns to a more promising future. Good points in a paper may be used to stress the attractive prospects of a writer, and the value of correcting a few faults or strengthening a few weaknesses may be presented to cheer a disheartened writer. Sometimes it is useful to turn from the paper on which the conference has been concentrated to a writer's next assignment and to help with that. At other times it may be better to summarize how a writer can improve a poor paper and then to turn the discussion away from writing to some work at which the employee is succeeding or to some work of special interest.

When writers resent without apparent reason all criticism given in conference, supervisors may give them their corrected papers to study, schedule other conferences, and trust to a session with fellow employees and the passage of time to make the writers more compliant. A suggestion that poor writers use a book, take a course, seek tutoring may be a good final note. To be effective, such a suggestion must be specific. What book, what course, what tutor? Will the company pay for the course or part of it? The best way to present such a suggestion is to lend the writers the book to look over, give them a catalogue with the course marked, or be ready with information about tutors. This help should be most specific for foreign-born writers who have come to the United States recently. They should not be left to flounder in a search for assistance they are poorly equipped to find.

A tactful editor yields a point now and then to a writer. Every educated person knows that some decisions about writing are matters of taste, and writers like to think that their taste will not be ignored. If a supervisor knows that a writer finds it hard to take criticism, the supervisor might occasionally consult the author about some debatable point. "Do you think that we might use a pronoun here instead of this synonym? Then we can use the synonym in these other sentences." If the writer stubbornly insists on retaining the synonym, the writing is not much worse, and the relations between supervisor and writer may be better. However, an editor or supervisor must use good judgment about yielding because some writers view any concession as a sign of weakness and try to take advantage of the supposed weakness.

## Special Problems

One of the most difficult editing problems is the writing of confused thinkers. They use correct words, appropriate style, and correct constructions, but they do not express the meaning they intend to convey. And the writers apparently do not notice this. Indeed, some supervisors report great difficulty in convincing such writers that they are not saying what they mean. What they intended to state is so firmly fixed in their minds that they cannot see what they have said. Such writers find it difficult to see that *a check on the progress of the experiment* is ambiguous; even with the help of a dictionary they will not understand that *check* has two meanings, both possible in the phrase. They are so intent upon one meaning that they refuse to admit another.

Some writers have difficulty whenever exactness is required. They write about "eliminating our present suppliers" and about "incredulous demands" and think readers should understand what they mean. When they plan to send such writing to readers who are not working for their company or government agency, an editor can insist that the reputation of the company or agent demands correct and precise writing. But when they write to their fellow employees, such writers may be difficult to convince. "He knows what I mean" is their defense. Sometimes a reader does know; sometimes he does not; certainly his substitute might not know. A more persuasive reason for care in writing is that writers who are careless with one kind of writing will make mistakes when they wish to be correct and accurate.

Time may help such writers. Although they are unaware of their error at the time that they write, do not notice it when they revise, and cannot see it when an editor or supervisor points it out, they may struggle with the misstatement, read and reread it, and discuss it with others until they finally see the error. A course may help confused or careless writers, particularly if the other members of a course equal or surpass them in rank. Even the stubbornest writer will not defend his or her clarity against a dozen readers who deny it. The complaints of readers force even the most obdurate writer to reconsider.

But writers who cannot see what is wrong with "Problems are encountered by changes in the conference hours" may prove difficult to convince of their weaknesses. And when a desire to hedge confuses their thinking further, they may be impossible. One doctor who wrote "There is every reason to expect that side effects will appear" was convinced he was not committing himself in any way. He did not expect any side effects but did not wish to say so. In attempting to evade a commitment, he made himself responsible for exactly the opposite of what he thought.

"If you say, 'There is every reason to believe that it will rain,'" he was told, "you are expressing the opinion that it will rain."

"No, I'm not," he insisted. "I'm saying that it probably won't rain, but I'm not committing myself. I'm hedging."

454

When a busy director summoned him to ask why the company was investing in a drug that would probably have side effects, he told the director that he had expected the tests to prove the absence of side effects. He was unable to understand how anyone could take any other meaning from his words.

If editors, or supervisors meet such a stupidly stubborn reaction, they must use their authority to change the wording. The writers may fret and fume, but when they are so wrong and so blind, someone must correct their writing.

## Explanations of Changes

Most writers are willing to correct writing if an editor can give them a sound reason for the change. Hence a good handbook of writing is useful to an editor or supervisor. After writers have been shown the rules for a few changes, they will, if they are reasonable, accept the editor's word for other rules. If a writer wants to be convinced of every point by reference to authority, a supervisor should explain some reasons and urge the writer to look up the rest. But supervisors should not expect writers to be convinced by such silly arguments as, "Well, this is the way I say it"; "I never heard it that way"; or "I like it this way and I'm the boss. When you're the boss, you can write it your way."

## Reasonableness and Competence in Editing

There is often more than one acceptable way of phrasing an idea. If a writer's expression is satisfactory, he should not have to change it. Good supervisors remember that editing can be too fussy; there is little point in caviling. Supervisors and editors should be reasonable and sensible. They should not press a writer to produce long papers just to convince readers that the division is working hard or insist that the writer use elegant diction or unnecessary jargon to convince readers that the subject is difficult. A supervisor who, despite the variety available, insists upon one organization, one beginning, one ending should be training robots not human beings. The best supervisors are those who permit their writers to depart from the set methods of a company or division when another method is better.

To make decisions and to give advice about writing, supervisors must be well informed. But government and industry often neglect the supervisors' training. Some writing courses are given for everyone but supervisors. The fact that supervisors realize this is foolish is shown by their requests to take or to audit these courses. When they have learned the principles of correct and effective writing, they find editing much easier. Many supervisors and editors need not a course in writing but a course in supervising and editing. Sharing problems, considering various approaches to writing and writers, becoming familiar with the editing of

455

other supervisors, studying the psychology of writing and of writers—all these help supervisors to improve. Some people may be born supervisors or editors, but most have to be trained and educated. Much writing in government and industry and much of the friction associated with editing are due to untrained, unprepared supervisors and editors. That the writing and friction are no worse than they are is due largely to the goodwill and common sense of those supervisors and editors. But common sense is not enough. Knowledge of the techniques of supervising and editing is essential. And by learning to understand some of a writer's difficulties and problems, a supervisor or editor can become more successful in other supervisory functions.

One difficulty in helping writers is that they need knowledge of their readers. It helps them to know roughly their readers' education, the extent of their understanding of the subject, and their purpose in reading the paper. To prepare good instructions, writers must know their readers well. And to compose an effective letter, they should know something of the personality of the person to whom they are writing. These are reasonable needs, and supervisors should supply what information they can. To insist that writers need not know anything about their readers is to insist on writing that succeeds only by accident.

Supervisors and editors work better with writers if they realize that good writing takes more time than poor writing and that to write something in a hundred words takes longer than to write it in a thousand. A person who is learning to write better needs time; corrections and changes require careful thought, especially at first.

Writers frequently interrupted or otherwise distracted are slower, poorer writers than they need be. At least while they are writing a first draft, writers should be free of interruptions and able to work in quiet surroundings that are not distracting (Chapter 1). If they cannot close their office doors or if there is no place for them to write except a noisy laboratory, it is an economy to send them to a library or the office of an absent executive. Many errors are avoided and time and effort are saved when writers have a quiet place to work.

Writers also need time to plan their writing. A period of mulling over a paper helps them find better ways to write. They need time to outline, time to write, time to put the paper aside after writing it, and time to revise (Chapter 2). Frequently their minds will be concerned with their writing while they are off the job. During dinner a phrase a writer wants may pop into her head; while reading his evening paper, a chemist may spot a type of ending that he can use; during a bowling game a better way of beginning a report may suddenly come to an engineer. And this costs the company or the government nothing but a little care in scheduling writing. If they can spread writing and revising, writers have time to work on papers in their minds. Therefore a supervisor's scheduling of writing is of major importance. It is not something for a supervisor to neglect or to take care of when there is nothing else to do; scheduling writing is an urgent task.

A good supervisor or editor is ready to help writers from the first plan to the last word of the final draft. If a writer organizes poorly or has a long, difficult report to organize, a conference on the outline may save many alterations later. Conferring on outlines is also useful for supervisors and writers who tend to disagree on the purpose or scope or organization of papers. And a supervisor can tell from a detailed outline whether a writer has understood the assignment (Chapter 4). Inexperienced workers and new workers benefit much from discussion of their outlines, for they gain a clearer idea of what is required. The old method of training professionals by letting them learn to write reports by reading those in the files is far inferior to guidance from a supervisor or editor. Imitating old reports may introduce writers to mistakes new to them, confirm bad habits, and dam originality.

A writer needs sensible answers to sensible questions. In some divisions of industry and government these are available. Patient, competent supervisors teach young administrators, scientisits, and engineers the principles of good writing. But in the many divisions where supervisors know no more about writing than new employees do, some of the supervisors are sensitive and try to cover their deficiency by pedantic insistence on their own expressions. The young people soon learn that the only way to get supervisory approval on a document is to write it in the supervisor's style. And that style may be antiquated, pompous, and verbose. Imitating such writing is not a salutary experience for a beginner. Some engineers and scientists have imitated the writing of supervisor after supervisor, and their own styles—though it may not be fair to call them that—are patchworks of the worst features of the writing of five or six supervisors. Imitators learn quickly that it is the poor characteristics of their style that supervisors like to see reproduced.

An editor or supervisor should keep informed of changes in English. To recommend style that was out of date twenty years ago, to punctuate according to rules current when the supervisor was in college, and to insist on old-fashioned business jargon is to court trouble. Writers lose respect for a supervisor who is out of date.

A supervisor or editor who is informed, who will look up a rule, who will listen to an intelligent point of view is a comfort to a writer. Professionals remember gratefully those executives who trained them to be better writers. Supervisors would be surprised at how often when we comment on a good point in a paper, a middle-aged or elderly writer will say, "Oh, I was trained by an excellent supervisor," or "My first supervisor was a great help to me. I was a terrible writer when I came here," or even, "I guess my boss finally taught me something about writing. You know, she's been pretty patient with me. I never learned anything about writing until I worked for her." Contentious, troublesome, irritating though writers may be, they are grateful ever after to the person who helps them to write better. That is a thought for the day when a supervisor feels that she does not want to see another writer—ever—or when an editor is certain he chose the wrong profession.

## The Extremes

The most difficult editor or supervisor of writing and the least effective is the one who goes to extremes. Zealotry, an unnecessary problem in any situation, particularly one that does not call for partisanship, is most troublesome in editing, where writers' sensitive feelings may be as mischievous as poltergeists. When editors go to extremes, the sensitivities of the writers may halt work and upset whole divisions.

Unfortunately, many supervisors and editors quickly reach extremes. Some common problems are (1) supervisors who defend the writing in their division like infatuated parents demanding high grades for their offspring; (2) supervisors and editors who would correct and change reports forever if that were possible; and (3) assistants and editors low on the corporate ladder who pander to their immediate supervisors by imposing on all the works they edit the favorite expressions, organizations, beginnings, and endings of their supervisors.

A typical example of the first group is the supervisor who was told by a vice president that a report prepared in the supervisor's division was well written on the whole but had a dull beginning. The supervisor's immediate reply, delivered as though it were irrefutable, was "*I* find it interesting."

"But you're not the reader for whom it's intended. Look at this first paragraph—five sentences of similar length each starting, 'This is the'— a weak sentence beginning."

"I like it that way. I wouldn't change a word."

The vice president changed the offending paragraph himself, secretly wondering whether the supervisor had written it and imposed it on the scientist's report. When he told the story to a consultant, she said, "You may be right. When parents complain about the low grades given their children's writing, one parent in the heat of anger will very often blurt out, 'I know it deserves a higher mark. I wrote it myself.'"

Equally effective in preventing the education and training of writers and in halting work is the manager who hates to release a paper because she wants to check it "just once more." After three rewritings and ten resubmissions with countless changes in each, the author rebels. But this does not deter the manager, who continues reading the paper and finding something to change each time. When the writer asks her to explain the changes, she "doesn't have time." Completed but unapproved reports pile up on her desk. Some members of her staff, weary of the hypercriticism, have locked in their desks reports never submitted to her. Report writing in the division is close to a standstill.

Supervisors or editors who insist on inserting their words or their supervisors' favorite words into every document annoy writers the most. Many writers will accept an occasional explanation, like "We have to write it this way for political reasons"; but only weaklings will accept the frequent insertion of passages that fawn on the next level of managers. A writer of spirit becomes rebellious under this kind of supervision.

One manager told his staff of scientists and engineers, "The report is yours only until you turn it in. After that it's mine to do what I want with." After his writers refused to lend their names to the rewritten reports, the supervisor was transferred to another country. The writers' only comment was, "We hope it's far away."

One of our students told us he had written a report months before, and then each person on the next higher rung of the corporate ladder rewrote it for the person on the next rung. One morning the first vice president requested the report for a trip to Washington. Hastily four more managers read the report and rewrote it. That made fourteen rewritings. The vice president read it and immediately rewrote it, then sent a copy of his version and of the fourteenth rewrite to our student, who, he apparently thought, had written the fourteenth version. The two versions were accompanied by a firm command to take the next report-writing course offered. When the engineer compared his report and the vice president's, they were so similar that either could have been a slightly edited version of the other.

The engineer sent the vice president a polite note and the two versions. The result for the manager was rather like a hurricane accompanied by tornadoes and waterspouts. Perhaps this was deserved, but it seems excessive when one realizes that in some companies fifteen rewritings are not unusual.

Careful editing is, to be sure, desirable, but not all writing is equally important or equally deserving of time and effort. Managers need the courage to send adequately edited reports to their readers in order to move on to other reports. Instead of wasting time on excessive editing, managers should spend it on teaching writers to understand the changes and to avoid repeating the errors and weaknesses the next time they write.

Many other forms of zealotry tempt a supervisor or editor of writing. The one we find most egregious is the defense of an editor's error by the statement "That's the way we do it here." An otherwise reasonable supervisor said, "Well, I looked it up, and I see the writing handbooks don't like it [a euphemism for 'label it incorrect']; but we've always done it this way at Sampson-Ulysses, we're used to it this way, and I'm not going to do it any other way." At the words "Sampson-Ulysses" she stood straight and almost touched her hand to her heart as though saluting a flag.

Akin to this manager are the ones who easily accept what they think management wants. One such manager hired a consultant to find the cause of poor writing in two divisions. The consultant noticed that poor writers ready for retirement in a year or two were assigned to edit the writing of new scientists and managers. These trainers were the failures of the company who had never climbed past their beginning jobs of thirty or more years earlier. "But I can't do anything about them," protested the manager. "Venus la Roseate never fires anyone."

The consultant showed the surprisingly large sums the system cost for interviews with disgruntled workers and for courses to correct what the company failures had taught the newcomers. "They are teaching new employees how to write poorly and incidentally how to fail, because that's what they know," said the consultant.

"I'll have to think about it," admitted the manager. "Venus-won't fire them. There are no other suitable jobs available. I can see that meanwhile we're losing the best results of three years of expensive college recruitment."

He found his answer—sending the trainers to take courses until their retirement, an answer that hastened several retirements. But the beginners they had taught took a long while under excellent supervision to recover. Perhaps they never learned or understood what had happened to them.


## Company Motivation

Many of the tribulations of supervisors and of editors disappear in a company that has an atmosphere conducive to the improvement of writing. When good writing is important in a company, employees are eager for help. But writing is not important because management says it is. Good writing is important when it is recognized as a part of an employee's work that counts toward advancement. As soon as advancement is based partly on writing skills and performance and as soon as this is known in a company, the supervision of writing becomes less of a problem and writers appreciate good editing.

Consultants on writing and teachers of writing immediately notice an atmosphere favorable to writers, because these consultants are welcomed with the same enthusiasm that a specialist who can solve a difficult scientific or engineering problem receives. When they return a second time, employees seek them out to ask questions and to discuss progress. The motivation for courses that is necessary elsewhere is superfluous in such an atmosphere. Writers are eager to improve, and unless a consultant or instructor is hopeless, the writers remain eager, interested, and alert. A consultant finds that employees could not be more intense about improving their writing if the company were expecting to make a monthly profit of several million dollars through writing. And good writing does profit such companies by saving expensive reading time, by keeping scientists informed, and by giving executives clear and complete information for decisions.

460

## BENEFITING FROM SUPERVISION OF WRITING
## AND FROM EDITING

Even excellent supervisors and editors cannot summon forth good writing by themselves. A poor supervisor and a cooperative writer can accomplish more than an excellent supervisor and an uncooperative writer. Some workers do not benefit from supervision and editing because they do not know how to benefit.

To improve, writers should pay attention, ask intelligent questions, and try to apply whatever seems useful. Obviously if, instead of trying to learn, they spend their time planning their next argument, they will not improve. A conference on writing is not an opportunity for writers to display their debating skills, to demonstrate that they know more than their supervisor, or to release their irritations. It is an opportunity to learn to improve writing.

A little reasonableness on the part of writers helps their editor or supervisor to instruct them more efficiently in the allotted time. Writers should try to think of their writing as impersonally as they think of their mathematics or science because touchy, resentful writers are difficult to help and therefore injure themselves more than any adverse criticism from an editor or supervisor can injure them.

Writers can also learn much from the corrections and changes made by an editor working for a publisher. Yet few writers realize this. Many approach an editor's changes antagonistically and fight bitterly although they can accomplish nothing by displaying bad temper and ignorance. But writers who examine changes, study the reasons for them, and try to improve their next papers can learn much about writing.

Some years ago we instructed a new journal editor by demonstrating on the two most poorly written papers awaiting publication. To save time we performed all the editing that would be desirable under the best circumstances, which are certainly not those that generally prevail in the editorial offices of most journals; and we indicated the editing that might be omitted in more normal circumstances. Then we discussed points that the new editor had noticed, and we marked one manuscript a little more. At the end of the day that paper had been edited more thoroughly than any ever published in the magazine. To spare the feelings of the writer we suggested that the editor erase the optional changes.

A few weeks later the editor told us that the edited manuscript with all the changes had been sent to the printer by mistake, and galleys had been mailed to the author. The editor was awaiting the writer's response with trepidation because of those optional corrections. The writer's galleys, however, were returned without comment, and we never thought about our zealous editing until more than a year later. We were reminded of it when the writer's next paper received an award as the best paper of the year.

We rushed to examine the manuscript of the prize paper and found that it contained not one of the errors or weaknesses, not even one of the wordy expressions, of the paper we had dissected. Instead of damning the editor who had made so many changes, this writer had learned everything that he could learn from the meticulous editing. And that, as shown by the subsequent award, was more than many writers learn during a lifetime.

Most of the editors and supervisors we meet are seriously concerned about improving writing and about learning how to help writers. Writers should not spurn this goodwill. If they realize that a supervisor or editor is trying to help them improve a skill important to their careers, they may be more tolerant of adverse criticism.

Nobody suffers more from the failure to improve than writers themselves. If they consciously or unconsciously attempt to revenge themselves upon a supervisor or a company by refusing to learn to write better, they damage their own careers. Their supervisors and their companies will succeed in spite of this lack of improvement. But each writer should ask, "Will I?"

Writers sometimes forget that they themselves are likely to supervise writers soon. In industry and government, the writers whose work is being edited today may be editing tomorrow. They can prepare themselves by taking advantage of advice, suggestions, information, and training. As supervisors, they will have to do more than express their own thoughts: they will have to know whether another writer's expression is correct and effective; they will have to explain, not just change; they will have to assume the writer's point of view and adapt instruction and advice to it; and they will need patience and understanding.

Since none of this will come to them with the title of editor or supervisor, writers should seize every opportunity to learn. In every conference about their writing, they should assimilate as much as they can of the principles of writing and the techniques of editing and supervising.

An intelligent professional can learn from poor as well as from good examples: one clever chemist who became the best supervisor of writers in his company told me that he had prepared for supervision by noting carefully in advance the supervisory mistakes that he intended to avoid.

That chemist is the antithesis of the complainers who spend their time fussing over every change in their writing, criticizing others sharply, and feeling unappreciated. Such complainers become the poorest of editors or supervisors not only because they have learned little about writing and supervising but also because they are likely to switch from destructive criticism of their supervisors to destructive criticism of their writers. Only by using their energies for learning rather than lamenting can they hope to become better writers and better prospects for promotion to a supervisory or editorial position. By examining carefully editors' changes, writers can evaluate them to determine whether to make them when they supervise or to avoid them. By learning good administrative techniques and by rejecting poor ones, particularly those to which they are prone, writers can prepare themselves to help others. And by taking a more broadminded view of others, even of supervisors, and a more critical view of themselves and their writing, they will become better prepared and better suited to help and develop others. Then when opportunities arise, when promotions come, they will be ready.

# APPENDICES

# APPENDIX A
# Fallacies to Forget

Fallacies about writing are destroyers. They lead to writing that is awkward, fuzzy, ambiguous, or unreadable. Those recently converted to belief in fallacies about writing bring to any discussion the worst features of arguments about politics and religion. As though it is not bad enough that their writing is being damaged by fallacies, they insist that everyone else practice the same mistakes. Thus error gains and holds the minds of writers.

Some fallacies place writers in straitjackets that make them hate to write. Some ruin the style of scientists and engineers whose writing was acceptable before they were forced to write in a poor, graceless style in order to have their theses approved. Other fallacies impose style so contrary to English idiom that writers who apply them long and faithfully may become handicapped for life.

The straitjacket fallacies concern mainly what writers must not do—split an infinitive, begin a sentence with some forbidden word, or end a sentence with a preposition. Some teachers rigidly enforce these fallacious rules. Many years later some of their students still believe that they must stop to check each sentence for the three fallacies and to rewrite immediately as though the forbidden word were obscene. Losing their train of thought blocks their thinking, and interrupting themselves in this way merely enforces their belief that they hate writing.

Writers lucky enough to escape lower-school emphasis on fallacious rules may meet powerful emphasis on fallacies in graduate work. Some professors in schools of science or engineering still insist that writers of theses use only the passive voice, string modifiers before nouns, and avoid the use of personal pronouns if they want their theses to pass. Fortunately, because of the good work of societies, associations, and the editing staffs in science and technology and especially because of effective leaders among the professors, the imposition of these fallacies is much less frequent than it was twenty years ago. Managers tell us, however, that some students still receive this erring instruction from a few graduate professors and emerge on the defensive with virulent cases of fallacy-itis.

But educators are not the only misleaders. Some supervisors in business, government, and industry recommend fallacies that are dangerous, for example, telegraphic style. The more faithfully a subordinate follows such advice and removes all articles and transitions from a report, the more likely it is that what is left will be unreadable. Later the writer will find that he has lost his idiomatic command of English articles altogether. When he writes for a new supervisor, he has to ask his co-writer, "Should I use *a* or *the* here?" "*The*, of course," she says. "I'm surprised you asked. I always thought English was your native language." Not any more. He is now a speaker and writer of the style of telegrams and cables and may never recover his command of the article in his mother tongue.

Another fallacy common in business, government, and industry is the file fallacy, which sends writers to the files when they need help with writing. A writer looking for a complimentary close for a letter examines her supervisor's files and copies a participial closing that was out of date in 1930. In fact, her colleagues advise her to write everything in her supervisor's style. If the absence of editing changes is an indication, the tip works. Her supervisor learned to write by copying the style of his supervisor, and that supervisor copied his supervisor, and so on back to Scrooge and Cratchit in their Victorian office.

Fallacies do not have to be old-fashioned to be dangerous. The modern advice given whole classes of writers or hundreds of convention listeners to shorten their sentences may work for those who write overlong sentences but will ruin the style of writers who have always used short sentences and now, in accordance with the advice, hack them to primer length.

The temptation or pressure to adopt fallacies being as great and as dangerous as these examples show, we have listed the fallacies most likely to cause problems for our readers and indicated next to each the chapter where it is discussed and illustrated.

Although these fallacies have demonstrated a remarkable ability to survive and to cause trouble, we are convinced that a determined writer can escape them and forget them. May all our readers note the falsity and forget the fallacies or, better still, never believe in them at all.

| Fallacy | Chapter |
|---|---|
| The first draft should be the final draft. | 1 and 2 |
| Put yourself in the reader's shoes (or place). | 2 |
| Grammar don't matter. | 6 |
| Good writing is just good grammar. | 6 |
| Punctuate the way you breathe: a comma for a short breath and a period for a pause. | 8 |
| Never change that precious first wording: That's the way it came to me; it must be the way to say it. | 9 |
| Write the way you talk. | 9 |
| Repeat ideas three times. | 9 |

# APPENDIX B

# Problem Words and Phrases

The recommended forms in this list are those now used by educated people for their writing and conversation in business and the professions. Although a few of the words or spellings labeled *substandard*, like *irregardless* and *alright*, may become generally accepted, we have not yet found them in prose of good quality.

We do not recommend colloquialisms, dialectal forms, or slang, because, however widespread their use on television, most fall into general disuse just as quickly as they appear. Slang that sparkles with newness today is forgotten next week. Who now says *skidoo, natch, palooka, jeepers*?

The acceptance of questionable words is difficult to assess, because there has been only one new edition of an unabridged dictionary in over twenty years. The trend in the current editions is toward permissiveness, many of the distinctions in earlier editions having been omitted. We have tried to maintain the standards of the classic guides and to indicate where current usage that we have observed in good prose has departed from these standards.

## A

**a, an**

The selection of **a** or **an** depends on the initial sound of the following word; **a** is used before a consonant sound and **an** before a vowel sound: "**a** byte, **a** heat exchanger, **a** toxin, **a** xylem ray, **an** anion, **an** intestinal problem, **an** ovenbird, **an** ultrasonic wave."

Note that the first sound, not the first letter, of the following word determines the selection: "**an** hour" but **a** "historian," "**an** F" but "**a** failure," "**an** s sound" but "**a** sibilant," "**an** unguent" but **a** uranium mine."

In using the indefinite article before abbreviations and acronyms, writers may have to choose between **a** and **an** on the basis of whether the first letter is usually read as part of an acronym, as a word, or as a letter.

Thus "**an** N.Y.U. team" indicates that *N* is read as the letter *en*, but "**a** N.Y. governor" indicates that *N* is read as the word *new*.

A writer who cannot find out how a new abbreviation is usually read should avoid using it with the indefinite article by changing the construction of the sentence or by spelling out the abbreviation.

### abbreviations, acronyms, clipped words

See Chapter 10.

### above

The use of **above** to refer to material previously mentioned is often vague or confusing and therefore irritating. If the reference is clearly to material that immediately precedes, writers should use **this, that, these, those**.

**Above** is used in legal briefs, as are **abovementioned, abovenamed, abovereferenced**—monstrosities best left to the legal profession.

> POOR: The **abovementioned** report is the problem.
> IMPROVED: **This** report [if mentioned close by] or **Salinger's** report [if it is too far away for pronoun reference] is the problem.

### absolute adjectives

Adjectives that do not logically lend themselves to comparison should not have it forced on them by the use of *more* or *most*. Some words do not permit comparison because they do not have degrees—*extinct, fatal, final, impossible, libelous, matchless, mortal, peerless, perfect, permanent, unique* (meaning *being the only one of its kind*), *universal*: "A very unique discovery, a rather peerless result, a very fatal illness, a quite impossible task"—all these phrases compare concepts that are incomparable, assign degrees to ideas that do not have degrees. Words should not be used loosely in informative prose, and therefore words such as those listed here should not be used in the comparative or superlative degree.

### accept, except

**Accept** means *to receive, to endure*; the verb **except** means *to leave out, to reject*; the preposition **except** means *with the exclusion of*.

> CORRECT: They **accepted** the proposal.
> The legislature **excepted** high schools from the proposed grants.
> Some members wished to apply the tax to all residents **except** those with annual incomes below ten thousand dollars.

## access, excess

The noun **access** means *ability to enter*. **Excess**, which is a noun and an adjective, means *an amount beyond a specified or understood limit*.

> CORRECT: He had **access** to the library at all hours.
> The **excess** was discarded with the trash.
> The **excess** baggage was refused.

## acoustics is/are (number)

**Acoustics** is singular when it means *the science of sound*; plural when it means *the qualities in an enclosure that reflect sound waves*.

> CORRECT: The engineer replied, "**Acoustics** interests me, and I read as much as I can about it."
> The **acoustics** of the auditorium were greatly improved by the new architect.

## acronyms

See Chapter 10.

## A.D., B.C.

The abbreviation for *anno Domini* ("in the year of our Lord") usually precedes the year, as in "**A.D.** 1611," but the abbreviation for *before Christ* usually follows the year, as in "3000 **B.C.**" **A.D.** should not be used for centuries. "In the tenth century **A.D.**" is incorrect and illogical. However, "in the third century **B.C.**" is logical and correct.

## adapt, adept, adopt

**Adapt** means *to change something for a reason*. **Adept** as a noun means *a skilled person* and as an adjective means *proficient, expert*. **Adopt** means (1) *to legally take the child of another*, (2) *to take and use as one's own* (the methods or habits of another, for example), (3) *to accept a law and put it into practice*.

## adjudicate, arbitrate, mediate

Negotiations require that writers know the differences in meaning of (1) **adjudicate**—*to hear evidence and settle a dispute judicially* (transitive verb) or *to act as judge* (intransitive verb), (2) **arbitrate**—*to hear evidence and give a binding decision*, and (3) **mediate**—*to listen to the arguments and to attempt to persuade the parties to accept an agreement*.

473

CORRECT: She did not think that the man who was **adjudicating** her case was impartial.

He did not like to **arbitrate** the strikes of that union because the documents submitted were incomplete and ambiguous.

The dean tried to **mediate** between instructors and students.

## adopt

See **adapt**.

## adopted, adoptive

A person who adopts a child is an **adoptive** parent; the child is an **adopted** child. (See **adapt**.)

## adverse, averse

**Adverse** is the stronger of these words. It means *antagonistic, opposed*. **Averse** means *disinclined, feeling distaste or repugnance*.

CORRECT: She is **averse** to noisy, angry arguments.

He is not **adverse** to your opinions; he does not like the way you express them.

This method often results in an **adverse** reaction.

## advice, advise

**Advice** is a noun; **advise**, a verb.

CORRECT: She gave him good **advice**.
She **advised** him well.

The words have different pronunciations, **advice** ending in an *s* sound and **advise**, in a *z* sound.

## affect, effect

To **affect**, a verb, is *to influence*; to **effect**, also a verb, is *to accomplish, to bring about*. An **effect**, a noun, is *that which is brought about*. **Affect** is used as a noun in psychology, where it means *the conscious, subjective characteristics of an emotion considered without the physical changes*. It is seldom used as a noun in writings on other subjects.

CORRECT: Deadlines **affected** her adversely.
She **effected** a change in his attitude toward female managers.
The **effect** of the change was immediately obvious.

**affix one's signature**

*Sign* is a good useful word that avoids false elegance.

**aforesaid**

Here is another pompous word: "The **aforesaid** charge was clear at the time." Why not "the charge," "this charge," "that charge," "the charge of theft"?

**agenda, agendas, agendum, agendums**

**Agenda**, a Latin plural, is a singular form in modern English. The plural is **agendas**. **Agenda** means *things to be done* or *things to be considered at a meeting*. These are usually listed, and one item on the list may be called an **agendum**; two or more items, **agendums**.

> CORRECT: The **agenda** for the meeting this week is long.
> The **agendas** for the meetings of the next two months are short.
> The members may vote to remove the less important **agendums** to shorten the meeting.
> The union members will not agree to removing **agendum** four, which is essential to their position.

**aggravate**

**Aggravate** means *to intensify* or *to make worse*. Used colloquially, particularly in the South, it means *to exasperate*.

**ain't**

This substandard contraction is avoided by most writers except in dialogue and humor. It is useful in conversation in the question "**ain't** I?" because suitable contractions like "isn't he?" and "aren't they?" are not available in the first person singular. "**Ain't** I" is far better than the ungrammatical, falsely elegant "aren't I?" And "amn't I?" is difficult to pronounce. "Am I not?" is a good substitute.

**all but one**

A noun and a verb following **all but one** are singular; however, if **one** is not followed by a noun, the verb is plural.

> CORRECT: **All but one** drug given to these patients *is* safe.
> **All but one** of the drugs *are* safe for those patients.
> **All but one** exit *was* blocked.
> **All but one** of the exits *were* blocked.

## All ready, already

**All ready**, an adjectival phrase, means *everyone* or *everything is ready*, but **already**, an adverb, means *by this time* or *previously*.

> CORRECT: The machines were **all ready**, but the operators were delayed at a prolonged meeting.
> They were **already** impatient.

## all right, alright

**All right** is the preferred spelling.

## All together, altogether

The phrase **all together** is applied to a group functioning as a unit; the adverb **altogether** means *entirely, wholly, thoroughly*.

> CORRECT: They were jammed **all together** because the meeting room was small.
> The patients were **altogether** bored during the tests.
> He forgot the answer **altogether**.

## allude, refer, elude

**Allude** is used for *indirect reference*; **refer** is used for *direct*.

> CORRECT: He **alluded** so subtly to his difficulties with his secretaries that no one noticed or remembered the problem.
> She **referred** to an article containing the evidence for her statement and cited the journal—date, volume, and page.
> He **referred** to his supervisor as "Old Stoneface."

**Elude** should not be confused with **allude** or **refer**. It means *to escape* or *to avoid with adroitness*: "For six months he **eluded** a close relationship with his supervisor."

## allusion, delusion, illusion, reference

An **allusion** is an *indirect* or *implied reference*. A **reference** mentions *directly* and identifies *specifically*, a **reference** in a scholarly journal, for example, being one of the most specific identifications. An **illusion** is a *false perception*; a **delusion** is a *false belief*.

> CORRECT: Her **allusions** to obscure scientists were difficult to trace.
> His **references** to scientists were helpful to others studying these early periods.

476

Many of the flying objects were **illusions**.

Some poor writers do not improve because they have **delusions** of their excellence.

## a lot of

**A lot of** is used colloquially in the sense of *many* or *much*: "A lot of people saw the comet that night." *Many* would be preferred by most writers.

## already

See **all ready**.

## alright

See **all right**.

## also

**Also** is not correct as a connective in place of *and*: "The committee discussed the length of coffee breaks, sick leave, also vacations." "And vacations" or "and also vacations" is preferable.

## alternate, alternative

**Alternate** means *every other one in a series* or *a substitute*. **Alternative** means *one of two possibilities* and suggests choice, which is not suggested by **alternate**.

> CORRECT: When the experiment failed, they tried the **alternate** method.
> As a student she complained that the only **alternative** to passing was failing.
> The necklace was a chain on which diamonds and emeralds appeared **alternately**.

**Alternate** is also a verb—"A wet season **alternates** with a dry on this island"—and a noun—"She left the decision to her **alternate** when she had to return home."

## alternative, choice

**Alternative** is used for *one of two*. For *three or more* the word is **choice**. Unfortunately, **choice** does not have the same meaning. **Alternative** implies a *compulsion to choose* or *an inevitability*: "the **alternatives** of life and death"; **choice** gives no indication that one must be selected or

477

that the choice is imperative. Therefore writers sometimes use **alternative** for one of more than two if they wish to stress the compulsion.

## although, though

**Although** and **though** have the same meaning but are usually found in different positions in a sentence. **Although** commonly introduces a clause that opens a sentence; **though** usually introduces a clause that follows a main clause.

> CORRECT: **Although** you may try, you cannot read from the corner of your eye.
> He referred to the article on mild high blood pressure **though** he did not agree with it.

**Though** is used in the expressions *as though* and *even though*.

## a.m., p.m., M, N

The expressions *ante meridiem* (**a.m.**) and *post meridiem* (**p.m.**) are always abbreviated. They used to be capitalized, but modern usage permits small letters: "11 **a.m.**, 5:30 **p.m.**" The specific time must be stated in numbers before the abbreviation.

The form for noon is 12 **M** (*meridies*) and for midnight 12 **p.m.** The use of **N** for noon is incorrect.

> CORRECT: He has an appointment tomorrow afternoon at 3:30 (not "at 3:30 **p.m.**," which is redundant).
> He has an appointment tomorrow at 3:30 **p.m.**
> She reminded him in the morning (not "in the **a.m.**").
> She reminded him at 11:00 **a.m.**
> The meeting begins at five-thirty in the afternoon (not "at five-thirty in the **p.m.**").

## among, between

**Among** should be used when more than two are involved; **between** is correct for two.

> CORRECT: The members of the basketball team discussed **among** themselves the likelihood of their winning.
> There had been deadly rivalry **between** the two chess champions for three years.

Some writers use *between* if more than two entities are mentioned but the relationships are one-to-one. Usually, however, these ideas can be rephrased. "There are hundreds of treaties **between** France, Great Britain, and the United States. (The treaties do not involve all three; they may be between France and Great Britain, between France and the United States, between Great Britain and the United States.)

### amount, number

**Amount** is used for *material in bulk or mass which does not have units that can be numbered*: "a small **amount** of sympathy, a large **amount** of money, a large **number** of chemicals, a small **amount** of the chemical, a small **amount** of sugar, a large **number** of sugar cubes."

### an

See **a**.

### and

**And** is sometimes carelessly replaced by other conjunctions that do not mean **and**. The following published sentence misuses *yet* for *and*: "There is nothing in Westchester like our 3 models with wood panelling, carpeting, post-and-beam construction, yet everyone loves them." Writers also confuse *however* with **and**: "Most stars are much larger than the earth; however a few are millions of times larger."

### and etc.

See **etc**.

### and/or

This legal shortcut is not necessary or desirable in other writing. Instead of "engineers **and/or** managers" a writer should choose "engineers *or* managers *or both*." This avoids the common misreading of **and/or** as "engineers and managers or both."

### and so, so that

**And so** means *as a result, therefore, consequently*; **so that** means *in order that, for that purpose*. They are commonly confused, as in the following example: "Titles and side headings are one of the most effective devices for stimulating reader interest and increasing reading ease. Newspaper and magazine publishers have long known this, **so that** today there is hardly a newspaper or magazine on the market that does not have an attractive format that invites the reader's interest." (Here **so that** should be **and so**.)

Using **so** for **so that** or **and so** is undesirable because it is often ambiguous. If a writer does use **so** to introduce a dependent clause, it should be preceded by a semicolon.

### anxious, eager

Careful writers distinguish between **anxious** (which suggests *worry* or *fear*) and **eager** (which suggests *enthusiasm* or sometimes *frustration at delay or restraint*).

> CORRECT: He was **eager** to lose ten pounds so that his new suit would fit better.
> He was **anxious** to lose weight because he had high blood pressure.
> She was **eager** to accept the Nobel Prize.

### anybody, any body

The single word is used for a *living person*. The two-word phrase is used for an *inanimate object*, such as a corpse; for an entity; or for a geographical unit.

> CORRECT: **Anybody** could see that the light at the crossing was out of order.
> Few libraries have complete collections of **any body** of literature.
> Some legislators insisted that the chemicals dumped near the factory did not pollute **any body** of water.

### anyone, any one

**Any one** is used for *a particular person or thing*: "He wrote five versions of the slogan but did not like **any one** of them." **Anyone** means *everyone* or *a person chosen indiscriminately*: "**Anyone** could have written a more effective slogan."

### any other

When someone or something is compared with others in its class, it is necessary to remove it from the class. This prevents the illogic of a statement that it is better than all in its class (including itself). The word *other* is useful in correcting such an illogical statement:

> INCORRECT: These new optical coatings are better than any coatings available today.
> CORRECT: . . . better than **any other** coatings. . . .

(See also **other**.)

## appraise, apprise

**Appraise** means *to estimate the value of*; **apprise** means *to inform, to give notice*.

> CORRECT: He **appraised** the estate at three million dollars.
> The supervisor said that she was not **apprised** of the agendas in time to study them.

## apropos

The adverb **apropos** is followed by the preposition *of*, not *to*: "**Apropos of** the article she cited, all chemists in this laboratory should use the safety checks described."

**Apropos** as an adjective means *relevant and opportune*: "He asked her to prepare an **apropos** speech for the rally."

## arbiter, arbitrator, mediator

An **arbiter** and an **arbitrator** hear evidence and make binding decisions, the **arbiter** often having some legal powers; **a mediator** attempts to bring accord between opposing parties by serving as an intermediary.

> CORRECT: The **arbitrator** fined the company severely for failure to provide safety equipment.
> The **mediator** encouraged both the union and the company to compromise in order to end the expensive and dangerous deadlock.
> The **arbiter** ruled in favor of the environmental group opposed to the development.

## arcane, archon

**Arcane**, an adjective, means *secret*; **archon**, a noun, means *a presiding officer*.

## as

The word **as** is an overworked conjunction that is often vague or completely confusing. It may mean *because, for, seeing that, since, in the manner that*. Sometimes by meaning too many of these, it means nothing: "It was easy to see that the patient was neither tired nor ill as he won the game of chess." This statement demands *while, because*, or some other conjunctive adverb. The explanatory **as** clause introducing a sentence is less objectionable, but it is not always unambiguous, and it has been overworked. Truly objectionable is the substitution of **as** for a relative pronoun: "Writers have no opportunity to discuss publication dates as is offered to supervisors" should be "Writers have no opportunity to discuss publication dates **such as** is offered to supervisors" or "Writers do not have the opportunity to discuss publication dates **that** is offered to supervisors."

### as, like

**As** introduces clauses of comparison and is the preferred usage: "The grid in Figure 11 confuses readers **as** does the grid in Figure 1.

**Like**, not **as**, introduces prepositional phrases (**like** followed by a noun, a pronoun, or an adjective and a noun or pronoun).

> CORRECT: This typewriter is **like** hers.
> The training book is **like** dozens of others.
> For this purpose drawings are **like** photographs.
> Bench models **like** this one seldom meet Federal standards.

### as . . . as, so . . . as

Formerly **as . . . as** was used for positive comparisons and **so . . . as** for negative.

> CORRECT: The communications industry and foreign users of rockets had **as** little choice in future launch plans **as** the Defense Department and United States scientists.
> The grounding of the shuttles may not prove to be **so** wise a move for the military space program **as** some advisers think.

Many writers still observe the difference, but in conversation **as . . . as** is widely used for both positive and negative comparisons.

### as far as . . . , as for . . .

Used alone the incomplete phrase **as far as** is incorrect. It must introduce a clause that completes it. **As for** is correct if followed by a noun or substantive.

> INCORRECT: **As far as** his honesty, I cannot recommend him.
> CORRECT: **As far as** his honesty **is concerned**, I cannot recommend him.
> **As for** his honesty, I would not rely on it.

### as follow, as follows

The correct form to introduce a list or statement is **as follows**. In this idiom the verb has no subject and does not change to agree with the nouns or pronouns in the sentence.

CORRECT: They listed the engineers under headings **as follows**: students, engineers with less than five years' training, engineers with more than five years' experience, engineers in sales divisions, engineers in management.

## as for

See **as far as**.

## assume, presume

Both words mean *to take for granted,* but **presume** indicates more certainty or confidence.

CORRECT: The committee **assumed** that the report would be accepted.
The vice president acted upon his suggested plan because he **presumed** that it would be accepted.

**Presume** also means *to exceed what is right or proper:* "She **presumed** on our relationship."

## assurance, insurance, ensurance

Some life insurance companies are named **assurance** companies. Companies that write policies for fire, theft, illness, etc., are **insurance** or casualty companies. **Insurance** guarantees against financial loss; **ensurance** is obsolete. The verb **to ensure** is used, however, for the meaning *to make certain.*

CORRECT: The agent of the **assurance** company told the widow that the payments arranged would support her and the children in their present style and also be adequate for the college education of the children.
His automobile **insurance** policy had no deductible clause.
Her budgeting **ensured** a comfortable life even though her social security and pension benefits were small.

## as the result of

**The result of** means that there is only one result. It is often used incorrectly for *a result of.*

INCORRECT: As **the result of** the landslide twelve people died. (One of the results of the landslide was twelve deaths; therefore *the result* is misleading. *A result* would be more accurate.)

483

**as to**

**As to** is substandard. **As for** is preferable. See also **whether or not**.

**average, mean, median**

An **average** in statistics is the sum of a series divided by the number of members in it; a **mean** is a midpoint between two extremes, and a **median** is the point in a series at which an equal number of members is above and below.

> CORRECT: If John's grade is 85, Ella's is 80, and Helen's is 90, their **average** grade is the sum 255 divided by three—85.
> If the high temperature is 90° and the low is 60°, the **mean** temperature is 75°.
> If the workers received the following salaries: $500, $450, $300, $250, and $150, the **median** salary is $300. (The **average** salary is $330, and the **mean** is $325.)

**averse**

See **adverse**.

**awhile, a while**

The adverb **awhile** means for a short time: "The committee discussion lasted **awhile**." The noun **while** means *a short time* or *an understood time*. **A while** is usually preceded by a preposition: "The chemist discussed the results of the experiment for **a while**." "The chairman returned in **a while**."

# B

**B.C.**

See **A.D.**

**because**

See **reason is because**.

**behalf: in behalf of, on behalf of**

**In behalf of** means *in the interest of*; **on behalf of** means *in place of* or *as the agent of*.

CORRECT: They pleaded for contributions **in behalf of** the homeless, starving poor.

The lawyer pleaded **on behalf of** the three defendants.

Because of the illness of the president, his wife attended the ceremony **on his behalf**.

## being as, being that

Use *since* or *because*.

INCORRECT: **Being as** we know the data are accurate, we can save time by using them.

CORRECT: **Because** we know that the data are accurate, we can save time by using them.

INCORRECT: **Being that** the conclusions in the report are long and cumbersome, he summarized them.

CORRECT: **Since** the conclusions in the report are long and cumbersome, he summarized them.

(See **since**.)

## beside, besides

**Beside** means *at the side of*; **besides** means *in addition to* or *other than*.

CORRECT: No chemicals **besides** those listed on the door are to be used.

**Beside** the books on the top shelf of this closet are pads and pencils.

## be sure and, be sure to

The correct form is **be sure to**.

CORRECT: Anyone using this equipment should **be sure to** read the safety instructions.

**Be sure to** lock the door when you leave.

## between, from . . . to

A noun following **between** is plural, but a noun after **from** and **to** is singular.

CORRECT: The elevator stopped **between** the third and fourth floors.

The building was dark **from** the second **to** the tenth floor.

485

## between you and I

Pronouns are in the objective case after **between**. **Between you and me** is always the correct form.

## biannual, biennial, bimonthly, bi-

**Biannual** means *twice a year*, or sometimes it means *once every two years*. **Biennial** means *every two years* or *lasting for two years*. Other words with the prefix **bi-**, such as *bimonthly* or *biweekly*, also have two meanings. *Biweekly*, for example, may mean *twice a week* or *every two weeks*. Because they may confuse, words with such a bi- prefix should be used carefully or avoided. If twice in a period is meant, the prefix *semi-* is unambiguous: *semiweekly, semiannually*.

## boss

**Boss** is generally considered unsuitable for use in writing, but it is widely used for political leaders or supervisors of laborers. It should not be used inappropriately: "the chemist's **boss**," "the assistant treasurer's **boss**."

## boys, girls

**Boys** or **girls** should not be used in writing about adults. In most instances **boy** is an insulting term for a grown man, and **girl** used for a grown woman is silly as well as demeaning. The acceptable terms are *men* and *women* or in certain contexts *ladies* and *gentlemen*. But the phrase *ladies and men* is not acceptable.

## bring, take

**Bring** is used for movement toward a speaker or writer; **take**, for movement away from a speaker or writer.

> CORRECT: She said that she would **bring** your check here tomorrow.
> He **took** our copier to the repair shop.

**Bring** is also used when accomplishment rather than movement is involved.

> CORRECT: She will increase the accuracy of our bookkeeping and **bring** our budgeting closer to reality.

## burglar, burglary, robber, robbery, thief, theft

A **burglar** breaks in with intent to steal; a **robber** steals after threatening force; a **thief** steals secretly or stealthily. The nouns **burglary**, **robbery**, and **theft** have the same differences in meaning.

> CORRECT: When the **burglar** found the house well protected electronically, he sawed a hole in the side of the house and entered without using windows or doors.
> My sister fainted because she was badly frightened by the **robber's** threats.
> Many hotel **thieves** like to steal in the early morning hours when most guests are sleeping soundly.

## but

The conjunction **but** is sometimes used carelessly and illogically, as in the sentence "He is not a Cornell graduate **but** he attended Harvard." The fact that he is not a graduate of Cornell does not contrast with the fact that he attended Harvard: "He is not a Cornell graduate; he attended Harvard." "He is not a Cornell **but** a Harvard graduate" is correct idiom.

**But** should not be used redundantly with *however*. "He admired the beautiful simplicity of Aristotle's system, **but** it does not follow, however, that he disliked the complexity of other systems." A contrast is stated adequately by **but** or *however*. In the illustrative sentence there is no need for either; a semicolon might be used between the two clauses.

A **but** contrast within a clause introduced by **but** is an awkward construction. One **but** should be removed in each of the following sentences: "He said that he would try to improve, **but** he did not improve **but** argued in the same way with his coworkers and his superintendent." "The results were not conclusive, but they were interesting but not suggestive." Fowler[1] aptly describes this use of **but** as "wheels within wheels."

## by means of

See **prepositions**.

# C

## callous, callus

**Callous** means *having calluses* but is most often used for *lack of sympathy for others* or *lack of feeling*. **Callus** is *a hard thickened area on skin*.

> CORRECT: The **callous** attitude of his coworkers made him feel like a punchcard.

The **callus** did not respond to treatment, and she became increasingly reluctant to have her hands seen in public.

### can, may

Careful writers distinguish between **can** and **may**, especially in informational prose. **Can** is used for *the ability or power to do*; **may**, *for permission to do*.

> CORRECT: **Can** new chemists use this procedure? (Are they able [knowledgeable enough] to use this procedure?)
> **May** new chemists use this procedure? (Are they permitted to use this procedure?)

### cancel out

This favorite of wordy writers should be avoided. Anything **canceled** is *out*. (See "Hiccups," Chapter 13.)

### canvas, canvass

**Canvas**, a noun, means *a kind of cloth*; **canvass**, a noun, means *a personal solicitation of votes* or *a survey of opinions*. **Canvass** is also a verb.

> CORRECT: **Canvas** was too heavy for the costumes; therefore we used all we had for backdrops.
> The **canvass** conducted by NOW was added to the papers being sent to Washington.
> He **canvassed** the entire plant in an attempt to find support.

### capital, capitol

A **capital** of a country or other geographic division is *the seat of government*; a **capitol** is *the building in which the government functions*. Thus a **capitol** is usually within a **capital**. A **capital** also is *the top of a column or shaft*, or it may mean *money or goods accumulated to provide income or an advantage*: "make **capital** of tax deductions."

### carat, caret, karat

**Carat** and **karat** are both units of measurement—**carat**, for precious stones and sometimes for gold; **karat**, for gold. **Caret** is a proofreading symbol indicating the insertion of something omitted from a proof.

## carp, cavil

Although often used interchangeably, these words carry different connotations for some people: **carp** suggesting *incessant complaining*; **cavil**, *disagreement over unimportant points.*

> CORRECT: She **carped** at his refusals to establish periodic reviews.
> The meetings were frequently prolonged because of his **caviling**.

## censor, censure

The verb **censor** means *to delete or forbid,* usually for moral or political reasons; **censure** means *to condemn or disapprove.*

The noun **censor** is *one who makes the deletions described;* the noun **censure** has a number of meanings, such as *a judgment involving condemnation, stern condemnation, an official reprimand.*

> CORRECT: The **censor** eliminated from the play all discussions of governmental tyranny.
> The striking students said that the dean's **censure** was lighter than they had expected.

## center around

A center is a point, and so the correct forms are *center at, center in, center on.* A writer who wishes to use *around* might replace *center* by *revolve* or *rotate.*

> CORRECT: His arguments **centered on** the psychology of those prone to violent acts.

## claim

**Claim** should not be used for such verbs as *assert, declare, state, say* unless some right is involved.

> CORRECT: On the basis of the patent, he **claimed** full ownership.
> She *said* that the tennis match had resulted in a number of injuries to both players.

## cliché

See Chapter 10.

## climactic, climatic

**Climactic** relates to *climax*; **climatic**, to *climate*.

> CORRECT: The action in space was so **climactic** that she compared it to the high points of several dramas.
> The **climatic** conditions in Florida do not affect the growth of this plant.

## clipped words

See Chapter 10.

## common, joint, mutual, reciprocal

These adjectives all suggest *sharing*. **Common** is used for *concepts or objects or actions applicable to several people or groups of people*: "the **common** good," "a **common** bank," "**common** effort," a **commonwealth**. **Joint** is a *combining of two or more things or ideas*: "a **joint** meeting of two associations," "a **joint** trip by several people," "**joint** ownership." **Mutual** usually refers to *intangibles shared by two people* and has the connotation of exchange: "**mutual** admiration," "**mutual** understanding" (admiration for one another, understanding of one another); an exception is the well-known phrase *mutual friend*, which many people would say should be *common friend* or *friend in common*. **Reciprocal** indicates *exchange or a weighing of one thing or idea against another*: "**reciprocal** trade agreement, **reciprocal** treaty, **reciprocal** hospitality."

## company, concern, firm, corporation

**Company** is used for *a group of people associated in a business or industrial enterprise*. It may also be used for *individuals whose names do not appear in the title of the business*: "Webb and Company." A **concern** means *a business or manufacturing unit*, not a professional organization. A **corporation** is *a legal entity* and must be chartered under the laws of incorporation. A **firm** is *a partnership of two or more persons*; it is usually unincorporated and is not a legal entity.

> CORRECT: The **company** is moving its headquarters next month. (For the use of the singular verb and singular pronoun, see "Company Names," Chapter 6.)
> Although he became president of his father's **concern**, he always regretted that he was in trade rather than in a profession.
> The **corporations** sued each other for more than twenty years.
> The union had to sue its lawyer, not his **firm**, because the **firm** was not a legal entity, but a partnership.

490

## compare, contrast

**Compare** is used when *similarities* or *similarities and differences* are presented. **Contrast** is used when *only differences* are presented.

> CORRECT: She **compared** the advantages and disadvantages of the equipment and found general similarities except in price.
> The report **compared** the precognition of children prone to accidents with that of children who had had no accidents.
> He **contrasted** the reactions under extreme heat and extreme cold.

## compare to, compare with

**Compare to** means *to note similarities in persons or objects of different categories*. **Compare with** means *to find or examine similarities and differences in objects or persons in the same category*.

> CORRECT: He **compared** the profits **to** a famine.
> The weatherman **compared** the spring storms **with** the summer storms.

## complementary, complimentary

**Complementary** means *serving to fill out* or *to complete* or *to supply each other's needs*. Complimentary means *favorable, flattering*, or *free as a courtesy*.

> CORRECT: The typists' manual and the guide to business letter writing are good **complementary** pamphlets.
> With the proposal she submitted **complimentary** letters written by well-known experts.
> He offered the visiting managers **complimentary** sightseeing tours of the city.
> Each luncheon guest received a **complimentary** bottle of perfume. (A *complimentary gift*, like a *free gift*, is redundant and should be avoided.)
> The **complimentary** closings of business letters come to us from a more formal and perhaps more hypocritical age.

## compulsion, compunction

Confusion of these nouns arises from the spelling; the meanings are completely different. **Compulsion** is *a strong impulse to act, coming from an external or an internal cause*: "Her **compulsion** to retire immediately was irrational." "Her **compulsion** to avoid parties was due to her parents' withdrawal from society." **Compunction** is *distress colored by a degree of guilt; it is felt for someone else, not for oneself*: "His **compunction** at her losing the promotion was easily offset by his glee at securing the position for himself."

## compulsory, mandatory

There is a difference in the strength of these two words: **mandatory**, the weaker of the two, is close to *obligatory*; **compulsory** means *enforced*.

> CORRECT: **Mandatory** arbitration never occurred between that company and the union until after the court had made two settlements **compulsory**.

## compunction

See **compulsion**.

## concave, convex

**Concave** means *arched* or *rounded inward like the interior of a bowl*; **convex** means *rounded like the exterior of a sphere*.

> CORRECT: The **concave** pavement caused the car to overturn.
> The **convex** addition to the building looked like a boil.

## concern

See **company**.

## condominium, condominiums

**Condominia** is an incorrect plural; the correct plural is **condominiums**. (See also "Foreign Singulars and Plurals," Chapter 6.)

## confer, conference

**Confer** is a verb; **conference** is a noun.

> CORRECT: The managers will **confer** at three o'clock (not "The managers will *conference* at three o'clock"). The **conference** will end at two o'clock.

## connotation, denotation

See Chapter 10.

## consul, council, counsel, councilor, counselor

A **consul** is an *official appointed by one country to represent it in another*; a **council** is a *group serving as an advisory or legislative body*; a

**counsel** is a *lawyer engaged in a court case* or a *lawyer advising an individual client or a corporate or public body*; **counsel** also is *advice*.

> CORRECT: The French **consul** discussed the new import taxes.
> The manager's **council** advised delay.
> Your **counsel** on my promotion was timely and helpful.
> The corporation's **counsel** is addressing the jury.

A **councilor** is a *member of a legislative or advisory body*; a **counselor** is *one who gives advice* or a *lawyer who manages a client's cases in court*.

> CORRECT: The **councilor** gave three suggestions to the president and promised to vote for any one of them.
> The **counselor** in the personnel department suggested that Joe consult his doctor about his problem.

### contact

The verb **contact**, which is vague, met with great opposition when it was introduced; but it is now accepted as colloquial. It is used mainly in business writing and speech. The more specific *write, call, cable* carry more meaning. (See "Common Use of Vague and Old-Fashioned Terms in Business," Chapter 7.)

### continual, continuous

**Continual** means **close recurrence**, as in "**continual** showers." **Continuous** means uninterrupted, as in "**continuous** rain."

### content, contents

**Content** and **contents** used as nouns have the same meaning but different degrees of formality, **content** being the more formal. **Content** usually connotes a *concept* or *idea*; **contents** may refer to *ideas* or to *something more tangible*.

> CORRECT: The **content** of the eulogy had been carefully selected.
> The **contents** of his briefcase were spread on the table.
> The **Table of Contents** was annoyingly incomplete, but it did lead us to expect some analysis of **content**.

### contrary, converse, opposite, reverse

**Reverse** means *contrary, converse, opposite*. **Converse** means *a statement reversed* or *a statement with subject and predicate interchanged*. (1) "Most excellent mathematicians are engineers" is the **converse** of (2)

"Most engineers are excellent mathematicians." (3) "Most engineers are not excellent mathematicians" is the **contrary** of the preceding sentence, and (4) "Most excellent mathematicians are not engineers" is the **contrary** of Sentence 1. Sentence 4 is the **reverse** of Sentence 1; and Sentence 3, the **reverse** of Sentence 2. **Contrary** and **opposite** denote a *definite contrast* or *irreconcilable ideas*.

### contrast

See **compare**.

### convex

See **concave, convex**.

### cop, officer, police

**Police** dislike being called **cops**, probably because the word has unfortunate associations, such as *copping* (*stealing*) and *copping a plea* (*plea bargaining*). **Police** is, therefore, a better term. For direct address **officer** is satisfactory because the **police** are **officers** of the law, and **officer** applies to either gender.

### corporation

See **company**.

### cupfuls, spoonfuls

Although an *s* indicating the plural was formerly inserted after the noun to which *ful* was added ("cupsful"), present usage places the *s* after *ful*.

The suffix *-ful* has one *l*; *full* as a separate word has two *l*'s: "a cupful, a full cup."

# D

### dais, lectern, podium

A **dais** is a *raised platform* on which speakers, officers of an organization, and others important to a meeting, such as guests of honor, may sit or stand during programs. A **podium** is a **dais** *that accommodates one person*, such as a lecturer or the conductor of an orchestra. A **lectern** is the *stand for a speaker's notes and for electronic equipment*.

CORRECT: Nervous as she was, she clung to the **lectern** with both hands.
The **dais** was too small to accommodate all the officers and speakers.
When the **podium** collapsed under his weight, the conductor good-naturedly
climbed on the step stool that the maintenance worker offered.

## data, datum

See "Foreign Singulars and Plurals," Chapter 6.

## deductive, inductive

See "Choosing Induction or Deduction," Chapter 3.

## definite, definitive

**Definite** and **definitive** mean *clear, unambiguous, having distinct limits*; but **definitive** also means *authoritative, not subject to change, apparently exhaustive.*

CORRECT: He demanded a **definite** answer.
He set **definite** standards for the new employees in his department and
detailed them clearly during employment interviews.
Her **definitive** study of that tribe helped new graduates to appreciate true
scholarship.
These **definitive** laws have lasted many centuries.

## degreed

**Degreed**, a favorite word of some personnel departments and of some
academics, should be replaced by simple, familiar terms.

POOR: She is **degreed** in chemistry.
IMPROVED: She has **a degree** in chemistry.
POOR: He is not **degreed** in educational methods and should not be allowed
to teach.
IMPROVED: He does not have **a degree** in educational methods. . . .

## denotation

See "Denotation," "Connotation," Chapter 10.

## deprecate, depreciate

To **deprecate** is *to register disapproval*; to **depreciate** is *to reduce in
value*. The distinction between the two words, however, is being ignored
by some educated writers.

495

CORRECT: He **deprecated** the time spent on the plans because there was no guarantee that they would be accepted.

The instability of the market **depreciated** the value of the stocks he had reserved for his retirement.


## depression, recession

These terms from economics have different meanings, but in times of **depression** or **recession** it is hard to get economists to agree on the exact difference between the two. A **depression** is a period of low general business activity accompanied by rising levels of unemployment. A **recession** is a brief period of slump in the economy during a time of generally prosperous business activity. **Recessions** have occurred during the recoveries from **depressions**.

CORRECT: She had to leave college when her father lost his job during the **depression**.

He said that she had suffered during the **depression** but had felt no effects during the **recession**.

Much depends on who suffers and how much; one person's depression may be another's recession.


## develop, discover, invent

To **develop** is *to change the state of something, usually to a better or more advanced condition.* To **discover** is *to find for the first time what was already there.* To **invent** is *to create or to produce for the first time something useful.*

CORRECT: Oil was **discovered** on the Texas property.

As soon as he had **invented** this design, his company **developed** it.

She **invented** two locking caps and applied for patents.


## diagnosis, prognosis

A **diagnosis** of a disease is *recognizing it from signs and symptoms*; a **prognosis** is a *forecast of the prospect of recovery* from the disease.

INCORRECT: The doctor **diagnosed** her patient as a case of measles. (A doctor **diagnoses** the disease, not the patient.)

CORRECT: After the doctor had received the report, she **diagnosed** the lesions on the neck as skin cancer.

She explained the type of cancer in detail, spoke of the patient's youth and strength, and offered an encouraging **prognosis**.

## die of, die from

The correct preposition is **of** when *die* means *to expire*.

> CORRECT: He **died of** starvation.
> She **died of** a cancer that had been neglected.

*Death*, however, may be followed by **from**.

> CORRECT: His **death from** pneumonia was unexpected.

## different from, different than

**Different from** is used to introduce a phrase: "I must speak carefully because on this point my manager's opinion is **different from** his supervisor's." **Different than** may be used before a clause: "After listening to my supervisor's pragmatic discussions, I found my manager's philosophical comments **different than** I had anticipated (or **different from what** I had anticipated).

## differ from, differ with

When **to differ** means *to be unlike*, the preposition **from** is used; when **to differ** means *to disagree*, the preposition **with** should be used.

> CORRECT: He **differs from** his sister so little that they seem to be identical twins, but they **differ with** each other so often and so angrily that one begins to suspect that like hates like.

## direct, directly

**Directly** is used ambiguously in many sentences; it means *immediately, straightforwardly,* or *person to person (no intermediary)*. These meanings frequently cause confusion in business writing:

> CORRECT: He will speak to the staff **directly** about that problem (*immediately* or *without an intermediary?*).
> He will go home **directly** after the meeting (*immediately* or *without stopping anywhere?*).
> Answer me **directly** (*immediately, frankly,* or *without involving anyone else?*).

## discomfit, discomfiture, discomfort

To **discomfit** is *to thwart* or *to disconcert*. (Its archaic meaning is *to rout in battle*.) To **discomfort** is *to make uncomfortable or uneasy*. The difference in meaning or degree is also true of the nouns.

497

CORRECT: He **discomfited** the rival candidate, leaving him speechless.
She was so **discomforted** by the vice president's remarks that she spilled her wine.

### discover

See **develop**.

### discreet, discrete

**Discreet** means *not revealing knowledge, careful, showing good judgment*; **discrete** means *constituting a separate entity; consisting of distinct, unconnected elements*.

CORRECT: The supervisor warned the guard to be **discreet** in watching for theft.
She compared the reproduction of the **discrete** sounds with the reproduction of the continuous sounds.

### disinterested, uninterested

A **disinterested** person is *unbiased* and *has no selfish interest in the matter being considered*. One who is **uninterested** does not have his or her mind or feelings concerned in the matter. **Disinterested** is also used, especially in the United States, to mean **uninterested**. (See "Changing Meanings," Chapter 10.)

CORRECT: The judge, who had invested most of his savings in the corporation on trial, could hardly be described as **disinterested**.
A completely **disinterested** jury is ideal but may be impossible to find for some trials.
Unfortunately the routine and delays of a court may soon leave even a promising jury **uninterested**.
Some **uninterested** students complain that their professors are boring.

### divorce, was divorced, was granted a divorce, obtained a divorce

"Joseph Andrews **was divorced** on July 1, 1980" implies that Andrews was the defendant in the case. "Joseph Andrews **was granted** or **obtained a divorce**" indicates that Andrews was the plaintiff. For divorces with no plaintiff or defendant relationship, sometimes called *uncontested* or *no-fault* divorces, the record or news may state, "The **divorce** of Helen and Joseph Andrews was granted on July 1, 1980.

## double negative

One type of **double negative** is substandard and should be avoided: "He doesn't have no book." Another type is the figure of speech called *litotes,* in which a negative term is negated to express a limited positive thought: "The data were not unimportant; the committee was not unfriendly." Litotes is useful for expressing certain shades of meaning that the negative or positive alone does not express, but the figure can easily become monotonous. (See "Litotes," Chapter 10.)

## dozen, dozens; hundred, hundreds; million, millions

If a number is stated, the singular (**dozen, hundred**) is used: "forty-nine **dozen** flasks, twenty **million** dollars." If no number is stated, the plural is correct: "**dozens** of flasks, **millions** of dollars, **hundreds** of people." An exception is "a **dozen** or the **million**," where the direct or indirect article replaces a number.

> CORRECT (exception): A **dozen** pencils is enough.

# E

## eager

See **anxious**.

## economics

**Economics** is a singular noun.

## effect

See **affect**.

## e.g., i.e.

These abbreviations for Latin phrases are frequently confused. Spelling them out in English (**e.g.**, "for example," and **i.e.**, "that is") is strongly recommended. (See also **that is**.)

## either, neither

**Either** and **neither** are used for only two choices. (See "Singular Indefinite Pronouns as Subjects," Chapter 6.)

> CORRECT: **Either** her interpretation or the tests themselves were at fault.

## elicit, illicit

**Elicit**, a verb, means *to draw forth*; **illicit**, an adjective, means *not permitted*.

> CORRECT: On the basis of the knowledge and training he offered and of the enthusiasm he **elicited**, they decided to employ him.
>
> His **illicit** sales of foreign currency led to the investigation.

## elude

See **allude**.

## emigrate, immigrate, migrate

To **emigrate** means *to leave one's country or the place where one lives in order to take up residence elsewhere*. To **immigrate** is *to enter a country of which one is not a native in order to live there*. To **migrate** is *to move from one country, place, or locality to another* (sometimes periodically).

> CORRECT: They were on a list of those waiting to **emigrate** from Armenia, but they never received permission to leave.
>
> It was a little easier for this group to **immigrate** to the United States because the members had relatives living in states with warm climates.
>
> In Old Testament times many tribes **migrated** for pasture for their animals.

## eminent, imminent, immanent

**Eminent** means *distinguished, outstanding*; **imminent** means *about to happen, threatening*; **immanent** is not a misspelling of imminent but a word meaning *having existence or effect only within the mind or consciousness, inherent*.

> CORRECT: The **eminent** scholar expected to win the Nobel Prize.
>
> The **imminent** storm caused the speaker to end her Commencement address abruptly.
>
> His busy life never even suggested his **immanent** plans.

## enormity, enormousness

**Enormity** means *monstrous wickedness* or *outrageousness*; **enormousness** means *immensity or vastness*.

> CORRECT: The **enormity** of substituting lying for reporting alienated the public.
>
> The **enormousness** of the proposed land grants surprised many business people.

**ensurance**

See **assurance**.

**equal**

See **more equal**.

**equally as good as**

The phrase **equally as good as** (or **equally as well as**) contains an unnecessary **equally**.

> CORRECT: Her editing is **as good as** her supervisor's.
> His reports were written **as well as** those of anyone in the department.

**Equally** is used when only one of the persons or things being compared is mentioned in the comparison.

> CORRECT: His father plays **equally** well.
> The new machine is **equally** slow.

**Esq., Esquire**

**Esquire** and its abbreviation **Esq.** are permissible only when no other titles are used. They follow the names of males and in the United States the names mainly of lawyers: "John Doe, Esq.; John Doe, Esquire," not "Mr. John Doe, Esq." The title, spelled out or abbreviated, is used mainly in addresses or in lists of names.

**etc., et cetera**

**Etc.**, the abbreviation of **et cetera** (*and others of the same kind* or *and so forth*) should be avoided; the American phrases *and so on, and others, and so forth* are suitable.

In expressions introduced by *for example, in addition*, and *such as*, **etc.** is not only redundant (at best) but silly (at worst): "Most tropical fruits— for example, mangoes, bananas, papayas, etc.—grow here" should be "Most tropical fruits—for example, mangoes, bananas, papayas—grow here" or "Mangoes, bananas, papayas, and other tropical fruits grow here." (See also "Tautology," Chapter 13.)

**everybody, everyone, somebody, someone**

These take singular verbs. (See Chapter 6.)

### every day, everyday

**Every day** is used for each day in succession: "It rained **every day** of her vacation." **Everyday**, an adjective, means *common, ordinary*: "The **everyday** tasks of a section head bored her."

### except

See **accept**.

### exhibit, exhibition

An **exhibition** is *a public showing*; an **exhibit** is *something shown in an exhibition*.

> CORRECT: The **exhibition** of early airplanes offered drawings, photographs, and mock-ups.
> These **exhibits** interested the adults, but the children preferred what they called "the button exhibit," where they pressed buttons to make **exhibits** rumble, screech, and bang.

### explicit, implicit

**Explicit** means *expressed directly or concisely or both*; **implicit** means *expressed indirectly or implied*.

> CORRECT: The eye surgeon's **explicit** discussion of macular degeneration was aided by simple drawings.
> The inadequacy of the known treatment was **implicit** in her discussion.

### explode, implode

The transitive verb **explode** means *to cause to burst*: "The gang threatens to **explode** a bomb." The transitive verb **implode** means *to cause to collapse inward*: "The blow **imploded** the tube." The intransitive verb **explode** means *to burst forth from internal energy*: "The plastic bomb **exploded**." The intransitive verb **implode** means *to collapse inward, to undergo violent compression, to become greatly reduced as if from collapsing*: "In this science fiction story hundreds of stars **implode**."

### exploit, explore

To **exploit** is *to use someone or something for gain*, often with the implication of selfishness; to **explore** is *to search for something* or *to examine*.

502

CORRECT: In an effort to attract tourists she hopefully **explored** the entire region and then **exploited** her findings, meager as they were.

## extinct

See **absolute adjectives**.

# F

## farther, further

A careful writer uses **farther** for spatial distance and **further** for other meanings: "She could throw a basketball **farther** than any other member of the team, but the coach kept insisting on **further** practice during her lunch hour." **Further** is also a transitive verb meaning *to promote* or *to advance*: "He **furthered** her career by introducing her to several influential people."

## fatal

See **absolute adjectives**.

## faze

See **phase**.

## feasible, possible

**Feasible** means *capable of being done*; **possible** means *capable of happening*.

CORRECT: Wrapping the pipes is a **feasible** solution to the problem of condensation.
It is **possible** that it will rain before he has wrapped all the pipes.

## feet, foot, hour, hours, inch, inches, mile, miles

It is correct as well as logical to write "a room twenty feet long," "a projection of ten inches," "an examination of six hours," "a marathon of thirty miles"; it is also correct to use the singular when the measurement precedes the noun modified: "a twenty-foot room," a "ten-inch projection," "a six-hour examination," "a thirty-mile marathon." (See "The Hyphen," Chapter 8.)

### feminist

Anyone who believes that women should have the same economic, political, and legal rights as men is a **feminist**; therefore the term may be used for males or females.

> CORRECT: She had always considered him a competent physicist, but when he rose at the NOW meeting to speak of women's rights, she considered him a **feminist** without peer.
> The women refused to be **feminists** for fear of ridicule.

### Feminist Movement, National Organization of Women, Women's Liberation Movement, Women's Movement

The generally acceptable terms for the groups that support equal rights for women follow: **NOW** is the largest official organization; **Women's Movement** is preferred to **Feminist Movement** by some feminists. *Women's Lib, Women's Libbers, Fem Lib, Fem Libbers* are slang terms of derogation and should be avoided.

### fever, temperature

A **fever** is a rise in body temperature above 98.6°F. **Temperature** is the degree of heat or cold measured on a definite scale, such as Fahrenheit or Celsius.

> CORRECT: His **fever** rose four degrees in two hours.
> According to this room thermometer, the **temperature** of the office is ten degrees too low.
> He measured the girl's **temperature** to see whether she had a **fever**.

### few

**Few**, which is singular, is preceded by *a* even though it requires a plural verb and a plural relative pronoun: "Only a **few** were irregular in their markings." The adjective **few** is followed by a plural substantive: "A **few** people accept her ideas."

### fewer, less

Although colloquially **fewer** and **less** are used to mean *not many*, in writing **fewer** is used for *items that can be numbered* and **less** for *quantity*: "**Fewer** than twenty examples," "**less** paper," "**fewer** pages," "**less** milk," "**fewer** customers," "**less** bureaucracy," "**fewer** offices."

## figuratively, literally

Some people err by using **literally** in attempts to intensify figures of speech or to prevent them from being taken literally: "When the manager saw the expense account, he **literally** exploded." (The sentence is intended to mean **figuratively** exploded. Removing **literally** will correct it.)

## finalize

Most writers reject **finalize** as substandard; furthermore the meaning is often misleading. For example, does it mean *to complete the work* or *to receive final approval*?

## firm

See **company**.

## first, firstly, second, secondly, third, thirdly . . .

American usage shows a marked preference for **first**, **second**, **third**. **Firstly**, in particular, sounds pretentious, too fancy, or too heavy.

> PREFERRED: **First**, we measured the ingredients; **second**, we prepared the slurry.

## fish, fishes

**Fish** is the usual plural; **fishes** is used when different species are meant.

> CORRECT: She will show us slides of the **fishes** we identified and studied in her course, which covered at least a hundred major species.
> The contaminated **fish** were removed from the shallows.

## flair, flare

Many writers confuse these words. **Flair** means *ability* or *inherent aptitude* or *aptness*; the noun **flare** is *a light signal* or *a bursting of flame or emotion* or *a widening at the bottom*; the verb **flare** has the same meaning as the noun.

> CORRECT: He had a **flair** for arriving just when he was wanted.
> Her **flair** for decorating developed into a career.
> The police placed **flares** at the scene of the accident.
> When he placed the steak on the grill, the coals **flared**.
> Her **flare** of annoyance lasted only a minute.
> Her **flared** skirt made it easy to stride.

## flammable, inflammable

Insurance companies have long urged the use of **flammable** for *likely to burn*, a use designed to prevent accidents to those who might mistake **inflammable** for *unable or unlikely to burn*. This is a clear case of safety first, but **inflammable** still appears with the meaning *likely to burn* and probably will continue to do so.

In factories, in instructions, and in laboratories it is important to use one form consistently so that workers will not be confused. (**Flammable** is now commonly seen on tankers, trucks, and other vehicles.)

## flaunt, flout

**Flaunt** means *to display boastfully or ostentatiously or contemptuously.* **Flout**, which is often misused, means *to scoff at* or *to treat with contempt.*

> CORRECT: He **flaunted** his lateness by entering noisily and asking questions that had already been answered.
> She **flouted** the rules so many times that there was no space on her record to list new examples of her contempt.

## flounder, founder

The verb **flounder** means *to thrash about, to move clumsily*; the verb **founder** means *to fail, to collapse,* or, *in the case of a ship, to become filled with water and then to sink.*

> CORRECT: Instead of answering the question, the witness **floundered** in his chair.
> The bank **foundered** because some of its officers were dishonest and the rest, ineffective.

## foot

See **feet**.

## forceful, forcible

**Forceful** means *effective, strong*; **forcible** means *characterized by force or power.*

> CORRECT: Her arguments for continuing the research grant were **forceful**.
> The police could find no signs of **forcible** entry.

## formally, formerly

**Formally** means *ceremonially, according to accepted practice, complying with conventional procedures.* **Formerly** means *just before* or *previously.*

CORRECT: He was **formerly** mayor of a small city.
She presented her resignation coldly and **formally**.

## former, latter

**Former** refers to the first of two; **latter**, to the second of two.

> CORRECT: Two criteria for selecting these designs are minimum aberration and optimal moments, but he relied entirely on the **former** because he did not completely understand the **latter**.

**Former** and **latter** are usually restricted to legal and other formal writing because they have a heavy effect. In all styles of writing, words like *the first* and *the second* or *one* and *the other* are appropriate.

> CORRECT: The speakers were Alyce Alexander and John Pope. The *first* discussed the background and defined the problem; the *second* reported the experimental procedure and the results.

When more than two are designated, **the former** or **the latter** should not be used. *The first named* (or *listed*) or *the last named* (or *listed*) is correct for a reference to members of a group.

## formerly

See **formally**.

## founder

See **flounder**.

## from . . . through, from . . . to

"She served as acting chair **from** January first **through** April first" means that she served on April first. "She served **from** January first **to** April first" means that she ended her service on March thirty-first. (See also **between**.)

## -ful

See **cupful**.

## further

See **farther**.

507

# G

### gender, sex

The term **gender** denotes whether a noun or pronoun is masculine, feminine, or neuter in grammatical use; it does not denote the sex of the person or thing referred to by the noun or pronoun. In German, for example, *das Mädchen* (*the girl*) is neuter gender; in French *la plume* (*the pen*) is feminine gender. English does not denote gender by the use of articles, like *das* or *la*, or by variations in case endings. In fact, gender lingers mainly in the personal pronouns *he, she, it.*

*Sex* is the word commonly used to refer to male and female. Insurance forms and medical questionnaires ask for one's sex, not gender.

### girls

See **boys**.

### got, gotten

Writers of British English use **got** as the past participle of *get*. In American English both **got** and **gotten** are used.

### graduated

The following examples use **graduated** correctly:

> He **was graduated** from college a year before his sister.
> She **graduated from** college a year before her sister.
> Wollmer College **graduated** him a year before his sister.

The incorrect form "He **graduated** Wollmer College" should be avoided.

### grateful, gratified

**Grateful** means *thankful*; **gratified** means *satisfied* and implies the fulfillment of an expectation or wish.

> CORRECT: He was **grateful** for her help with his report.
> She was **gratified** that her report led to a change in a faulty company policy.

# H

### hanged, hung

In careful writing **hanged** is used for suspending a being by the neck until dead. **Hung** is used for inanimate objects.

CORRECT: In the seventeenth century people were **hanged** for many crimes that today would merit only brief imprisonment.

When she **hung** her first picture in her new apartment, the nail went through the wall into her neighbor's apartment.

## harbor, port

**Harbor** and **port** are nearly synonymous, but a **port** must have facilities for loading and unloading.

CORRECT: It was a safe **harbor**, but as it had no **port** facilities, the ship had to go farther.

## hardly, scarcely

**Hardly** and **scarcely** have negative meanings; they should not be allowed to attract unnecessary negatives. The following constructions are often heard but are incorrect: "They couldn't **hardly** see the slide." "**Hardly** nothing appealed to his appetite." "The slide was prepared without **scarcely** any checking of the data." "He spoke without **scarcely** any preparation."

CORRECT: They could **hardly** see the slide.
**Hardly** anything appealed to his appetite.
The slide was prepared with **scarcely** any checking of the data.
He spoke with **scarcely** any preparation.

## he, his

See **its**.

## hereby, herein, hereto, herewith

None of these are necessary; they add legal pomposity and stuffiness.

## hopefully

**Hopefully**, an adverb, means *in a hopeful way*: "She looked **hopefully** at the doctor." Recently **hopefully** has been used and overused in such senses as *I hope, we hope, it is hoped that*: "Hopefully he will come for the holidays." Purists and many other writers reject this use, sometimes forcefully. The sensible recommendation is to avoid the second use in writing and to avoid its overuse in familiar speech. (See also "Vague Words," Chapter 10.)

509

**hour, hours**

See **feet**.

**hundred, hundreds**

See **dozen.**

# I

**I, me**

Courtesy requires that pronouns in the first person be placed last when used with other nouns or pronouns designating people.

> CORRECT: He gave the reports to my assistant and **me**.
> Drs. Kornfeld, Smith, Appleton, and Jones and **I** visited the plant.
> The degrees were awarded to Mary Lambert, James Sullivan, James White, and **me**.

(See also "The Evasive I" Chapter 9.)

**I, my, mine**

See **its**; "The Evasive *I*," Chapter 9; and "The Pronouns *I, We, You*," Chapter 14.

**i.e.**

See **that is** and **e.g.**

**if**

See **provided**.

**if, whether**

**If** introduces one possibility; **whether** introduces two. **Whether** is sometimes confused with **if**. "Let me know **if** you are coming" means that you need not let me know if you are not coming. "Therefore he could not decide **if** the new regulations were in effect" is poor because the writer meant **whether**. The sentence "Please notify us **if** this statement of your expenses is incorrect" means that notification is unnecessary if the statement is correct. "Please let us know **whether** this statement of your expenses is incorrect" is a request for an answer in either case.

## illicit

See **elicit**.

## immanent

See **eminent**.

## immigrate

See **emigrate**.

## imminent

See **eminent**.

## impeach, impeachment

**Impeachment** does not mean *trial, conviction,* or *removal from office.* It does mean *the bringing of charges* or *a challenge to credibility or validity.*

> CORRECT: Andrew Johnson was **impeached** but was cleared of charges and remained in office.
>
> The **impeachment** of the witness by the prosecution did not result in a conviction or a trial.

(The adjective **unimpeachable**, meaning *free of discredit, unassailable,* is commonly used: "He could not be attacked because his conduct was **unimpeachable**.")

## implicit

See **explicit**.

## implode

See **explode**.

## impossible

See **absolute adjectives; feasible**.

## imply, infer

**Imply** means *to suggest* or *to say indirectly.* **Infer** means *to deduce from statements or other information.*

511

CORRECT: The manager **implied** that the chemistry project was costly and impractical.
The chairman of the chemistry project **inferred** that the manager wanted to use the remaining project funds for other purposes.

## impracticable, impractical

See **practicable**.

## inch, inches

See **feet**.

## incredible, incredulous

**Incredible** means *unbelievable*; **incredulous** means *skeptical* or *unbelieving*: "Her teaching was **incredible**" does not state whether the teaching was too good or too poor to be believed. "He was **incredulous** when they reported seeing an unidentified flying object."

## inductive

See "Choice of Induction or Deduction," Chapter 3.

## infer

See **imply**.

## inflammable

See **flammable**.

## ingenious, ingenuous

**Ingenious**, a praiseful word, means *having special ability to discover, to invent, to contrive* or *characterized by resourcefulness and cleverness, often in design*. **Ingenuous**, although it once had an equally favorable connotation, now means *having childlike simplicity, lacking subtlety.*

CORRECT: Every one of his twenty-nine patent ideas was **ingenious**.
Her response to praise of her report was **ingenuous**: "The whole department helped me; that's why it's good."

## in one's behalf, on one's behalf

See **behalf**.

512

**in re**

See **re**.

**in regard to, in regards to, as regards, with regard to**

**In regard to** and **with regard to** (meaning *in reference to*) are correct. Adding an *s* to *regard* in these phrases is incorrect. But the *s* is used in **as regards** (concerning), **best regards, regards**. See **prepositions**.

**inside of, outside of**

The **of** should be omitted for the meaning *within* or *beyond a boundary*. "The information he requested was outside of the scope of his investigation" should be "The information . . . was **outside** the scope of his investigation" and "The mixture is inside of the tank" should be "The mixture is **inside** the tank."

**insurance**

See **assurance**.

**in the matter of**

See **prepositions**.

**invent**

See **develop**.

**irregardless, regardless**

**Irregardless** is a double negative: **ir-** and **-less** both negate. **Regardless** is the correct word.

**is when, is where**

**Is when** and **is where** should be avoided. (See "Is When, Is Where, Is Because," Chapter 6.)

> INCORRECT: Noon **is when** he felt the pain and collapsed.
> IMPROVED: He felt the pain and collapsed at noon.
> INCORRECT: The office **is where** his strange behavior was noticed first.
> IMPROVED: His strange behavior was noticed first in the office.
> INCORRECT: A violation **is where** you tear pages from library books.
> IMPROVED: Tearing pages from library books is a violation.
> INCORRECT: Silent gallstones **is when** they have never caused symptoms.

IMPROVED: Silent gallstones are those that have never caused symptoms.
Gallstones that have never caused symptoms are called "silent gallstones."

### its, it's

**Its** is the possessive case of the pronoun *it*. **It's** is the contraction of *it is*. Writers should be careful not to confuse them. English personal and relative pronouns do not take an apostrophe to show possession:

PERSONAL PRONOUNS

| Nominative | Possessive |
|:---:|:---:|
| I | my, mine |
| you | your, yours |
| he | his |
| she | her, hers |
| it | its |
| we | our, ours |
| they | their, theirs |

# J

### job, position, post, situation, occupation

These words have essentially the same meaning, the difference being mainly in connotation. **Job** was once considered colloquial but now is generally used to describe any work or the specific duties of an employed person. **Position** is a more formal term and is generally used in applying for employment: "I am interested in the **position** of programmer." **Situation** is rather old-fashioned, although it remains in classified advertisements: "**Situations** Wanted." **Post** meaning position is used mainly in British English. It is used in the United States to designate a diplomatic assignment: "His first **post** was Assistant Consul in Bern." **Occupation** signifies an activity that may or may not be compensated: doctor, student, homemaker, carpenter. Information on one's **occupation** is often requested on forms for personal data.

### judicial, judicious

**Judicial** means *related to the law*; **judicious** means *wise, just.*

CORRECT: She explained that the change in permissible lead levels stemmed from a **judicial** decision.
His **judicious** examination of the proposal spared us many mistaken conclusions.

514

# K

## karat

See **carat**.

## kind of, rather, somewhat

**Rather** and **somewhat** are the words to use when they are meant.

> CORRECT: The patient was **rather** tired (not *kind of*).
> The test was **somewhat** tiring (not *kind of*).

## kind of a, sort of a

The *a* in **kind of a, sort of a** should be removed: "that **kind of** pill, this **sort of** answer."

# L

## last, latest

**Last** suggests *final*; **latest** means *most recent*.

> CORRECT: The **latest** biographer suggests that Doc's powers were waning during his long terminal illness.
> His **last** wish was to face death alone.

## latter

See **former**.

## lay, lie

**Lay** is a transitive verb; **lie**, an intransitive verb. **Lay** is generally used by itself; **lie** nearly always is followed by a preposition or an adverb: "**lie** *on the bed*, **lie** *in the grave*, **lie** *at the door*, **lie** *down*." Unfortunately, the past tense of **lie** is spelled *lay*; therefore a writer must be careful not to confuse readers.

> CORRECT: Each morning he **lays** the equipment on the benches as soon as he arrives.
> She **lay** on the lawn for two hours yesterday in spite of the work waiting for her inside.

The equipment **had been lying** on the bench for two days before he put it away (participle of **lie**).

The treasure **had lain** under the sea for three hundred years before it was located (participle of **lie**).

They **are laying** a pipeline from the port to the plant (participle of **lay**).

They **laid** the first pipeline fifty-three years ago (past tense of **lay**).

They **had laid** the equipment on all the benches before he decided to change the work order (participle of **lay**).

## lead, led

**Lead** (pronounced with the long *e* of *see*) is the present tense, and **led** is the past tense of the verb meaning *to guide, to direct, to go in advance*. **Lead**, the metal, although pronounced like the past tense *led*, should not be confused with it. The noun and adjective **lead** (pronounced like the present tense) have a number of meanings; those most commonly used relate to the verb.

> CORRECT: His imagination often **leads** him to places that he has never visited before.
> She **led** the research team.
> **Lead** shields against radioactivity.
> He was the **lead** in the play that failed.
> The **lead** in this news article is intended to shock.
> The **lead** horse is lame.

## leave, let

**Leave** and **let** do not have the same meaning. "**Leave** me alone" is a request for solitude; "**let** me alone" means *do not disturb or interfere with me*. **Leave** also means to *deposit*.

> CORRECT: She advised him to **leave** the report on the desk and **to leave** the busy writer alone in the office.
> The baby-sitter finally insisted, "You must **let** me alone while I am feeding the baby."

## lectern

See **dais**.

## led

See **lead**.

## less

See **fewer**.

**let**

See **leave**.

**libel, slander**

Although in American usage *false defamatory written statements or presentation in pictures* is always called **libel, slander** refers to such statements made *orally*. **Libel**, however, may be used for oral statements, also, but **slander** does not refer to written material. In common use in Britain **libel** and **slander** are, according to Fowler, synonyms meaning "a deliberate, untrue, derogatory statement, usually about a person."[2]

**libelous**

See **absolute adjectives**.

**lie**

See **lay**.

**lightening, lightning**

**Lightening** is the present participle of the verb *to lighten*, meaning *to remove a burden* or *to cheer or gladden* or *to make or become brighter*. **Lightning** is *the flashing light produced by atmospheric electricity discharging*; the adjective **lightning** means *with the speed and suddenness of* **lightning**.

> CORRECT: After a depressing meeting his antics were aimed at **lightening** our spirits.
> His **lightning** decisions about hiring and firing were criticized by the other supervisors.
> The **lightening** sky promised a beautiful day.

**like**

See **as**.

**literally**

See **figuratively**.

**loath, loathe**

**Loath**, the adjective, means *reluctant because acting would conflict with one's predilections, opinions, or liking*. **Loathe**, the verb, means *to detest*.

517

CORRECT: Because of his experience in the clinic, he was **loath** to take the drug that his doctor had prescribed.

She **loathed** everyone who smuggled or sold drugs, and she despaired when such an offender was given a light sentence.

### lot, lots

Although **lot** and **lots** (meaning *much* or *a great amount*) are marked *standard* in some dictionaries, many readers object to their use except in informal speech: The boy said, "There are **lots of** mistakes in my paper. I should have spent **a lot** more time on it."

INCORRECT: Among the **lots** of treatments for arthritis, the two under discussion are used **a lot**.

IMPROVED: Among the many treatments for arthritis . . . are used frequently.

### lunch, luncheon

**Lunch** and **luncheon** are two names for a midday meal. **Lunch** suggests more informality; even the sandwich eaten at a desk may be called **lunch**. **Luncheon** is used for *a more elaborate or elegant meal with some advance arrangements*. It may include speakers or other program features, such as fashion shows, raffles, or short films. Even though in this age of dieting a guest may eat just a chef's salad, this may be a **lunch** or a **luncheon** according to the atmosphere.

# M

### M

See **a.m.**

### majority, plurality

A **majority** is *a number greater than half of a total*. It is also *the amount by which this number exceeds the total remainder*.

CORRECT: He received three hundred votes, well over the simple **majority** needed. (There were 500 votes, and so 251 would have been enough to win the election.)

He won the election by a **majority** of 100.

A **majority** of the employees agreed (more than half).

A **plurality** is *the largest number of votes received by one of three or more*

518

*contestants if that number is less than a majority.* It also is *the amount exceeding the nearest number of votes.*

> CORRECT: She won the election with a **plurality** of 61. (There were three candidates and five hundred voters. She won 240 votes—not enough to be a **majority**: 251, but well over those of the next candidate, who had 179 votes.)

## make a motion, move

The correct word for proposing a measure in a deliberative body is **move**: "I **move** that the club adjourn." The proposal that is moved is known as a **motion**.

> CORRECT: (at the meeting): I **move** that the meeting be adjourned.
> (in the minutes): At five o'clock Mr. Walker **moved** the adjournment of the meeting.
> The **motion** to adjourn was passed.

## mandatory

See **compulsory**.

## mania

See **phobia**.

## marginal

When **marginal** is used to mean *small* or *little*, the writer might better use *small* or *little*. In economics **marginal** means *hardly enough for a profit* and should be used for that sense in writing on finance and related subjects.

> CORRECT: The meteorologist said that the cold front would have *little* effect in the Southeast.
> They could not continue to run the toiletries division on a **marginal** basis.

## matchless

See **absolute adjectives**.

## material, matériel

**Material** is the *substance from which something is made*; **matériel** is *supplies or equipment for an organization, usually military.*

CORRECT: The **material** of the chairs and tables was plastic of poor quality.
The **matériel** was rotting and rusting in the sheds.

## may

See **can**.

## me, myself

Some writers in the United States are so shy of the pronoun *I* that
they use **me** or **myself** where it is grammatically incorrect: "Dr. Jones,
Mr. Smith, and **me** will reorganize the records." (*I* is correct.) "The
speaker praised the community work that Tom Jones, Barbara Green,
and **myself** had done." (*I* is correct.) "He questioned the laboratory assistants and **myself**" (should be . . . *me*). (See also "The Evasive I," Chapter 9.)

## mean

See **average**.

## means (to, of, for)

**Means** in the sense of *resources* (*money or property*) is plural. **Means**
in the sense of *an agency useful to an end* may be singular or plural.

CORRECT: His **means** were not sufficient to meet the uncertainties and risks
of his business.
Every **means** for testing the drugs was considered.
Three **means** of holding attention were demonstrated.

## media, medium, mediums

**Medium** is the singular and **media** is the plural for a *channel of communication*. **Medium** is also a *person who claims to be a link of communication between the spiritual world and the earthly*; the plural of this
**medium** is **mediums**.

CORRECT: If discussion is restricted to the **medium** of radio, the candidates
must omit the **media** of television and print.
Both **mediums** wore filmy white gowns with long wide sleeves and moved
their arms languorously.

(See "Foreign Singulars and Plurals," Chapter 6.)

## median

See **average**.

## memoranda, memorandum, memorandums

The singular is **memorandum**. The Latin plural is **memoranda**; the plural formed according to the rules for English is **memorandums**.

> CORRECT: His **memorandum** on cafeteria hours was ignored (not **memoranda** unless he wrote more than one).
> She sent dozens of **memoranda** to trainees (not **memorandas**).
> His **memorandums** to his assistants never had fewer than three colors of ink.
> Her **memorandum** is ready for typing (not **memoranda**, which is not a singular form).

(See "Foreign Singulars and Plurals," Chapter 6.)

## meticulous

**Meticulous** means *overcareful* or *marked by extreme care*. It does not mean *careful* or *very careful*.

> CORRECT: He was so **meticulous** in his research that he never completed it, and all his reports were years overdue.
> His notes were so **meticulous** that they were useful to all the researchers who followed him.

## migrate

See **emigrate**.

## mile, miles

See **feet**.

## militate, mitigate

**Militate**, an intransitive verb, means *to have weight or effect against or for*. **Mitigate**, a transitive verb, means *to mollify, to alleviate, to relieve*; it suggests a moderating of something violent or painful.

> CORRECT: His unwillingness to hear the ideas of others **militated** against arbitration.
> She **mitigated** the writer's suffering by reducing her editing of his work.

521

## million, millions

See **dozen**.

## minimal, minimum

**Minimal**, an adjective, means *the least possible in size, number, or degree*; **minimum**, as noun or adjective, refers to *the smallest amount possible*.

> CORRECT: The supervisor suggested that they pay **minimal** attention to slight changes in color.
> Five hundred samples is the **minimum** required for the tests.
> She suggested that the **minimal** requirement for the tests be doubled.
> The **minimum** wage is fixed by law.

## minimize

**Minimize** means *to reduce to the least possible amount* (to a minimum); therefore **minimize** *greatly* or *considerably* is not logically possible. Verbs that may be qualified are *reduce, minify, diminish, decrease*.

> CORRECT: He was delighted that the consultant was able to decrease the dangers greatly (not "**minimize** them greatly").

## mitigate

See **militate**.

## minus

See **plus**.

## moral, morale

**Moral**, noun or adjective, refers to *right conduct or character*. **Morale**, a noun, refers to *mental or emotional condition*.

> CORRECT: His **morals** provided strength in the discussions of our competitors.
> Her **moral** tone, though never too conspicuous, did affect many of the decisions.
> The **morale** of the workers was never so low as during his presidency.

## more equal

**Equal** is an absolute adjective; **equitable** may be compared. **Equal** means *of the same measure, quality, amount, or number as another*; **equitable** means *fair dealing with all*.

INCORRECT: The union wants **more-equal** distribution of overtime work.
CORRECT: The union wants **equal** distribution of overtime work.
The union wants **more-equitable** distribution of overtime work.

## more than one

The phrase **more than one** is singular, but if **than one** is separated from **more**, a plural verb may be preferred.

CORRECT: **More than one** of the infections is serious.
He said that **more** infections **than one** were due to poor hygiene.

## mortal

See **absolute adjectives**.

## myself

See **me, -self**.

# N

## N

See **a.m.**

## namely

**Namely** should be reserved for scholarly notes and should not be used in the text of a paper.

WEAK: Two members, **namely** Joy Hill and James Hill, have volunteered.
IMPROVED: Two members, Joy Hill and James Hill, have volunteered.

## need, needs

The verb **need** is used for all persons in a negative construction. **Needs** is used for the third person singular in an affirmative construction.

CORRECT: For an absence of less than a week, an employee **need** not submit a medical certificate.
I **need** not attend the meeting.
She **needs** three million dollars.
He **needs** approval for three grants.

## neither

See **either**.

## none

**None** is correct as a singular or plural pronoun. The plural use is more common.

> CORRECT: **None** of the students have (or has) reported on a return to modernism in architecture as a reaction to postmodernism.
> None of the ideas were acceptable.

## no such a

The **a** should be omitted: "**No such** compound was known."

## nothing

**Nothing** is always singular: "**Nothing** about adhesions was mentioned until the seminar."

## nouns used as verbs

There has been a vogue lately for turning nouns into verbs even when satisfactory verbs are already in use. Among such monstrosities are *to author, to module, to chef, to Hollywood, to janitor, to gallery, to conference.* Even though the practice is standard in English (*to pilot, to doctor, to table, to man, to flower*), this is usage that writers in business and industry should not be the first to try. The practice is disliked intensely by sensitive readers. Our recommendation is to resist it.

## NOW

See **Feminist Movement**.

# O

## oculist, ophthalmologist, optician, optometrist

**Oculist** and **ophthalmologist** have the same meaning, but physicians practicing ophthalmology prefer **ophthalmologist**, which is not so easily confused with **optician** and **optometrist**. An **optometrist** is trained to test eyes and prescribe corrective lenses. An **optician** makes and sells optical supplies.

**officer**

See **cop**.

**OK, O.K., okay**

Although it began as slang, the expression **OK** is accepted now by many informal writers, who should not overuse it. Formal writers and speakers avoid it. The expression is seldom appropriate in informational prose. **OK** may be a noun, an adjective, or an adverb. All of the spellings shown are acceptable if used consistently.

> ACCEPTABLE: She refused to give her **OK** (**O.K., okay**) to the report.
> He considered increasing the budget of his division, an **okay** goal even when his company was suffering losses.
> The computer has been running **O.K.** since he repaired it.

**one (the pronoun)**

See Chapter 6.

**one or more, one or two**

These phrases require plural verbs and pronouns.

> CORRECT: **One or more** of the speeches were embarrassing to the corporation.
> **One or two** biologists working in Laboratory A were still complaining about their irritated eyes even though the Safety Committee had announced that there were no irritating substances in the air.

**only**

See "Single Words and Phrases," Chapter 6.

**on one's behalf**

See **behalf**.

**on to, onto**

**On to** is used for *as far as*; **onto** is used for *to a position on*.

> CORRECT: The expedition went **on to** Cairo.
> She was always the first to climb **onto** the wagon.

525

## opposite

See **contrary, converse, opposite, reverse**.

## optimum

**Optimum**, a noun, means the best that can be achieved within specified limitations. As an adjective **optimum** means *best* or *best possible*, either of which is to be preferred to **optimum**.

> CORRECT: In the choice of raw materials, the **optimum** often depends on a balance of quality and price.
>
> This feed will give the *best* product (or . . . *best possible product* if that is meant).

## oral, verbal

**Verbal** means *in words*, and **oral** means *in spoken words*. **Oral** (when that is meant) is more helpful to understanding than is the general **verbal**. Using *written* and *oral* is strongly recommended. **Verbal** thus is left for such communication as "a **verbal** attack," as distinguished from a physical attack, and "a **verbal** interpretation," one of words, not of substance and meaning.

> CORRECT: His **oral** report seemed better than the written one because his personality was highly persuasive.
>
> The **verbal** and photographic displays did not conceal the lack of substance in the proposal.

## ordinance, ordnance

**Ordinance** is a *decree* or *law*; **ordnance** is *military supplies* or a *divison of the army in charge of such supplies*.

> CORRECT: The **ordinance** against smoking is strictly enforced.
>
> They planned to examine the **ordnance** as soon as the audit of the books was complete.

## other

When comparison is expressed, writers should take care to use **other** so that they will not seem to be including in the comparison the object compared.

> INCORRECT: The first vice president has *more* secretarial assistants *than* any administrative officer. (This implies that the first vice president is not an administrative officer.)

CORRECT: The first vice president has *more* secretarial assistants *than* any **other** administrative officer.

When the superlative degree is expressed, **other** is not needed: "The first vice president has the *most* secretarial assistants of any administrative officer."

### outside of

**Outside of** is used colloquially for the meaning *except*. For the meaning *beyond a boundary,* **outside** is used without the preposition.

COLLOQUIAL: **Outside of** his secretary, no one in the company knew that he was leaving.
CORRECT: He stood **outside** his office to say his goodbyes.

### overlay, overlie, underlay, underlie

These verbs follow the rule for **lay** and **lie** (see **lay**). A writer uncertain of whether to use **overlay** or **overlie** can find the answer by rephrasing the sentence so that the prefix follows the verb **lay** or **lie**: "Knowing the cause that (**underlays/underlies**) the spread of the disease is very helpful in treatment" may be changed to "Knowing the cause that (**lays/lies**) under the spread of. . . ." The choice clearly is **lies** and thus **underlies**. In the same way changing "He (**overlays/overlies**) the enamel with a finish he invented" to "He (**lays/lies**) a finish over the enamel . . ." shows that the choice is obviously **overlays**. Avoid misspelling these words as –**ly** or –**lys**, instead of the correct –**lie** and –**lies**.

## P

### pair, pairs

The plural form has the *s* ending: two **pairs** of gloves, four **pairs** of pants, six **pairs** of socks. The verb that is governed by **pair** is singular if the objects are mates, plural if they are individual items:

CORRECT: This **pair** of scissors needs sharpening.
We are all leaving at five o'clock except that **pair**, who are going to their homes. (Here **pair** is considered plural because it refers to individuals. The fact that they are acting jointly does not change the number.)

### party, person

The use of **party** for **person** is a vulgarism popularized by early telephone operators: "Your **party** is ready now"; "Your **party** is on the line."

The restaurant use of **party** for one or more persons should be avoided. The word *group* is suitable in many instances. In legal documents, however, **party** for **person** is customary.

> CORRECT: Which **person** (or better, *who*) ordered the whole wheat bread and the decorated cake (not which **party** ordered . . .)?
> The funeral group is in Dining Room A (not the funeral **party** . . . ).
> Mrs. Smith's guests are in Dining Room A.
> The committee (or those investigating the accident) will meet on Tuesday (not the **parties** investigating . . . ).

### peerless

See **absolute adjectives**.

### people, peoples, persons

**People**, a collective noun, is used for more than one person. **Persons**, however, is often used when an exact number is given or when the components are thought of as individuals: "the 2,687 **persons** injured," "the two **persons** dying painfully," but "**people** converging from all directions," "about forty **people** waiting on line."

**People** is used for a single race or nation, but the verb that it governs is plural; **peoples** means more than one race or nation: "The French **people** do not respond well to this kind of advertising"; "The **peoples** of Asia share our difficulties in understanding the point of view of the **peoples** of Africa."

### per

**Per** is overworked in commercial writing and therefore is rejected by many writers and editors. It is acceptable when it means *for each* in writing on statistics, economics, accounting, and so forth.

> POOR: Write the letters as **per** my instructions (*according to* my instructions, *in accordance with* my instructions, *as I instructed* are preferable).
> The tickets were three dollars **per** each adult (. . . *for each adult* is preferable).
> More than five dollars **per** dozen is very expensive (. . . *a dozen* is preferable).

### percent, percentage

**Percent** and **percentage** mean *a part of the whole*. **Percent** *requires* a specific amount: "two **percent**," "85 **percent**." **Percentage** often is preceded by an adjective: "a small **percentage**," "the usual **percentage**"; also a preceding article, noun, or pronoun requires **percentage** rather than **percent**: "a **percentage**," "his **percentage**."

> CORRECT: Twenty **percent** of the tourists were late and missed the bus.
> He invested 25 **percent** in metals, 15 **percent** in bonds, and 60 **percent** in CD's.
> The discouraging **percentage** of unsold tickets for the boat rides worried the treasurer.
> My profits have decreased markedly even though my **percentage** of the business has remained unchanged.

### perfect

See **absolute adjectives**.

### permanent

See **absolute adjectives**.

### personnel

**Personnel**, a collective noun, means *a body of persons, usually employees, or of soldiers*, and it also means *the division that is concerned with those persons*. It is not a substitute for *persons* or *people*.

> CORRECT: The Advertising Department was increased by three **persons** (not three *personnel*).

Being a collective noun, **personnel** takes a singular verb for collective action and a plural verb for individual action.

> CORRECT: The **personnel** of this division has good morale.
> The **personnel** of this company enjoy social meetings, picnics, and many sports in their hometowns.

### persons

See **people**.

### phase, faze

**Phase** is a part in a cycle or development: "the early **phase** of the egg," "the **phases** of his life after fifty."
**Faze** means *daunt*: "The threat of an examination did not **faze** him."

## phobia, mania

A **phobia** is an exaggerated fear: **acrophobia**—fear of heights, **cataphobia**—fear of falling, **dentophobia**—fear of dentists. One meaning of **mania** is *extreme and unreasonable enthusiasm, a craze*. It also describes the manic phase of *manic depressive* psychosis. There is no obvious reason to confuse exaggerated fear with extreme, unreasonable enthusiasm; yet incorrect uses like the following are not uncommon: "The new restaurant is fast becoming a **phobia** with sophisticated diners." (**Mania** conveys the desired meaning.) "He would not join the hiking club because of his **mania** about heights." ("He would not join the hiking club because of his **acrophobia**" is correct.)

## plus, minus

**Plus** as a substitute for *and* should not be used in writing: "Jute is expensive; plus it is flammable." **Plus** may, of course, be used in its mathematical meaning. Writers should note that it does not change the number of a singular noun that precedes it.

> CORRECT: The deposit of $4,950 **plus** the amount in the savings account is enough to cover this check.

**Minus** in the sense of *deprived of* is best confined to informal speech and humorous contexts: "**Minus** his podium and gavel he seemed diminished, as though someone had let the air out of a balloon figure."

## p.m.

See **a.m.**

## podium

See **dais**.

## policeman

See **cop**.

## port

See **harbor**.

## possessive pronouns and the apostrophe

See **its**.

530

**possible**

See **feasible**.

**practicable, practical, impracticable, impractical**

**Practicable** means *capable of being done, feasible*; **practical** means *useful, exhibiting usefulness*.

> CORRECT: His plans were **practicable** but much too involved and expensive for our small company.
> Having common sense, the assistant treasurer designed a **practical** system of record keeping for the checking and storage rooms.
> In contrast to his dreamy brother, he was sensible and **practical**.

**Impracticable** and **impractical** are the negative forms, the denials of **practicable** and **practical**.

> CORRECT: **Impracticable** plans and wild fantasies occupied him during the working day.
> She clung to her **impractical** methods in spite of orders to change.
> The vice president's objections made any suggestion of change **impracticable**.

**precede, proceed, procedure**

Spelling is the first problem in the use of these words; the second is misusing **proceed**. A fourth-grade teacher's injunction, heard years ago, still applies: "You and I *go* to our seats; the commencement speakers and other honored guests **proceed** to the platform." **Proceed** is an elegant word that should not be used for *go, travel, move*; nor should it be added unnecessarily: "She **proceeded** to explain at unnecessary length all the reasons for her precipitate departure" (She *explained* . . . ). "He **proceeded** to repair the lock" (He *repaired* the lock or *began to* . . . or *started to* . . . ). "Walking backward, he **proceeded** to bump into the desk, knock down the telephone, and collide with the excited dog" (He *bumped* . . . *knocked down* . . . *and collided with*). *Pro-* indicates forward; therefore **proceed** should not be used if the action is backward: "The train **proceeded** back to the station it had just left" (The train *went back* or *returned* . . . ).

**precipitant, precipitate, precipitous**

As an adjective **precipitant** means *hasty, speedy*; as an adjective **precipitate** means *rash, without deliberate care, excessively hasty, sudden, abrupt*. **Precipitous**, an adjective, means *very steep, like a prec-*

531

*ipice*. **Precipitate** usually modifies actions; **precipitous**, physical characteristics.

> CORRECT: Those who had heart trouble were told to avoid the **precipitous** climb.
> Her **precipitate** attempt to flout the warning by climbing the steepest section caused an accident.
> **Precipitant** dismissal from the program was the first result she met.

The verb **precipitate** means *to provoke* or *to start quickly*.

> CORRECT: His abrupt refusal **precipitated** a quarrel.
> Her announcement **precipitated** a rush for the train.

In science **precipitate** may be a noun or a verb, and **precipitant** is a noun. Anything precipitated from a solution is the **precipitate**; anything causing the precipitation is the **precipitant**.

> CORRECT: The **precipitate** obtained from the first experiment was more satisfactory than that from the second, but the **precipitant** used in the second experiment was less expensive.
> The meteorologist predicted that a cold wind arriving late in the day would cause the clouds to **precipitate** snow or sleet before morning.

### preclude, prevent

To **preclude** is *to rule out beforehand*: "His motion **precluded** any discussion of military matters." To **prevent** is *to keep from happening*: "She would talk of nothing but how to **prevent** war."

### prepositions

Poor style results from cumbersome prepositional phrases. They should be replaced by simpler expressions because writing that contains many such groups seems heavy and awkward. The following sentences illustrate that the phrases *in regard to, by means of*, and *in the matter of* are usually easy to omit or to replace:

> POOR: **In regard to** scientists, they are addicted to detective fiction.
> IMPROVED: Scientists are addicted to detective fiction.
> POOR: **By means of** raising the temperature he solved the problem.
> IMPROVED: **By** raising the temperature, he solved the problem.
> POOR: **In the matter of** professional ethics it may be said that such action violates the code.
> IMPROVED: Such action violates the code of professional ethics.

Wordy prepositional phrases tend to attract other unnecessary words, and the total effect is fuzzy. When the unnecessary words are removed, the sentences become clear-cut. A writer, however, should not omit necessary prepositions. Omitting necessary prepositions and conjunctions is characteristic of telegraphic style, and telegraphic style is neither clear nor readable. (See Chapter 13 and "Telegraphic Style," Chapter 9.)

### prescribe, proscribe

**Prescribe** means *to write or give medical prescriptions, to lay down as a guide or rule*, or *to name as a remedy*. **Proscribe** means *to forbid as harmful or unlawful, to prohibit*.

> CORRECT: She **prescribed** a low-salt diet.
> The doctor **proscribed** visitors, TV, and other excitement.

### presently

**Presently** means *before long, without undue delay*. It is a somewhat pretentious substitute for *now*, a good, simple useful word.

> CORRECT: The director will see you **presently** (meaning *before long*).
> The director will see you *now* (meaning *immediately*).
> **Presently** the cafeteria will be out of salads. (*Before long* or *soon* is better suited to the simple meaning.)

### presume

See **assume**.

### presumptive, presumptuous

**Presumptive** means *offering grounds for a reason, opinion, or belief*. Lawyers use it more often than other writers. **Presumptuous** means *taking liberties, overstepping the limits of propriety*.

> CORRECT: His tendency toward **presumptive** analysis made conferences difficult for his illogical and fanciful clients.
> She thought that it would be **presumptuous** to ask the judge for further explanation.

### prevent

See **preclude**.

### preventative, preventive

**Preventive** is preferable.

### principal, principle

As an adjective **principal** means *chief, main*; as a noun it means *head of a school or of a group* or *an amount of money considered apart from any interest it may command or any profit it may engender*. **Principle**, a noun, means a *fundamental truth, a basic rule*, or a *moral standard*.

> CORRECT: Our **principal** negotiator became ill, and the contract was delayed.
> The teachers asked the **principal** for six more courses in expository writing.
> Lowering the interest rate and shortening the term of the loan would enable us to repay the **principal** much sooner.
> Fifty years ago many textbooks were entitled ***Principles*** *of Mathematics,* ***Principles*** *of Chemistry,* ***Principles*** *of English,* and so forth.
> Her **principles**, if any, were hard to judge from her actions.

### prior to, before

The phrase **prior to**, like repetition of the pronoun *one*, seems stilted to American readers. The good simple word **before** is an ideal substitute.

### procedure

See **precede**.

### proceed

See **precede**.

### professional

**Professionals** are *those engaged in learned occupations or in occupations that require advanced training and proficiency in the sciences or liberal arts*. The adjective **professional** means, in addition, *characterized by and conforming to the technical or ethical standards of a profession*. In sports **professionals** are those who play for money.

> CORRECT: There are few **professionals** in the company, because the entire emphasis is on sales.
> She was **professional** in all her work and in her relationships with all the members of the staff.

## prognosis

See **diagnosis**.

## prone, supine

**Prone** means that *the front, or ventral, surface is downward*; **supine** means *lying on the back with the face upward*.

> CORRECT: She could not identify him while he was **prone**, but when the workers placed him in a **supine** position, she saw that he was her neighbor.

## proscribe

See **prescribe**.

## proved, proven

Both are acceptable past participles of *to prove*.

> CORRECT: The workers thought that they had **proved** (or **proven**) the case before the jury went to lunch.

## provided, providing, if

For one possibility **if** is better. **Provided that** and **providing that** are correct when conditions complicate the thought of the clause.

> CORRECT: **If** the dyed sample lightens, increase the time the fabric is left in the dye.
> **Provided that** (or **Providing that**) Sections 44C and 129A apply and the client has been notified of the problem, stop production of the dyes for cotton and for silk.

## purposefully, purposely

These words are close in meaning: **purposely** (*intentionally*) suggests less determination than **purposefully**, which means *with a purpose or aim*.

> CORRECT: He **purposely** spoke only of the analysis and research design of optical fiber sensors.
> Knowing that he was avoiding mention of performance evaluation, she **purposefully** directed the discussion to that subject.

**pursuant to our agreement**

Good writers prefer simple expressions like *as we agreed.*

# Q

### questionable, questioning

**Questionable** means *giving reason for challenge or doubt* or *arousing suspicions of immorality, falseness, dishonesty.* **Questioning** means *doubting, interrogating, examining.*

> CORRECT: The union's interpretation of the ruling was **questionable**.
> The reporter cast a **questioning** eye at the irate witness.

# R

### re, in re

The forms **re** and **in re** above the body of business letters and memorandums are used to mean *in the matter of.* These terms are permissible in legal papers, but we do not recommend their use elsewhere.

### reason is because, reason is that

The **reason is that** should be preferred; the **reason is because** is tautological, cause being stated twice, and ungrammatical because the verb *is* should not be followed by an adverbial clause.

> INCORRECT: The **reason** this is called "objective tinnitus" **is because** it can be heard by others besides the person suffering from the defect.
> CORRECTED: This is called "objective tinnitus" *because* it can be heard. . . .
> The **reason** this is called "objective tinnitus" **is that** it can be heard. . . .

Another reason for avoiding the redundancy is that some uses of the incorrect expression sound childish and awkward.

### rebut, refute

**Rebut** is *to argue against, to contradict*; **refute** is *to prove wrong.*

> CORRECT: She **rebutted** the point skillfully, but the audience was not swayed.
> She **refuted** the point, and everyone agreed with her.

536

**recession**

See **depression**.

**recourse, resort, resource**

Similar in sound and sometimes in meaning, the words **recourse, resort, resource** offer distinctions in denotation and connotation to writers. **Recourse** means both *turning to someone for help or refuge* and *the source of the help*. **Resort** also means *the source of help*. As a verb it means *to turn to someone or something*. **Resource** is *a source of help, a source of expertise*, and (usually in the plural) *a source of revenue* or *one's total wealth*. **Resort** often conveys the finality of "last resort"; **resource**, some of the optimism of *resourceful*, or at least no feeling of finality.

> CORRECT: Eager to purchase the new equipment, she decided that her best **recourse** was the finance department.
> He had a valuable **resource** in his old professor.
> Her **resources** were adequate for financing her own business for several years.
> They wisely refused to use every other **resort** before consulting a financial adviser.
> She **resorted** to evasion when she was questioned.

**refer**

See **allude**.

**refute**

See **rebut**.

**regretful, regrettable**

The verb **regret** means *to be very sorry for*. **Regretful**, an adjective, means *filled with sorrow*; **regrettable**, an adjective, means *deserving of regret*.

> CORRECT: The condition of the neglected laboratories was **regrettable** (not **regretful**, which is a condition of beings, not of objects).
> His attitude was most **regretful** when he spoke of the accident.

**renown, renowned**

**Renown** is *fame*; **renowned** is *celebrated, famous, having renown*.

> INCORRECT: The **renown** speaker was applauded loudly (should be **renowned**).

CORRECT: Her **renown** gave weight to her suggestions about the education of women.

## respective, respectively

**Respective** and **respectively** are used when two series or lists are being matched.

> CORRECT: She conferred with Boulang, Wetekost, and Domicelli in Paris, Vienna, and Rome **respectively**. (The sentence means that she conferred with Boulang in Paris, with Wetekost in Vienna, and with Domicelli in Rome.)
>
> Boulang, Wetekost, and Domicelli will report next month to their **respective** assignments in Vienna, Rome, and Paris. (This sentence means that Boulang will report to Vienna, Wetekost to Rome, and Domicelli to Paris.)

Careful writers avoid unnecessary use of **respective** and **respectively** because these words make style seem heavy or pretentious. In a complicated listing they force the reader to go back to match the partners in the series or lists. It is better to rewrite such sentences.

> IMPROVED: She conferred with Boulang in Paris, Wetekost in Vienna, and Domicelli in Rome. Next month Boulang will report to Vienna; Wetekost, to Rome; and Domicelli, to Paris.

When **respective** or **respectively** is used unnecessarily, it should be removed: "Boulang, Domicelli, and Wetekost will leave for their assignments."

## result of

See **as the result of**.

## reverse

See **contrary**.

## rob, steal

One **robs** a person or establishment and **steals** money or goods.

> CORRECT: Three men **robbed** Brownsley Bank last night. They **stole** more than seventy thousand dollars.

**robber, robbery**

See **burglar**.

# S

## said

**Said** as an adjective referring to preceding material occurs mainly in business and legal writing. The best business and legal writers avoid the word. "**Said** witness was excused." (*This* witness . . . or *That* witness. . . .)

## scan

**Scan** means *to glance over* ("to **scan** the want ads"), but it also means *to examine minutely* as well as *to produce an image of a part of the body by combining the data from different viewpoints or angles* ("to **scan** the patient's liver"). A writer must be certain that the context indicates clearly either a hasty, casual viewing or a scrutinizing examination.

## scissors is, are

**Scissors** is a plural noun which may be either singular or plural in a construction.

> CORRECT: The **scissors** are dull.
> The **scissors** is engraved with a bird.

## -self, -selves

Reflexive pronouns—**myself, yourself, herself, himself, itself, ourselves, yourselves, themselves**—refer to the person or persons expressed in the sentence.

> CORRECT: She gave **herself** an interesting assignment.
> INCORRECT: They gave **myself** the answer.
> CORRECT: They gave **me** the answer.

Pronouns with the **-self** suffix are used for reflexive action ("She hit **herself** on the door," "He dressed **himself** this morning," "Mary allowed **herself** a second helping of dessert") and for emphasis ("I'll repair it **myself**," "They **themselves** started the quarrel," "You **yourself** should offer to help"). (See also Chapter 6.)

Writers should not use **myself** to avoid *I*.

> INCORRECT: **Myself** thinks him foolish for not checking his expense account.
> CORRECT: **I** think him foolish. . . .

539

When used as intensives, pronouns with the **-self** suffix may follow immediately the noun or pronoun they reflect or may end the clause: "He **himself** took no action," "He took no action **himself**."

## set, sit

**Set** is a transitive verb meaning *to place, to put* ("He **set** the package on the floor," "They **set** the chairs on the desks before cleaning the floor") and *to adjust, to establish* ("I have **set** this watch twice today," "He **set** the gears before starting the machine," "She **set** the date for the next meeting and then left").

**Sit** is an intransitive verb meaning *to rest on a chair or a bench or other location*: "She always **sat** on the desk although her employer disliked the habit." "Please **sit** down until I call you." **Sit** is often followed by *down* although *down* does not change the meaning of the verb.

A number of informal English idioms employ these verbs: "**Set** foot in" (*to enter*), "**sitting** pretty" (*successful or favorably placed*). (See also "Principal Parts of Verbs," Chapter 6.)

## sex

See **gender**.

## sexist nouns

One of the first attacks by a government bureau on sexist nouns was directed at changing those that named occupations such as the following:

| Sexist Noun | Suggested Replacement |
|---|---|
| cleaning woman | cleaner |
| fireman | fire fighter |
| foreman | supervisor |
| mailman | mail carrier, letter carrier |
| policeman, policewoman | police officer |
| salesman | salesperson, sales clerk |
| steward, stewardess | flight attendant |

These and similar changes are simple, desirable substitutions; unfortunately the government agency that distributed the list had not yet changed its name—the Manpower Commission.

The word *man*, as in *Manpower Commission*, leads to consideration of the use of *man* to describe people. Today *man* is so closely connected with one sex that other words should be used for the concept of men and women. Some suggestions are *people, persons, population, populace*. For *man-*

*power*, then, a writer might use *population resources*; and for *man-made: artificial, synthetic, not natural.*

The term *lady*, which has had wide application from *lady of the court* to *scrub lady*, should be used only when *gentleman* is or would be used for a man. Instead of *lady*, writers should use *woman* or *female*, as in "He did not know that George Eliot is a woman author (not *authoress*)"; "She was the first *female* professor on our faculty." The word *girl* is also inappropriate for an adult woman. The term is particularly demeaning for a secretary or assistant and should be replaced by the name of the person or of the position: *Mrs. Jones, my secretary, my assistant.* (See **boys**.)

Writers should also avoid diminutives and feminized forms such as *suffragette* for *suffragist*, *usherette* for *usher*, *poetess* for *poet*, *manageress* for *manager*. *Actress* and *waitress* are exceptions.

## sexist pronouns

See "The Sexist Pronouns: He, She; His, Her, Hers; Himself, Herself," Chapter 6.

## shall/will, should/would

See "Shall and Will, Should and Would," Chapter 6.

## simplistic

**Simplistic** means *oversimplified, made simple by the omission of complications.*

> CORRECT: Many a **simplistic** discussion of violence sees violence only in distant countries and omits the violence in the place where the discussion is being held.
> His explanations of grammar are so **simplistic** that even when his students recognize the correct and incorrect forms, they do not understand why one form is right and the other wrong.

## since

The conjunction **since** means (1) *in the period following the time stated* ("**Since** he was elected, he has spoken to me only once"); (2) *continuously from the time* or *ever since* ("**Since** the publication of his findings the world of physics has been changing rapidly"); and (3) *because* or *inasmuch as* ("**Since** the readings are inaccurate, the plant must repeat the runs").

**Since** frequently is used ambiguously by careless writers.

> POOR: **Since** the department was reorganized, the work has been more evenly distributed

**Since** the FDA announced this ruling, the experiments have been reported.

**Since** storms have become severe, the plant is closed four or five days every winter.

**Since** he learned that ambiguity may cause accidents, he has corrected every safety publication carefully.

The writers of these sentences meant "Ever since the department was organized . . . ," "Because the FDA announced . . . ," "Because storms have become severe . . . ," "Ever since he learned . . . ," but each sentence conveys at least two meanings.

### sinus

The **sinus** is a cavity in the body, not a disease.

INCORRECT: He missed the meeting because of a **sinus** attack.

CORRECT: He missed the meeting because of a **sinus** infection (or because of sinusitis).

### sit

See **set**.

### slander

See **libel**.

### so, very

**So** should not be substituted for **very**: "She is not **very** accurate"; "The equations are **very** long." But a writer may use **so** if it is followed by a clause: "She is not **so** accurate that I dare depend on her"; "The equations are **so** long that the editor must use more than one page for one equation"; "He is not **so** reliable as you think."

### so . . . as

See **as . . . as**.

### somebody

See **everybody**.

### somewhere, somewheres

**Somewheres** is substandard for **somewhere**.

CORRECT: He will find the entry **somewhere** if he looks through these books long enough.

## so that

See **and so**.

## specie, species

**Specie** means *money in coin*; **in specie** means *in the same or similar form* and *in coin*. **Species**, both the singular and plural form of the noun, means *a class with common attributes* or *a division of a class or category*. The word also has other meanings, especially in biology.

> CORRECT: It is difficult to assess the **specie** found in ships that sank hundreds of years ago.
> The effect of aerobic activity on the heart and muscle of **species** other than the human should be studied further.
> Her study showed that at least one **species** of gorilla has made remarkable progress in communicating.

## split infinitive

See "Never Splitting an Infinitive," Chapter 10.

## stationary, stationery

**Stationary** means *immobile*; **stationery** is *materials for writing or typing, including matching letter paper and envelopes.*

> CORRECT: The wheel was **stationary** when the accident occurred.
> He complained that workers were taking home company **stationery**.

## such a

**Such a** is correctly used if the word or phrase it introduces is followed by a clause starting with *that*. It is not correct as an intensive if the clause is omitted.

> CORRECT: They had **such a** problem with shipboard operation of the laser that they postponed the experiment.
> The children were in **such a** noisy, destructive mood today that we had to assign two more teachers to their bus.
> INCORRECT: He made **such a** mistake. (He made a *bad* or a *serious* mistake.)

## supine

See **prone**.

# T

**take**

See **bring**.

**temperature**

See **fever**.

**thanking in advance**

Admonitions against thanking in advance and explanations of the impression it makes have been dwindling since Strunk and many other advisers warned writers against it. Use of the phrase **thanking you in advance** suggests that the writer is too lazy or too stingy to thank someone for a favor after receiving it and therefore combines the thanks with the request. It also suggests somewhat high-handedly that the reader of the letter has no choice but to grant the favor or to answer the request. Since this is seldom true (many readers fill their wastepaper baskets with these letters), perhaps the writers might more accurately write, "thanking or damning you in advance." Then they will not have to telephone or write to complain when their letter does not achieve its goal.

Courteous writers will avoid the poor taste of thanking in advance by requesting politely and then, if the favor is granted, thanking the grantor in another letter or in a telephone call.

This acknowledgment frequently is a good opening for the next letter one writes to the grantor, according to our clients, many of whom say good letter openings are hard to find.

**than me, than I**

The nominative case, **I**, should be used if **I** is the subject of the clause understood after **than**.

> CORRECT: He can estimate sales more accurately **than I**. (*Can* is understood after *I*.)

**that**

Writers should be careful to include **that** whenever the sense requires it and to place it correctly. The sentence "The reporter maintained during the conference **that** the public relations director had given him the figures" differs from "The reporter maintained **that** during the conference the public relations director had given him the figures." If the conjunction **that** is not expressed, temporary misreading may occur: "The reporter said he had heard the information to be given at the press conference was incomplete." A reader who understands that the reporter said he had heard the information to be given at the press conference is pulled up sharply by the *was incomplete* and has to reassemble the sentence in his mind or reread it. "The reporter said he had heard **that** the information to be given at the press conference was incomplete" is immediately clear.

If omission of **that** does not cause a misreading, may **that** be omitted? The conjunction **that** is expressed more frequently in writing than in informal speech. Conversational sentences are usually shorter and less intricate than written sentences and therefore require fewer guides. But even in formal writing some verbs are not always followed by **that**, for example, *believe, presume, suppose, think*. **That** is usually expressed after such verbs as *agree, assert, assume, calculate, conceive, learn, maintain, state*, and *suggest*. The decision whether to express **that** after a verb may depend on the purpose of the writing, the sound of the sentence, and the writer's opinion of the readers' needs. Unsure writers should remember that omitting the conjunction **that** may be confusing. Unless the major aim is conversational informality, writers who are unsure of their judgment should express **that** after verbs which are sometimes followed by **that**. Careful observation of the usage of good contemporary writers will guide those who feel more certain of their perception.

But writers should take pains to avoid unnecessary repetitions. For example, the second **that** should be omitted from the following sentence: "They were told **that** although the convention would not recess until Monday **that** there was not time for new business." And strings of interdependent **that** clauses, like strings of other interdependent clauses, should be avoided.

## that, this, which

Sentences in which **that, this**, or **which** refer to entire preceding sentences or to clauses in preceding sentences should always be examined carefully. If an intervening noun is misread as the antecedent, the result may be confusing.

> AMBIGUOUS: Some of the work was delayed by a hitherto unknown side effect of the drug. **This** resulted in the loss of the contract (or ". . . **which** resulted in the loss of the contract" or ". . . **that** resulted in the loss of the contract).
>
> CLEAR: The contract was lost because a hitherto unknown side effect of the drug delayed some of the work.
>
> A hitherto unknown side effect, which delayed some of the work, resulted in the loss of the contract.
>
> A hitherto unknown side effect of the drug delayed some of the work, and the contract was lost.
>
> A delay in some of the work, caused by a hitherto unknown side effect of the drug, resulted in the loss of the contract.

### that, which

The question of whether to use **that** only for restrictive clauses and **which** only for nonrestrictive clauses is much discussed. In such sentences as the following the distinction between restriction and nonrestriction depends on commas and on the context:

> The formulas **that** are dangerous to birds are not to be used without authorization.
> These two formulas, **which** are dangerous to birds, should be stored separately.

Fowler[3] states, "If writers would agree to regard *that* as the defining restrictive relative pronoun, and *which* as the non-defining [nonrestrictive], there would be much gain both in lucidity and in ease. Some there are who follow this principle now; but it would be idle to pretend that it is the practice either of most or of the best writers." Some writers, for example, would use **that** to indicate restriction in the first sentence above, but others would use **which**. We doubt that handbook writers can change this, desirable though such a change may be. Both rely on commas to indicate nonrestriction. (See "The Comma," Chapter 8.)

### that is

**That is** introduces a statement equivalent to the preceding statement, not one that corrects the preceding statement. In the following sentence the statements connected by **that is** are equivalents: "The entire research staff, **that is**, administrators, scientists, engineers, laboratory technicians, laboratory assistants, secretaries, and typists, will move on Monday." But "There are twenty-nine papers in the symposium; **that is**, three of them are abstracts" should be "There are twenty-six papers and three abstracts in the symposium" or "There are twenty-nine papers, three of them abstracts, in the symposium" or "There are twenty-nine papers in the symposium, but three of them are abstracts."

### their, there, they're

**Their** is the third person plural pronoun in the possessive case. **There** is an adverb meaning *in that place*, an expletive introducing a construction, an interjection. **They're** is a contraction of *they are*.

> CORRECT: They submitted **their** most recent reports to show how much **their** writing had improved.
> The secretary pointed to the file and told them to leave the reports **there**.
> **There** were twelve reports.
> **There**! I have done it.
> The writers say that **they're** glad to have additional editorial help.

546

## these kinds, this kind

A writer who uses **this** with singular **kind** and **these** with plural **kinds** will have no trouble about agreement in number, but some writers make the common error of using **these kind**.

> INCORRECT: He discussed the weekly management meetings and said he had no confidence in **these kind** of time wasters (should be **this kind** of time waster).
>
> **These kind** of solar heaters are too expensive for these homes (should be **these kinds** . . . are or **this kind** of solar heater is).

## they, their, theirs

See **its**.

## thief, theft

See **burglar**.

## this

See **that, this, which**.

## though

See **although**.

## thus, thusly

**Thusly** is substandard.

## till, until, 'til, til, 'till

**Till** and **until** are both standard words and may be interchanged. The contraction **'til** is sometimes used, although unnecessarily. **Til** and **'till** should not be used at all.

> CORRECT: **Till** (or **until**) she had tested each component separately, she did not have the necessary data.
>
> He telephoned to say that he would wait for you **until** (or **till**) one o'clock.

## to, too, two

**To** is a preposition; **too** is an adverb; and **two** is a noun, pronoun, or adjective.

CORRECT: They went **to** Key West.
She was **too** tired to walk along Duval Street.
**Two** of them went fishing; **two** stayed at home.
At the beach she read **two** stories by Hemingway.

### tortuous, torturous

**Tortuous** means *devious; having twists, bends, or turns*; **torturous** means *cruelly painful* or *producing pain cruelly*.

CORRECT: The **tortuous** mountain road was the scene of many accidents.
The violent group used many **torturous** punishments on their victims and threatened to use others.
His **tortuous** schemes alienated many fellow workers.

### try and, try to

The correct form is **try to**.

INCORRECT: They will **try and** repair the furnace.
CORRECT: They will **try to** repair the furnace.

### turbid, turgid

**Turbid** means *thick, opaque, or muddy*; **turgid** means *bombastic, pompous*.

CORRECT: The **turbid** stream was examined as a possible source of pollution.
The **turgid** report wasted the committee's funds because few could understand it, and those few said that it lacked credibility.

# U

### uninterested

See **disinterested**.

### unique

See **absolute adjectives**.

### universal

See **absolute adjectives**.

## usage, utilization

**Usage** is *traditional practice* or *the manner of employing words*. **Utilization** means *making a new use of* or *turning to practical use*.

> CORRECT: **Usage** reflects many characteristics of the society in which it developed.
> Her **utilization** of a parachute to protect her car fascinated her neighbors.

## use to, used to

The past participle or the past tense is **used**. **Use to** may be the result of slovenly speech.

> CORRECT: They **used to** have more time to write.

## utilize

See **usage**.

# V

## verbal

See **oral**.

## very

See "Hedging and Intensifying," Chapter 13.

## via

**Via** refers to location or layout; it means *by way of*, not *by means of*.

> CORRECT: They went to Niagara Falls **via** Buffalo.
> Pipe A carries the effluent to Pipe C **via** Pipe B.
> INCORRECT: They tested the result **via** a spectrometer.
> CORRECT: They tested the result *by means of* (or *by*) a spectrometer.

## visit, visitation

**Visit** should be used unless the visitor is supernatural.

# W

## we

See **I**.

## we, our, ours

See **its**.

## when, where

Time and place are frequently interchanged by mistake. In the sentence "**Where** colleges do not schedule these conferences, meeting candidates takes two or three days instead of one" the writer meant **when** or *if*, not **where**. And the sentence "In the instances **where** men neglect the methods of science and concentrate their thinking on the results of science, this also occurs" needs **when** or *if* in place of the wordy *in the instances where*. The writer of "The richest ore was found **when** the vein was deepest" meant **where**, not **when**. The writer of "There are many examples **when** the computers would be more accurate than the methods now used" was not referring to time but to divisions of a company. Confusion of time and place is likely to occur in sentences containing unnecessary words like *in the instances, in cases, in fields*.

## where, when, that

**Where** and **when** are sometimes carelessly misused for **that**.

> INCORRECT: I read in the paper **where** he was convicted of embezzlement.
> CORRECT: I read in the paper **that** he was convicted of embezzlement.
> INCORRECT: It was right after her vacation **when** she received her promotion.
> CORRECT: It was right after her vacation **that** she received her promotion
> (or "She received her promotion right after her vacation").

## whether, whether or not, as to whether

The conjunction **whether** may be followed by **or not** for emphasis or sense, but **whether** is customarily used without the stated negative; however, a sentence such as "the date must be recorded **whether** the patients cooperate" demands **or not**. Unnecessary restatement of the alternatives should be avoided. It is not emphatic. Thus "**whether or not** the experiments do or do not succeed is the question" repeats ineffectively. Any one of the following sentences is better: "**Whether or not** the experiments succeed is the question"; "**Whether** the experiments succeed is the question"; "Will the experiments succeed? is the question."

**As to** preceding **whether** is unnecessary. In "The question **as to whether** the experiment should be continued was discussed but was not voted upon" and in "They could not decide **as to whether** to call the meeting," **as to** should be omitted.

### which

For the use of **that** or **which** for restriction, see **that, which**. For **which** with a clause or sentence as antecedent, see **that, this, which**.

### which, who, whom, whose

The relative pronouns **who, whose, whom** are used for persons, personified objects, and sometimes animals. **Which** is used for things, for animals, and for collective nouns (even when the collective nouns designate groups of people).

The pronoun **whose** may be used instead of an awkward **of which**: "The letter was sent to the department **whose** chairman had requested it." When it is possible to avoid this substitution, it is better to do so, as the following sentences illustrate:

> He examined the reports **whose** bindings were damaged.
> He examined the reports **with** damaged bindings.
> He went to the library **whose** founder was his grandfather.
> He went to the library **founded by** his grandfather.

Awkward use of the relative pronoun **which** as an adjective should be avoided. "He did not present his vouchers on time, **which** negligence caused him trouble for weeks" might be rephrased in a number of ways: ". . . on time, negligence which caused him trouble . . ."; "Neglecting to present his vouchers on time caused him trouble for weeks"; "His negligence in failing to present his vouchers. . . ."

### while

**While** should be used carefully so that it will not be ambiguous. **While** means *as long as* and *during the time that*, but **while** also means *although*. In the sentence "**While** he wrote, he paid no attention to anything else," **while** means *as long as* or *during the time that*. In "**While** his reports are brief and well organized, his speeches are long and discursive," **while** means *whereas*. In "**While** he never seemed anxious about the deadline, he did press very hard to complete the experiments," **while** means *although*. But what does **while** mean in "**While** the animals were quiet, they did not seem sick"? Does it mean *as long as* or *although*? In sentences where **while** is ambiguous, a careful writer uses another conjunction, even though the ambiguity may be temporary, as in "**While** he did not

551

like the defendant, his behavior was impeccably disinterested." In the following sentence **while** is used incorrectly for *and*: "However, difficulties were encountered on a large scale because the oxime could not be filtered, **while** centrifugation proved long and tedious." Writers should avoid carelessly using **while** for *and*, a use that is always incorrect.

Some advisers recommend that **while** be restricted to constructions related to time, but with care a writer may use **while** in other constructions.

### who, whoever, whom, whomever

**Who** and **whoever** are in the nominative case and therefore subjects of the verb or verbs that they govern. **Whom** and **whomever** are in the objective case and therefore are objects of the verb or preposition to which they relate. Errors often arise when writers confuse the subject of a dependent clause with the object in the preceding clause or when the subject pronoun in the objective case appears before its verb. "The foundation will award the grant to **whomever** submits the best proposal within a given length of time" should be " . . . to **whoever** submits." The whole clause beginning with **whoever** is the object of *to*; **whoever** is the subject of that clause and therefore is in the nominative case, but it has been attracted into the objective case because it follows the preposition *to*. "**Whomever** the committee selects will represent the department at the meeting" should be "**Whoever** the committee selects will represent. . . ." The subject of the main clause must be in the nominative case. "**Who** did John expect to see?" should be "**Whom** did John expect to see?" (The fact that the pronoun is separated from its verb and appears before it does not mean that the pronoun is the subject of the sentence.)

There are several ways to be certain of selecting the correct case. One method is to substitute a noun and relative pronoun or a personal pronoun and relative pronoun for **who** or **whom** or **whoever** or **whomever**: "The foundation will award the grant to *the person who* submits. . . ." Clearly, then, **who**, in the nominative case, is needed. "He **whom** the committee selects will represent. . . . *Whom the committee selects* is a relative clause modifying the subject of *will represent*. If the sentence is a question, the case usually may be determined by rephrasing it as a statement: "John expected to see *him* or *her*"; therefore the objective case, **whom**, should be used in the interrogative form also.

Some editors and writers think that differentiating between **who** and **whom** is a lost cause, but writers should be aware that if they misuse **who** and **whom**, many readers will consider them illiterate.

### -wise

See "Vogue Words," Chapter 10.

**Women's Liberation**

See **Feminist Movement**.

**worth, worthwhile**

To avoid false elegance, writers do not use **worthwhile** when simple **worth** serves the purpose: "The convention is **worth** attending" (not **worthwhile** attending).

# X

**Xmas, Christmas**

**Christmas** is appropriate. **Xmas** appears in commercial writing, but many readers find it objectionable.

# Y

**yet**

See **and**.

## NOTES

1. Fowler, H. W., *A Dictionary of Modern English Usage*, 2nd ed. (New York and Oxford: Oxford UP, 1965) 69.
2. Fowler 333.
3. Fowler 626.

# APPENDIX C

# Documentation

Borrowed material—whether direct quotations, paraphrases, or concepts—and information about the work of others must be unmistakably attributed to the originator. While the style of the list of references and of the text references may vary from publisher to publisher or company to company, the information to be included is largely the same.

Writers who plan to publish should consult their editor or the appropriate journal for instructions on style. The suggestions in this appendix are general and by no means exhaustive.

In collecting references, writers must be thorough. They should be certain that the material is copied accurately and completely. Copying machines are a great help. The author's or authors' names, the title of the work and if needed of the article or section, the editor or translator if one is named, the publisher, the place of publication, the date, and the volume and page number should be obtained when the reference is consulted; then there is no last-minute scramble for missing information.

## NOTES

References to the work of others must be complete the first time they appear. Subsequent references are as brief as possible. The references are footnotes at the bottom of the appropriate page or notes at the end of an article, chapter, section, or sometimes a book. They are identified in the text by numbers or by a brief form of the reference, or citation (page 559).

End notes have been preferred over footnotes recently for the sake of economy, but as computer composition replaces labor-intensive hand typesetting, footnotes may come back into favor.

One advantage of footnotes is that only the information necessary to identify the citation need appear. If the author or the title or both are part of the text, the first footnote may contain only the publication data and the page. Subsequent footnotes may need only the page number. The same logically holds true for numbered entries in end notes, but some publishers consider that the separation of text and reference requires repetition of at least the authors' names. If the material cited is important to the subject of the work and is not merely an acknowledgment of examples borrowed, use of this form requires also a complete bibliography so that readers may judge the thoroughness of the research.

In typical footnotes or end notes, the first reference to a work must be complete, so that a reader may easily identify it. References begin with the author's name. The name is always inverted in an alphabetical list for ease of reference. Many publishers prefer normal order for names in numbered references. In an alphabetical list, for two or three authors of a work the name of only the first author is inverted. For more than three authors, *and others* or *et al.* may follow the name of the first author.

### Numerical or Alphabetical Order

Footnotes are nearly always numbered. End notes may be numbered or arranged in alphabetical order according to authors' names. The choice depends on the form used for citations in the text.

In alphabetized notes the author's name is not repeated when there is more than one entry for an author. Instead, a long dash replaces the name in the second and subsequent entries. Works by the author alone are listed first, then works by the author and a second author, then works by the author and two others. Each work appears only once.

> Brown, Y. Z., *Politics* 251 (April 2014a):1575.
> ———, *Statesmanship* 3 (November 2014b):974.
> Smith, A. B., *Business as Usual* (New York: Wiley, 2002a)14.
> ———, *Marketing for the Next Decade* (Boston: Sunflower, 2002b)249.

Where numbers are used, the works are numbered consecutively as they are referred to in the text. Each reference, then, must be numbered. The first time a work is mentioned in the text, a complete reference appears in the notes. Subsequent citations are also numbered, and the reference is repeated in the notes in shortened form.

### Published Material

#### Books

For books the author's name is followed by the title of the book, the publication data (place of publication, publisher, date), and where appropriate the volume and page:

> Jones, F. L., *Coins and Paper* (New York: Wiley, 2087) 86.

If the book is a collection with an editor, the editor's name appears in place of the author's and is followed by *ed.*, or the author's name and work may appear and then the title of the collection and the editor's name.

If the book is a translation, the author's name appears first, then the title, then *trans. John Johnson.* Anonymous books are entered in the list by title.

Publishers' names may be shortened as long as they are identifiable. Abbreviations like *Inc., Ltd., Co.* and the words they represent are omitted. (A list of shortened forms of publishers' names is in the annual *Books in Print.*)

### Journals

For articles authors' names are followed by the title of the article in quotation marks and then by the title of the journal in italics (underlined), the volume of the publication, the date (month and year, or season and year, or in the case of a weekly, bimonthly, or newspaper the day of publication), and the page:

> Smith, J. L., and C. T. Jones, "Footnotes and End Notes," *Journal of National Bibliographers* 45 (March 2083): 17.

In references to scientific and technical journals the title of the article is often omitted, the article being identified only by its location in the periodical. The reason is perhaps that the titles, being descriptive, tend to be long. Some journals that omit the titles of articles use Roman type for the title of the periodicals. Some publishers use standard abbreviations for the titles of periodicals, especially in the sciences:

> White, J. L., and A. B. Brown, *Q. Cer. Soc.* 105 (March 1993): 42.

(A list of accepted abbreviations for journal titles may be found in ANSI Z39.5 1969 [New York: American National Standards Institute] and in lists for specific disciplines.)

### Unpublished Material and Other References

Letters, manuscripts, personal communications may also appear in the notes. If the notes are numbered, the reference may give brief information or may be a sentence:

> 1. Telephone conversation between D. L. Jones and the author March 1987.
> *or*
> 2. At a meeting of the XYZ Society in September 1986 D. L. Jones told the author that he considered the theory of expanding egos "very shaky."

If there are no numbers, the information must be inserted under the name of the person concerned:

> Jones, D.L., Personal communication March 1987.

Forms for documenting tapes, filmstrips, video cassettes, computer printouts, and other sources of information may be found in the books mentioned under "References for Style" at the end of this appendix.

## Subsequent References

Subsequent references in numbered lists contain the last name of the author or authors and the page. In the case of more than one work by an author or authors, the author's name, the title or a shortened title of the book or article, and, if needed, the page are given:

> 6. White and Brown 42.
> 7. Jones, *Coins* 54.
> 8. Jones, "Gold Standard" 16.

> NOTE: If article titles are not used, the article is identified by the year. (Some publishers use the year instead of the short title for books also.) If the author or authors have more than one publication in a year, letters may be used after the year.

Subsequent references to a work are unnecessary in alphabetized end notes, because short-form or author-date references are used in the text (page 559).

## Latin Abbreviations

As with other Latin abbreviations, bibliographic terms are being discarded. Instead, the brief reference form is used. For *Ibid.* (*ibidem*), "in the same place"; for *op. cit.* (*opera citato*), "in the work cited"; and for *loc. cit.* (*loco citato*), "in the place cited," the author's name and the page or, for several works by the author, the author's name, a short form of the title or the year of publication, and the page are used in numbered references.

## Comments, Examples, Supplementary Information

Clarifications should be incorporated in the text. If excessively long, they may be given in an appendix. Very brief material that does not fit into the text may appear in a footnote or an end note with, as a rule, an identifying symbol, rather than a consecutive number.

## TEXT REFERENCES

A number or other reference in the text to a footnote or end note closely follows the material cited.

If numbered references are used, the number may be a superior figure or a figure in text type or italics enclosed in parentheses or brackets. It corresponds to the number in the notes:

Jones[2] maintains. . . .
*or*
Jones (2) maintains. . . .
*or*
Several authorities (*1,5,7*) concur, some with minor reservations.

Instead of numbers, some publishers, especially in science and technology, prefer to include in the text a short reference enclosed in parentheses. The short form is the last name of the author or authors and the year of publication or a short form of the title, followed where appropriate by the page:

This theory had been offered before (Jones 1983).
. . . (Jones and Brown 1979, 67).
. . . (Jones, Brown, and Black 1986a, 104) [The authors published more than one article in 1986].
. . . (Jones, *Coins* 7).

If this short form is used in the text, the end notes are alphabetized according to the authors' names, and no numbers are necessary. Each work appears in the notes only once.

## BIBLIOGRAPHIES

A bibliography, which usually appears at the end of a paper or book, is an alphabetical listing, by author, of works consulted. Each work appears only once, and the information is the same as that in the complete reference in the end notes but with slight differences in punctuation. The page references are omitted for books. For articles the first page or the inclusive pages are used. The last name of the sole author or of the first of several authors of a work is inverted. Entries are not numbered.

Many writers dispense with a bibliography and depend on end notes or footnotes. There are, however, some advantages to adding a bibliography, particularly for long papers.

End notes, which may be called *References, Literature Cited,* or any descriptive title, and footnotes contain only works cited in the text. Also, if the references are numbered, the reader has to search through the notes to see what authors were cited. This chore is an argument for inverting authors' names, as they are thus easier to spot.

A bibliography enables a reader to see quickly what works were used. Furthermore, a bibliography may contain works consulted but not cited. These may be woven in with the cited material, or there may be a section of cited works and one of works consulted but not cited. A bibliography also may be divided into subjects, the works in each division being alphabetized. Some writers prefer to separate the bibliography into several sections, one for each type of reference: books, periodicals, unpublished papers, personal communications, government reports, video and audio cassettes, and other classifications.

## EXAMPLE OF NOTES WITH NUMBERED ENTRIES

1. John Smith and George R. Brown, *A History of Finance* (New York: Wiley, 2089) 10.
2. Joseph Jones, "What about the Gold Standard?" *Proceedings of World Finance Society* 3 (2091): 49.
3. Detailed discussion of notes and bibliographies appears in the style guides of many professional societies. [Such remarks should preferably be in the text. They do not fit into unnumbered notes.]
4. Smith and Brown 306.
5. Joseph Jones, *Coins and Paper* (New York: Wiley, 2091) 618.
6. Jones, *Coins* 401.
7. U. J. Black, Jr., letter to the author, 3 Aug. 1990.
8. Jones, *Coins* 23.
9. Jones, "Gold Standard" 41.
10. Henry Miller et al., *Credit* (New York: Wiley, 2078) 18. [Four authors]

## EXAMPLE OF NOTES WITHOUT NUMBERING

(The text citations relating to this list are in the short form.)

Black, J. R., Jr., letter to the author, 3 Aug. 1990.

Jones, Joseph, *Coins and Paper* (New York: Wiley, 2091a) 618.

———, *Proc. World Finance Soc.* 3 (2091b): 49. [The title of the article might have been included and the name of the journal spelled out.]

Smith, John, and George R. Brown, *A History of Finance* (New York: Wiley, 2089) 48.

## EXAMPLE OF A BIBLIOGRAPHY

Black, J. R., Jr. Letter to the author. 3 Aug. 1990.

Jones, Joseph. *Coins and Paper.* New York: Wiley, 2091.

———. "Bond Prices." *Proceedings of World Finance Society.* 3 (2091): 40–52.

Smith, John, and George R. Brown. *A History of Finance.* New York: Wiley, 2089

## REFERENCES FOR STYLE

Further information on documentation is available in *The Chicago Manual of Style*, 13 ed. (Chicago: Chicago P, 1982); Joseph Gibaldi and Walter S. Achtert, *MLA Handbook for Writers of Research Papers*, 2nd ed. (New York: MLA, 1984); United States Government Printing Office, *Style Manual* (Washington, D.C.: GPO, 1984); and the style manuals of various associations and publishers.

# Index

Congratulation, letters of, 417
Conjunctions:
  ambiguous:
    *as*, 481
    *since*, 541
    *when, where*, 550
    *while*, 551
    cause of fuzzy writing, 551
  beginning sentences with short, 321
  coherence achieved by, 362–363
  confusion of time and place, 550
  correlative, 186
    used to connect subjects, 186
  expressing addition, comparison and
    contrast, exemplification, place,
    purpose, repetition, result, summary,
    time, 362–363
  misuse of:
    *and so*, 479
    *but*, 487
    *however* for *and*, 479
    *if* for *whether*, 510
    *so that*, 479
    *that*, 544–546
    *yet* for *and*, 479
  omission of *that*, 544–545
  position of, for transition between
    paragraphs, 365
  unnecessary repetition of, 544–545
  used:
    in argumentation, 53
    in chronological order, 37
    in spatial order, 39–40
  *whether* or *whether or not*, 550–551
Connectives, *see* Coherence; Conjunctions
Connotation, 283–284
Constructions:
  confusion in, 197–198
  incomplete in comparisons, 196–197
  misplaced modifiers, 200–202
  of paragraphs, 335–338, 361–368
  parallel, 198–199, 328–331
  roundabout, 392–393
  of sentences, 313–331
  split, 286–287
Contact, ambiguous use of, 493
Contents, *see also* Material; Subject Matter
  of instructions, 422
  of letters of condolence, 418–419
  of reports, 437–438
  technical information, 413
Contract, letters of, 413
Contrast:
  beginning with, 88–89
  device for sentence emphasis, 326–327
  organization of, 53–57
  paragraph development by, 346–347

Correction, *see* Revision
Correctness:
  annual check for, 29
  goal of revision, 28–29
Correlative conjunctions, 186
Courtesy in letters, 410
Co-writers, advantage of outline to, 65

Dangling modifiers, 200–202
Dashes, 237–240
  for abrupt shift or break in thought, 238
  commas and, 238
  with parenthetical material, 240
  with supplementary material containing
    commas, 240
*Data,* 182–185
Data in reports, 437–438
Deduction, choice of induction or, 57–58
Definite article, 468
Definition:
  beginning with, 91–94
  confusion in, 298
  developing paragraphs by, 347–350
  emphasis in, 298–299
  necessary and unnecessary, 299–300
  separate section of report for, 299
  subordination of, 298–299
Demonstrative adjectives, coherence and,
  361
Denotation, 283
Details:
  beginning with, 94–99
  developing paragraphs by, 350–352
  interest of reader in, 21, 27, 436–437
Development of paragraphs, methods of,
  339–361
  abstract and summary, 339–342
  analogy, 342–344
  analysis, 344–345
  cause and effect, 345–346
  combination of methods, 359–361
  comparison and contrast, 346–347
  definition, 347–350
  details and particulars, 350–352
  examples and illustrations, 352–353
  questions and answers, 353–354
  reiteration, 354–356
  scope and qualification, 356–358
  straw man, 358–359
  summary, 339–342
Diagonal, *see* Slashes
Dictation:
  aid to informal style, 14–15
  disadvantages of impromptu, 35, 66
  value of outline for, 35, 65–66
Diction, *see also* Brevity; Words
  business, 217–218